PROCEEDINGS OF THE BERKELEY CONFERENCE IN HONOR OF JERZY NEYMAN AND JACK KIEFER

Volume I

THE WADSWORTH STATISTICS/PROBABILITY SERIES

Series Editors

Peter J. Bickel, University of California, Berkeley
William S. Cleveland, AT&T Bell Laboratories
Richard M. Dudley, Massachusetts Institute of Technology

Richard A. Becker and John M. Chambers
Extending the S System

Richard A. Becker and John M. Chambers
S: An Interactive Environment for Data Analysis and Graphics

Peter J. Bickel, Kjell Doksum, and J. L. Hodges, Jr.
Festschrift for Erich L. Lehmann

George E. P. Box
The Collected Works of George E. P. Box
Volumes I and II, edited by George C. Tiao

Leo Breiman, Jerome H. Friedman, Richard A. Olshen, and Charles J. Stone
Classification and Regression Trees

John M. Chambers, William S. Cleveland, Beat Kleiner, and Paul A. Tukey
Graphical Methods for Data Analysis

Franklin A. Graybill
Matrices with Applications in Statistics, Second Edition

Lucien M. Le Cam and Richard A. Olshen eds.
*Proceedings of the Berkeley Conference in Honor
of Jerzy Neyman and Jack Kiefer*, Volumes I and II

John W. Tukey
The Collected Works of John W. Tukey
Volume I: *Time Series, 1949-1964*, edited by David R. Brillinger
Volume II: *Time Series, 1965-1984*, edited by David R. Brillinger

PROCEEDINGS OF THE BERKELEY CONFERENCE IN HONOR OF JERZY NEYMAN AND JACK KIEFER

Volume I

edited by

LUCIEN M. LE CAM
University of California, Berkeley

RICHARD A. OLSHEN
University of California, San Diego

with the assistance of
Ching-Shui Cheng, Richard M. Dudley, Morris L. Eaton,
James A. Koziol, Elizabeth L. Scott, and Michael Woodroofe

a copublication of
Wadsworth Advanced Books & Software
Monterey, California
A Division of Wadsworth, Inc.
and
The Institute of Mathematical Statistics
Hayward, California

Sponsoring Editor: *John Kimmel*
Production Editor: *Andrea Cava*
Copy Editor: *Yvonne Howell*
Compositor: *Jove Statistical Typing Service*

Printed in the United States of America
1 2 3 4 5 6 7 8 9 10

ISBN 0-534-03312-1

Wadsworth Advanced Books & Software
A Division of Wadsworth, Inc.

Library of Congress Cataloging-in-Publication Data
(Revised for vol. 1)
Main entry under title:

Proceedings of the Berkeley conference in honor of
    Jerzy Neyman and Jack Kiefer.

    (The Wadsworth statistics/probability series)
    Proceedings of a conference held in Berkeley, Calif.,
June 20-July 1, 1983.
    1. Mathematical statistics--Congresses. I. Le Cam,
Lucien M. (Lucien Marie), 1924-   . II. Olshen,
Richard A., 1942-   . III. Cheng, Ching-Shiu.
IV. Neyman, Jerzy, 1894-1981. V. Kiefer, Jack,
1924-1981. VI. Series.
QA276.A1P75 1985   519.5   85-13750
ISBN 0-534-03312-1 (v. 1)
ISBN 0-534-03357-1 (v. 2)

PREFACE

On August 5, 1981, Jerzy Neyman died of a heart attack. Then on August 10, 1981, Jack Kiefer died of a heart attack. Their deaths left the Berkeley Statistics Department in a state of confusion and shock. In less than a week, we had lost two colleagues, two of the most eminent statisticians of our century. We also had lost two friends.

Memorial sessions were held in their honor at various meetings of learned societies, as is fitting for scientists of their rank. However, for reasons of heart as well as science, we felt that a special tribute should be payed to them at Berkeley, where they had so many friends. A meeting was eventually organized. It took place from June 20 to July 1, 1983. Jack's friends at Cornell felt similarly. They organized a meeting that took place immediately after the Berkeley conference.

The Berkeley meeting started on June 20, 1983. A formal opening ceremony was held with dedication of room 1011, Evans Hall, to Jerzy Neyman. Professor Harald Cramér, who came from Sweden for the occasion, gave the main address. Then, Chancellor I. Michael Heyman formally declared room 1011 *The Jerzy Neyman Room* and unveiled the bronze plaque that carries an unintended variation of one of Neyman's favorite sayings: Life is complicated, but not uninteresting. (The unintended variation is an addition of *it is* after the *but*.) It was the start of a busy, exciting two-week conference.

The idea of publishing proceedings that would constitute a more permanent memorial grew during the conference itself. We are very grateful to our eminent colleagues who agreed to participate, present recent advances in their fields, and write them out for publication. Their contributions fill two volumes. Their names are well known; no introductions are necessary. However, some explanations of the choice of topics are in order. We concentrated on subjects close to the preoccupations of Neyman and Kiefer. This means that, although not all branches of statistics are covered, a wide variety of domains in theoretical and applied statistics are represented.

The reader will find many interesting ideas and many problems in Volume I. Some are treated through sophisticated mathematics, and some by sheer logic and a bit of arithmetic. The areas represented can be classified roughly as foundations, applications of statistics to important societal problems, and stochastic modeling in biological and medical studies.

Volume II collects papers in domains that include empirical measures, stochastic approximations procedures and sequential analysis, design of experiments, and multivariate analysis and asymptotics.

The Neyman-Pearson theory of testing hypotheses was born half a century ago. It set the foundations of what became known as the theory of statistical decision functions and has been the basis of most theoretical developments since then. Its present status is recorded by E. L. Lehmann in the first article. The nagging problems that arise from the choice of the reference system for probabilistic analysis or conditioning were attacked magistrally by Kiefer. The work of Kiefer is followed by J. O. Berger. There are other problems in that not everything can be described adequately in terms of risk functions, as we are reminded by C. Blyth. One of the visible phenomena of the past thirty years has been the resurgence of Bayesian analysis. Persi Diaconis tells us how to do it honestly.

With the advent of the computer age, data analysis also has seen a resurgence. Peter Huber describes the problems that arise if one tries to do both statistics and data analysis.

Leaving foundations, we have a fundamental essay by J. Tukey on how to randomize effectively and economically. The presentation gives special emphasis to randomization in weather modification experiments, a subject on which Neyman was still writing on the eve of his death. With this emphasis, the paper also serves as a transition to the next group of papers, which are concerned with problems of special societal importance.

Both Kiefer and Neyman were very concerned with such problems and with the necessity for statisticians to take stands on them. In this group, S. M. Dawkins and E. L. Scott continue their studies with Neyman on weather modification. R. S. Daggett and D. A. Freedman mix statistics with antitrust suits. T. A. Speed shows how unreliable the assessment of reliability of nuclear plants can be, and L. Breiman explains how committees decide things that cannot be decided. With C. Heyde we enter into stochastic modeling and its use in another of society's problems: population growth.

One of the longstanding interests of Neyman was the use of stochastic modeling and mathematical methods, especially in problems of biology and health. His work on such matters started in Poland and was barely interrupted by the time of his residence in England. He became deeply involved in models of carcinogenesis and models of the effects of radiation on cells. In fact, a few weeks before his death, he had organized a special two-day conference to try to bring together scientists with different areas of expertise in that difficult subject. The proceedings of that conference were published under the title *Probability Models and Cancer* (North-Holland, 1982). This explains the selection of topics in our next group of papers.

The paper by Karlin and Ost is not about cancer but about matching sequences of genes. This has to be done for many purposes. The methods introduced are invaluable.

Next, Olshen, Gilpin, Henning, LeWinter, Collins, and Ross apply several techniques to making twelve-month prognoses for patients who have suffered myocardial infarction. Following this, Fredkin, Montal, and Rice model acetylcholine receptors. Two papers, one by Baldwin and Byers and one by Dillman and Koziol describe the status of very promising techniques based on the use of monoclonal antibodies.

H. Rubin's paper on cell diversity is a bit lonely here. It was part of a session involving H. Smith, V. Ling, and M. Harris, who will publish elsewhere. It was also supposed to be accompanied by papers presenting more reductionist views based on the observations of oncogenes. However, the oncogene experts were meeting elsewhere.

For the effects of radiation on cells, B. Palcik describes what could and should be done to obtain information on the effects of low doses of radiation. Then we have a trilogy of papers by N. Albright, C. Tobias, and T. C. H. Yang about the stochastic models used to describe mechanisms of damage and repair in irradiated cells. As already mentioned, this was a subject of major interest to Neyman, who published his own models with P. S. Puri.

Besides the biological papers already mentioned, some other papers are missing. The conference included a special session on the spatial distribution of galaxies with the participation of astronomers M. Davis, S. White, and G. Abell. This was because of the long involvement of J. Neyman and E. L. Scott in that subject.

Unfortunately Professor George Abell did not get to finish his paper on "The Largest Structure in the Universe." He was working on it shortly before his death, also of a heart attack. To his family and friends we express our deepest regrets.

Volume II is somewhat more technical than Volume I in the mathematical sense of the word. It starts with a short appreciation of Kiefer's contributions, written by J. Sacks. This is followed by a group of papers on empirical cumulatives and empirical measures. Empirical cumulatives are a favorite tool of statisticians. A very substantial part of their theory is due to Kiefer, some jointly with J. Wolfowitz or A. Dvoretzky. R. Pyke gives us a survey of the present state of affairs. Empirical cumulatives are given by the empirical measure of certain sets such as quadrants. The theory for other classes of sets forming Vapnik-Červonenkis classes has made leaps and bounds in the last few years. The papers by K. Alexander,

R. M. Dudley, and D. Pollard record recent progress. C. Stone's
paper is about a related subject: the optimal selection of histograms.

From this we go to sequential subjects with a paper by H. Chernoff and J. Petkau on the sequential design of classical trials.
P. Groeneboom discusses the estimation of a monotone density; properties of certain jump processes generated by Brownian motion are key
tools.

A subject in which the names of pioneers are Robbins and Munro
and Kiefer and Wolfowtiz is that of stochastic approximation methods,
described here by T. L. Lai. Optimal shapes of stopping barriers
for sequential tests is the subject of two papers, one by R. Lerche
and one by D. Siegmund.

Kiefer's contribution to the present-day theory of design of
experiments is covered in papers by C. S. Cheng, Z. Galil, A. Gio-
vagnoli and H. Wynn, and by J. Sacks and Ylvisaker.

In multivariate analysis, M. Eaton handles random matrices.
Klonecki and Zontek look at the admissibility of estimates of variance components. Following a paper by P. S. Puri on nonidentifiabil-
ity in quality control situations, there is a group of papers classified under the general category of *asymptotics*. One branch of that
large field has to do with higher-order efficiencies. It is a bit
frightening to the uninitiated because of its reliance on enormous
formulas. P. Bickel, F. Götze, and W. van Zwet give us a simpler
view. J. Pfanzagl describes his own method of attack, using dif-
ferentiability with rates of convergence. J. K. Ghosh and P. K. Sen
attack problems in which the distributions are mixtures. Hartigan
deals with a failure of likelihood asymptotics for normal mixtures. I.
Basawa gives an account of developments around Neyman's $C(\alpha)$ tests and
related objects. C. Klaassen and W. van Zwet estimate score functions to see how well one can do in the presence of nuisance parameters. R. Davies deals with situations in which the second deriva-
tives of the log likelihood remain very random.

A paper by R. Beran and P. W. Millar ends Volume II with a solid
foothold in the future. Using resampling techniques, they avoid all
classical assumptions and still get high-quality confidence balls.

A substantial number of contributed papers were presented at the conference but not reproduced here. Their authors are deserving of many thanks for adding to the content and spirit of the conference.

Most of the work involved in collecting and processing the papers was done by Richard Olshen. He was helped by C. S. Cheng, R. M. Dudley, M. L. Eaton, J. A. Koziol, E. L. Scott, M. Woodroofe, and a bevy of anonymous referees. To all, we extend our sincerest thanks.

The conference was organized by Le Cam and Elizabeth L. Scott with the help and advice of an advisory committee appointed by the Institute of Mathematical Statistics. It consisted of the chairman, Mark Kac, Samuel Karlin, Frederick Mosteller, Richard Olshen, David Brillinger, and Lucien Le Cam. We regret that Mark Kac is no longer with us, but we still express our thanks to him and to the other members of his committee.

The conference was organized using the facilities of the University of California at Berkeley, whose administration was most cooperative. We wish to express our appreciation to its members and most particularly to Chancellor I. Michael Heyman. For the actual workings of the meeting, we drew on the goodwill and expertise of many members of our staff including Kate Caldwell, who ran the show, Sheila Gerber, Debora Haaxman, Sharon Howard, and Jo Dee Koller and Diana O'Reilly. Other tasks were performed by our students, especially Demissie Alemayehu, Yo-Lin Chang, Victor de la Pena, Ross Ihaka, and Pao Kuei Wu. Joan Pappas did excellent work handling manuscripts, keeping track of correspondence, and typing some revisions of papers. Their help is greatly appreciated.

Finally, we are grateful for the funds provided by the National Science Foundation, the Air Force Office of Scientific Research, the Army Research Office, the National Institutes of Health, the Office of Naval Research, the University of California Cancer Research Coordinating Committee, and the Faculty of the Department of Statistics. The *Proceedings* themselves were produced by Wadsworth Advanced Books & Software. We are very grateful for the infinite patience of Andrea Cava and John Kimmel.

Lucien M. Le Cam
Richard A. Olshen

CONTENTS OF VOLUME I

CONTENTS OF VOLUME II

PROCEEDINGS OF THE BERKELEY CONFERENCE IN HONOR OF JERZY NEYMAN AND JACK KIEFER

Volume I

THE NEYMAN-PEARSON THEORY AFTER FIFTY YEARS

E. L. LEHMANN   University of California, Berkeley

ABSTRACT: To commemorate the 50th anniversary of the Neyman-Pearson paradigm, this paper sketches some aspects of its development during the past half-century. In particular, the relevance of this approach to data analysis and Bayesian statistics is discussed.

1. INTRODUCTION

The Neyman-Pearson theory of hypothesis testing was developed by its two authors during the years 1926-1934.[1] However, its official birthday may be assigned to 1933, which saw the publication of the seminal paper "On the Problem of the Most Efficient Tests of Statistical Hypotheses." In this paper, the authors provided both the first complete statement of their new approach and some of its most important applications.

Research partially supported by National Science Foundation Grant MCS82-01498.

From *Proceedings of the Berkeley Conference in Honor of Jerzy Neyman and Jack Kiefer*, Volume I, Lucien M. Le Cam and Richard A. Olshen, eds., copyright © 1985 by Wadsworth, Inc. All rights reserved.

[1]Accounts of the collaboration can be found in Pearson's recollections (1966) and in the Neyman biography by Reid (1983).

The theory contains three principal ingredients:

   (i)   a *parametric model* specifying the assumed family of distribu-
         tions of the observable random variables;

   (ii) a *criterion* measuring the performance of a given statistical
         procedure, namely, the power function of the test in question;

   (iii) the advocacy of *optimality*: that the most desirable test is
         obtained by maximizing the power.

Despite the fact that these ideas had their forerunners in the
work of Fisher,[2] the Neyman-Pearson paradigm formulated for the first
time a clear program and provided a completely novel approach to
hypothesis testing, the first "exact" small-sample theory of its kind.

It is the purpose of this paper to commemorate the 50th anniversary of the NP theory by considering some aspects of its development
during the past half-century and to inquire into its relevance for
statistics today. A particular concern will be the contribution the
NP approach can make to data analysis and Bayesian statistics. All
three of these paradigms have had extensive developments, generalizations, and elaborations during the period in question.

Data analysis had its origin in descriptive statistics, which
dealt with the description and summary of data without making any
assumptions, in particular, without assuming an underlying stochastic
model. Under the leadership of Tukey (see, for example, his paper of
1962), many new techniques were developed and new tasks added such as
the discovery of underlying structure and of transformations to simplify such structure. Recent treatments of the subject can be found in
the books by Tukey (1977), Mosteller and Tukey (1977), and Hoaglin,
Mosteller, and Tukey (1983). For critiques, see, for example, Dempster
(1983) and Mallows (1983).

The Neyman-Pearson theory, which was originally concerned only
with hypothesis testing, was adapted by Neyman (1937) to deal also

---

[2]Particularly in his fundamental paper of 1922, which dealt with point
estimation but also emphasized parametric models and the fact that
maximum likelihood estimators minimize the asymptotic variance. Fisher
was preceded by Edgeworth.

with estimation by confidence sets and was generalized by Wald to his general decision theory, in which the power function was replaced by a general risk function, new optimality criteria were introduced, and the restriction to parametric models was jettisoned (although in Wald's final exposition of 1950 the examples were all parametric).

Bayesian statistics had come to a first flowering in the work of Laplace. It had been pushed aside in deliberate efforts by Fisher, and Neyman and Pearson, to build an "objective" theory and was reawakened by L. J. Savage[3] (particularly in his 1954 book) and earlier, from a somewhat different point of view, by Jeffreys (1939). One of its principal features is the assumption that the parameters of the parametric model should themselves be viewed as random variables whose distribution is assumed to be known.

The needs that lie behind the development of data analysis and Bayesian statistics had their impact also on the NP theory. In particular, the desire to free statistical analysis from the stringent and unreliable assumptions of parametric models moved the NP theory in the direction of data analysis, and the need to incorporate prior information and subjective impressions brought it closer to Bayesian statistics. The two trends will be sketched in Sections 2-4.

2. ROBUSTNESS AND ADAPTATION

A typical example of the classical NP approach is the normal two-sample problem. If one assumes that $X_1, \ldots, X_m$ and $Y_1, \ldots, Y_n$ are independent with normal distributions

$$X_1, \ldots, X_m : N(\xi, \sigma^2)$$
$$Y_1, \ldots, Y_n : N(\eta, \sigma^2),$$

(2.1)

then the optimal test of the hypothesis $\eta = \xi$ against the one-sided alternatives $\eta > \xi$ under a variety of optimality criteria (UMP

---

[3]Savage acknowledged the profound influence on his thinking of de Finetti, as presented, for example, in de Finetti (1937).

unbiased, UMP invariant, most stringent, and so forth) is the two-sample t-test.

However, the validity of this model is frequently doubtful: the distribution of the errors may not be normal, the two variances not be equal, the observations not be independent. We shall here consider only the effect of non-normality, and hence the model

$$X_1, \ldots, X_m : F(x - \xi)$$
$$Y_1, \ldots, Y_n : F(y - \eta),$$

(2.2)

where F is unknown. Before embarking on this program, we should mention a type of model that is intermediate between (2.1) and (2.2). In this formulation, it is assumed that the distribution F in (2.2) is approximately normal, for example, that

$$F(x) = (1 - \varepsilon)\Phi(x) + \varepsilon H(x),$$

(2.3)

where ε is given and H is an arbitrary distribution function. Such models, which have been studied by Huber [see, for example, Huber (1981)], will not be considered in the present paper.

The first question that was asked about the robustness of the t-test concerned the actual level $\alpha_{m,n}(F)$ of the t-test carried out at nominal level α when the model is given by (2.2). This question was raised quite early; for a review of some of the relevant literature, see Tan (1982). It immediately follows from the central limit theorem and Slutsky's theorem that

$$\alpha_{m,n}(F) \to \alpha \quad \text{as } m, n \to \infty$$

(2.4)

for any F with finite second moment. The same type of argument shows further that, for large m and n, not only the level but also the power of the t-test is approximately equal to the power predicted under the normal model.

If the value of the significance level is taken very seriously, the approximate nature of the level under (2.2) may be considered inadequate. The difficulty can be overcome by a modification of the t-test, the *permutation* (or *randomization*) t test due to R. A. Fisher (1935). In this test the t-statistic is compared not with all the values it takes on as the X's and Y's range over the sample space but only with the values obtained by permuting the $m + n$ observed values $(x_1, \ldots, x_m, y_1, \ldots, y_n)$. The resulting test has a level that is independent of F in the model (2.2). In addition, it has asymptotically the same power as the t-test (Hoeffding, 1952). It thus appears to be an ideal solution of the robustness problem.

Unfortunately, this conclusion overlooks the fact that, if the normal model is wrong, the t-test loses its optimality properties and may actually not be a particularly desirable test to use. The distinction is that between robustness of validity (or, more generally, performance robustness) and robustness of efficiency, introduced by Tukey and McLaughlin (1963). It is closely related to the concepts of criterion and inference robustness proposed by Box and Tiao (1964). The t-test and its permutation version are performance-robust in that their performance is approximately that asserted by the normal theory even under model (2.2). However, since the efficiency of these tests does not continue to hold even approximately in (2.2) when F is not normal, performance robustness is not as strong a recommendation as appears at first sight.

There exist, in fact, tests (based on the rank of the observations) whose significance level is independent of F and whose power for many realistic distributions F is much higher than that of the t-test. A particularly striking example is the normal-scores test N, whose asymptotic relative efficiency (ARE) $e_{N, t}(F)$ with respect to the t-test satisfies (Chernoff and Savage, 1958)

$$e_{N, t}(F) \geqslant 1 \quad \text{for all } F, \tag{2.5}$$

with equality holding only when F is normal, and with $e_{N, t}(F)$ becoming arbitrarily large when the tail of F is sufficiently heavy. The test N has the asymptotic optimality property of maximizing $\inf e_{N, t}(F)$ in model (2.2).

In this solution to the robustness problem the original model (2.1) still plays a central (though implicit) role because it sets up the t-test as the standard with which other tests are being compared. If we want to treat (2.2) without any special relevance to the normal case, a more natural efficiency measure of a test is its *absolute*

efficiency $e(F)$, that is, its efficiency relative to the best parametric test of $H : \Delta = \eta - \xi = 0$ when F is known (for example, the UMP
invariant test). Stein (1956) suggested the existence of a test that
would satisfy

$$e(F) = 1 \tag{2.6}$$

for all sufficiently smooth F. The program of working out such *adaptive* procedures (which adapt themselves to the unknown true F) has
been carried out for a number of problems. For the estimation of the
center of a symmetric distribution on the basis of a sample X_1, ...,
X_n, for example, fully efficient adaptive estimators were obtained by
Beran (1974) and Stone (1975). The latter shows that the small-sample
efficiency of his estimator is of the order of .9 for a variety of
distributions of quite different shapes (such as normal, Cauchy, and
double-exponential) for sample sizes as small as $n = 40$. The conditions under which adaptation is possible have been investigated by
Bickel (1982).

3. RELATION WITH DATA ANALYSIS

For removing the assumption of normality from model (2.1), adaptive
tests and estimators provide the ultimate answer. However, they seem
to be far from the methods and concerns of data analysis as an
approach to statistics without any modeling. In fact, they are closer
than appears at first sight.

Adaptive procedures begin by estimating the unknown distribution
F or its density f; they then apply the procedure that would be best
if the estimated f were the true one. However, the problem of estimating an unknown density f is quite close to that of describing a
data set by means, say, of a histogram. Exactly the same issues arise
in both (interval width, smoothing, etc.); see, for example, Scott
(1979), Freedman and Diaconis (1981), and Hoaglin, Mosteller, and
Tukey (1983).

There is, of course, a difference between the two methods of attack. The modeling approach can measure and compare the performance of different procedures; it can ask how close the estimated density is likely to be to the true one and obtain the interval width that provides the best convergence rate. However, this requires making some assumptions: for example, that the observations are n independent realizations of a random variable whose distribution has a smooth density.

The Neyman-Pearson program as represented by (ii) and (iii) of the Introduction can be applied not only to data description but also to other aspects of data analysis. We shall briefly consider its contribution to two of those: data summaries and exploration of underlying structure.

To summarize a simple one-dimensional batch of data, one calculates statistics that indicate, for example, the center and the spread of the batch. If one is willing to assume that the observations X_1, ..., X_n are a sample from a distribution F, one will be concerned instead with measures $\mu(F)$ of the center and spread of F (or of skewness, Kurtosis, etc.), and the summary statistics become estimators of the unknown $\mu(F)$.[4] Some of the concerns about such summaries can be shared by the two approaches; resistance to outliers is an important example. However, the assumption of an underlying probability model provides new possibilities. In particular, one can now again measure and compare the performance of different estimators. The choice of suitable measures $\mu(F)$ (and their estimators) from this point of view is discussed in Bickel and Lehmann (1975/1976).

Methods for discovering underlying structure (including the choice of transformations to simplify the structure) can again be strengthened by making some assumptions. As an example, we mention recent work concerned with model selection that assumes a class of possible models of varying complexity; see, for example, Stone (1981).

---

[4]Typically $\mu(F)$ is estimated by $\mu(F_n)$, the functional μ evaluated at the empirical distribution function F_n.

The data are used to choose a suitable model within this class.   An interesting new feature of this application of the Neyman-Pearson-Wald approach is that the loss function must include a component representing the complexity of the model.

It thus appears that many of the problems considered in data analysis can also be studied from the NPW point of view.   The latter will always require some (although possibly rather weak) assumptions, and this can never quite match the freedom of a totally unrestrained examination of the data.   On the other hand, it has the advantage of allowing evaluation and comparison of different methods.   In addition, it may permit the extrapolation of findings to other data sets, whereas a purely data-analytic approach is bound to the data set under consideration.

4. RELATION WITH BAYESIAN INFERENCE

The move toward data analysis has as its aim a weakening of assumptions that are frequently over-detailed and unrealistic.   In the other direction, the Bayesian approach tries to take account of additional information, not represented in the classical models, that affects our expectations regarding the parameter values likely to obtain.   This is achieved by postulating that the parameters themselves are random variables (though unobservable) with a known probability distribution.

Non-Bayesians and most Bayesians appear to be in agreement on two basic facts:

(i)   The procedures one wants to use are Bayes' procedures with respect to some prior distribution (or at least limits of such procedures).   For Bayesians this follows from first principles. From a decision theoretic point of view it is a consequence of the fact that only Bayes' procedures and their limits are admissible.

(ii) Obtaining the appropriate prior distribution is usually a difficult task.   Different Bayesian ideas of how to go about this are represented, for example, by the books of Raiffa and Schlaifer (1961), Box and Tiao (1964), and Lindley (1965).   In the present section, we shall briefly discuss the contribution of decision theory to this problem.

Since a large-sample approach often brings great simplification, let us begin by considering the asymptotic situation of a large number of i.i.d. random variables having density $f(x, \theta)$, where θ is real-valued and has prior density $\pi(\theta)$ with $\pi(\theta) > 0$ for all θ and $E|\theta| < \infty$. Then, under suitable regularity conditions on $f(x, \theta)$, the Bayes' estimator $\delta_{\pi,n}$ of θ with respect to squared error loss satisfies[5]

$$\sqrt{n}(\delta_{\pi,n} - \theta) \to N(0, 1/I(\theta)) \quad \text{in law,} \tag{4.1}$$

where $I(\theta)$ is the amount of information each X_i contains about θ. See, for example, Lehmann (1983), Section 6.7.

Thus, to the accuracy of this approximation, the Bayes' procedures share the asymptotic optimality of maximum likelihood estimators: they are efficient. However, it is a consequence of this result that the theory cannot distinguish between them. In fact, the Bayes' estimators corresponding to two different prior densities π and π' satisfy

$$\sqrt{n}(\delta_{\pi,n} - \delta_{\pi',n}) \to 0 \quad \text{in probability.} \tag{4.2}$$

Thus, large-sample theory does indeed greatly simplify the situation. Unfortunately it simplifies it too much.

A crucial assumption in these considerations is that $\pi(\theta) > 0$ for all θ, because a Bayes' procedure pays no attention to parameter values that are impossible a priori. This suggests that optimal risk can be attained most rapidly (in a minimax sense) if π is not too uneven, i.e., if there are no θ-values for which $\pi(\theta)$ is relatively very low. Actually, such an argument should also take into account the amount of information the sample contains about different values of θ, which then points to an uninformative prior, and results in an approach that some authors call *objective* Bayesian. Furthermore, in problems in which invariance considerations are applicable, the decision theoretic invariance principle leads directly, even for small

---

[5]However, see Diaconis and Freedman (1983) and Freedman and Diaconis (1983).

samples, to the Bayes' solution corresponding to an invariant (and hence noninformative) prior.

If, based on previous experience and perhaps other less tangible and more subjective impressions, one has a preferred prior distribution, decision theory suggests an alternative approach. Compare the behavior of procedures corresponding to a number of representative priors, including the prior of one's choice, and adopt the latter if its performance is satisfactory. Essentially the same strategy is recommended in a book, written from a Bayesian point of view, by J. Berger (1980). In the contrary case, one may want to tone down one's prior opinions in the direction of a noninformative prior.

A compromise of this kind is offered by the theory of restricted Bayes solutions. Suppose one has arrived at a preferred prior Λ, which leads to very high values of the risk function for some parameter values and about which one does not feel very confident. Consider for a moment the minimax solution as a competitor that has a much smaller maximum risk M but which corresponds to a prior very different from Λ. As a compromise between these alternatives, let us limit the allowable maximum risk not to its minimum value M but to $M(1 + \varepsilon)$, so that the risk function is required to satisfy

$$R(\theta, \delta) \leqslant M(1 + \varepsilon) \quad \text{for all } \theta. \tag{4.3}$$

The *restricted Bayes' procedure* $\delta_{\Lambda, \varepsilon}$ is then obtained by minimizing the Bayes' risk $\int R(\theta, \delta)d\Lambda(\theta)$ subject to (4.3); see Hodges and Lehmann (1953). The choice of ε in (4.3) will reflect both the confidence one places on Λ and the cost of excessively high risk. Applications of this approach have been developed by Efron and Morris (1971), Bickel (1983), and Jelihovschi (1984).

5. CONCLUSIONS AND ACKNOWLEDGMENTS

The Neyman-Pearson theory makes two principal recommendations: Evaluate the behavior of statistical procedures in terms of some performance criterion, and then opt for the procedure that performs best. The application of these recommendations requires a probabilistic model for the observations, although such models can be much broader than those originally envisaged by Neyman and Pearson.

As has been indicated in the preceding sections, the first of these recommendations can be usefully applied to a wide spectrum of statistical techniques ranging from procedures inspired by data analysis to those obtained from Bayesian principles. It is difficult to imagine statistics as a subject in which this idea would not play a central role.

Complete reliance on optimality, on the other hand, suffers from some drawbacks. In particular,

(i)   procedures that minimize the risk may be quite unsatisfactory in other respects such as robustness or interpretability;

(ii) most optimality criteria lead to solutions only in fairly simple problems.

As a consequence, an optimal solution (if it exists) should not be considered the ultimate answer, but its performance should be studied, just as that of any other proposed procedure, from all relevant points of view. On the other hand, a procedure that is best in some way is of value as a benchmark with which competing procedures can be compared. It shows how well one can do in this particular direction and how much is lost when other criteria have to be taken into account.

It is suggested in Sections 3 and 4 that the procedures obtained from a decision theoretic point of view may not differ as much from those reached via a data analytic or Bayesian philosophy as is frequently claimed and that all three ways of treating problems can have contributions to make. This point has been made by other authors.

The relationship between decision theory and data analysis is explored by Thisted (1981) and in the book by Hoaglin, Mosteller, and Tukey (1983). Regarding the uses of decision theory in a Bayesian setting, the work of Berger (1980) has already been mentioned. A different combination of Bayesian and Neyman-Pearson theory has been recommended by Box (1980), who proposes a Bayesian approach to the estimation of the parameters of the model and a test based on sampling considerations for testing the model.

Finally, it was shown how all three lines of attack can be used in a substantive study by Mosteller and Wallace (1964), and a plea for "ecumenism" in statistics is put forth by Box (1983).

REFERENCES

Andrews, D. F. (1978). Article on "Data Analysis, Exploratory." In *Internat. Encycl. of Statist.* Kruskal and Tanur, eds. Free Press, New York.

Beran, R. (1974). Asymptotically efficient adaptive rank estimates in location models. *Ann. Statist. 2*, 63–74.

Berger, J. (1980). *Statistical Decision Theory.* Springer, New York.

Bickel, P. J. (1981). Minimax estimation of the mean of a normal distribution when the parameter space is restricted. *Ann. Statist. 9*, 1301–1309.

Bickel, P. J. (1982). On adaptive estimation. *Ann. Statist. 10*, 647–671.

Bickel, P. J. (1984). Parametric robustness. *Ann. Statist. 12.*

Bickel, P. J., and Lehmann, E. L. (1975/1976). Descriptive statistics for nonparametric models. *Ann. Statist. 3*, 1038–1069; *4*, 1139–1158.

Box, G. E. P. (1980). Sampling and Bayes' inference in scientific modelling and robustness. *J. Roy. Statist. Soc.* (A) *143*, 383–430.

Box, G. E. P. (1983). An apology for ecumenism in statistics. In *Scientific Inference, Data Analysis, and Robustness.* Box, Leonard, and Wu, eds. John Wiley, New York.

Box, G. E. P., and Tiao, G. C. (1964). A note on criterion robustness and inference robustness. *Biometrika 51*, 169–173.

Box, G. E. P., and Tiao, G. C. (1973). *Bayesian Inference in Statistical Analysis.* Addison-Wesley, Reading.

Chernoff, H., and Savage, I. R. (1958). Asymptotic normality and efficiency of some nonparametric competitors of the t-test. *Ann. Math. Statist. 29*, 972–994.

de Finetti, B. (1937). La prévision: ses lois logiques, ses sources sujectives. *Ann. de l'Inst. Henri Poincaré 7*, 1–68. [Trans. in Kyburg, H., and Smokler, H. (1964), *Studies in Subjective Probability*, John Wiley, New York.]

Dempster, A. P. (1983). Purposes and limitations of data Analysis. In *Scientific Inference, Data Analysis, and Robustness*. Box, Leonard, and Wu, eds. John Wiley, New York.

Diaconis, P., and Freedman, D. (1983). Frequency properties of Bayes rules. In *Scientific Inference, Data Analysis, and Robustness*. Box, Leonard, and Wu, eds. John Wiley, New York.

Efron, B., and Morris, C. (1973). Stein's estimation rule and its competitors—an empirical Bayes approach. *J. Amer. Statist. Assoc. 68*, 117-130.

Fisher, R. A. (1922). On the mathematical foundations of theoretical statistics. *Phil. Trans. Roy. Soc.* (A) *222*, 309-368.

Fisher, R. A. (1935). *The Design of Experiments*. Oliver and Boyd, Edinburgh.

Freedman, D., and Diaconis, P. (1981). On the histogram as a density estimator: L_2 theory. *Zeitsch. Wahrsch. 57*, 453-476.

Freedman, D., and Diaconis, P. (1983). On inconsistent Bayes estimates in the discrete case. *Ann. Statist. 11*, 1109-1118.

Hoaglin, D. C., Mosteller, F., and Tukey, J. W. (1983). *Understanding Robust and Exploratory Data Analysis*. John Wiley, New York.

Hodges, J. L., Jr., and Lehmann, E. L. (1953). The use of previous experience in reaching statistical decisions. *Ann. Math. Statist. 23*, 396-407.

Hoeffding, W. (1952). The large-sample power of tests based on permutations of observations. *Ann. Math. Statist. 23*, 169-192.

Huber, P. J. (1981). *Robust Statistics*. John Wiley, New York.

Jeffreys, H. (1939). *The Theory of Probability*. Oxford University Press, Oxford.

Jelihovschi, E. (1984). Estimation of Poisson parameters, subject to constraint. Ph.D. dissertation, University of California, Berkeley.

Lehmann, E. L. (1983). *Theory of Point Estimation*. John Wiley, New York.

Lindley, D. (1965). *Probability and Statistics*. Vol. 2: *Inference*. Cambridge University Press, Cambridge.

Mallows, C. L. (1983). Data description. In *Scientific Inference, Data Analysis, and Robustness*. Box, Leonard, and Wu, eds. John Wiley, New York.

Mosteller, F., and Tukey, J. W. (1977). *Data Analysis and Regression*. John Wiley, New York.

Mosteller, F., and Wallace, D. (1964). *Inference and Disputed Authorship: The Federalist*. Addison-Wesley, Reading.

Neyman, J. (1937). Outline of a theory of statistical estimation based on the classical theory of probability. *Phil. Trans. Roy. Soc.* (A) *236*, 333-380.

Neyman, J., and Pearson, E. S. (1933). On the problem of the most efficient tests of statistical hypotheses. *Phil. Trans. Roy. Soc.* (A) *231*, 289-337.

Pearson, E. S. (1966). The Neyman-Pearson story: 1926-1934. In *Research Papers in Statistics: Festschrift for J. Neyman.* F. N. David, ed. John Wiley, New York.

Raiffa, H., and Schlaifer, R. (1961). *Applied Statistical Decision Theory.* Harvard University Pres, Cambridge.

Reid, C. (1982). *Neyman—From Life.* Springer, New York.

Savage, L. J. (1954). *Foundations of Statistics.* John Wiley, New York. 2nd revised ed., Dover (1972).

Scott, D. W. (1979). On optimal and data-based histograms. *Biometrika 66*, 605-610.

Stein, C. (1956). Efficient nonparametric testing and estimation. *Proc. Third Berkeley Symp. Math. Statist. and Prob. 1*, 187-196.

Stone, C. J. (1975). Adaptive maximum likelihood estimators of a location parameter. *Ann. Statist. 3*, 267-284.

Stone, C. J. (1981). Admissible selection of an accurate and parsimonious normal linear regression model. *Ann. Statist. 9*, 475-485.

Student [W. S. Gosset] (1908). On the probable error of a mean. *Biometrika 6*, 1-25.

Tan, W. Y. (1982). Sampling distributions and robustness of t, F and variance-ratio in two samples and ANOVA models with respect to departure from normality. *Communications in Statistics, Theory and Methods 11* (No. 22), 2485-2511.

Thisbed, R. (1982). Decision theoretic regression diagnostics. In *Statistical Decision Theory and Related Topics.* Gupta and Berger, eds. Academic Press, New York. Vol. II, pp. 363-382.

Tukey, J. W., and McLaughlin, D. H. (1963). Less vulnerable confidence and significance procedures for location based on a single sample: Trimming/Winsorization 1. *Sankhya 25*, 331-352.

Tukey, J. W. (1977). *Exploratory Data Analysis.* Addison-Wesley, Reading.

Wald, A. (1950). *Statistical Decision Functions.* John Wiley, New York.

THE FREQUENTIST VIEWPOINT AND CONDITIONING

JAMES BERGER Purdue University

ABSTRACT: The relationship of the frequentist viewpoint and the conditional viewpoints in statistics is discussed. After an attempt to define the frequentist viewpoint, it is shown that Kiefer's conditional and estimated confidence theories, particularly the latter, are very successful in dealing with many of the conditional difficulties of standard frequentist theory, while maintaining frequentist validity. Not all conflicts with the conditional viewpoint can be avoided, however, as shown by the likelihood principle and the stopping rule principle. These principles and the conflicts are briefly reviewed. Also, a justification for the stopping rule principle is given in terms of frequentist admissibility.

1. INTRODUCTION

This paper is based on a talk given in the memorial session for Jack Kiefer at the Berkeley Conference in honor of Keifer and Neyman. The bulk of the talk was devoted to a review of Kiefer's work on conditional confidence and estimated confidence. The talk also touched

Research supported by the National Science Foundation under Grant MCS 81-01670.

on broader issues, such as what it means to be a frequentist and what
the basic issues in conditioning are. This paper will mainly be
devoted to a discussion of these broader issues, partly because a
review of Kiefer's work on the subject will appear in the volume of
his collected works, partly because Kiefer's work can best be appreciated in a general discussion of the issues, and partly because the
broader issues themselves deserve more exposure than they are commonly accorded.

Section 2 of the paper begins with a discussion of what it means
to be a frequentist, based on the original views of the developer of
the frequentist school, Jerzy Neyman. This issue deserves attention
because Neyman's original justification for the frequentist position
(which strikes us as the best justification that has been given) is
not the justification most commonly taught. Furthermore, the issue
is important in determining the type of conditional frequentist
theory that is most justifiable. Kiefer and Brownie discussed, in a
series of papers [Kiefer (1975, 1976, 1977) and Brownie and Kiefer
(1977)], two possible approaches to such a conditional frequentist
theory, the *conditional confidence* and *estimated confidence*
approaches. These are briefly reviewed in Section 2.2, followed, in
Section 2.3, by a discussion of the very interesting problem of
selection of a conditional frequentist procedure.

Section 3 turns to a discussion of conditioning from a more
philosophical viewpoint, concentrating on such issues as the likelihood principle and the stopping rule principle. Formal developments
are eschewed in preference for a more intuitive presentation of the
issues and discussion of conflicts and paradoxes. For instance, a
little known paradox, between the frequentist notion of admissibility and the idea that decisions should depend on the stopping rule,
is discussed. Section 4 presents some conclusions.

This paper is in no sense meant to be a thorough review of any
of the topics discussed. Instead, it is intended to serve as an
introduction to a number of interesting and too often ignored issues.
Extensive references are not given, and indeed no attempt has been
made to trace ideas to sources. More scholarly reviews of and

references for these topics can be found in (for example) Kiefer (1977), Berger and Wolpert (1984), and Berger (1984a).

Before proceeding, some notation will be introduced. Also, a series of simple examples will be presented to give a feeling for the conditioning issues and to provide a background for later discussion.

It will be assumed that an "experiment" E is performed, which consists of observing a random quantity X having distribution P_θ on a sample space $\mathfrak{X}$, $\theta \in \Theta$ being unknown. (For the most part θ will be taken to be a parameter, so that $\{P_\theta\}$ is a parametric family, but this need not necessarily be the case; θ could just index some non-parametric family.) When $\{P_\theta\}$ is a dominated family with respect to a measure ν, we shall denote the density of X by

$$f(x|\theta) = dP_\theta(x)/d\nu(x). \tag{1.1}$$

Expectation over the distribution P_θ will be denoted by E_θ. Finally, the actual data from the experiment (i.e., the realization of X) will be denoted by x.

The following examples are well-known illustrations of the conditioning problem. For historical references and other examples of conditioning problems see Kiefer (1977), Berger (1984a), and Berger and Wolpert (1984).

Example 1: Suppose $X = (X_1, X_2)$, where X_1 and X_2 are independently distributed according to the distribution

$$P_\theta(X_i = \theta - 1) = P_\theta(X_i = \theta + 1) = \tfrac{1}{2},$$

where $-\infty < \theta < \infty$. Consider the confidence procedure defined by

$$C(x) = \begin{cases} \text{the point } \tfrac{1}{2}(x_1 + x_2) & \text{if } |x_1 - x_2| = 2 \\ \text{the point } x_1 - 1 & \text{if } |x_1 - x_2| = 0 \end{cases}.$$

Since (by an easy calculation)

$$P_\theta(C(X) \text{ contains } \theta) = .75 \text{ for all } \theta,$$

the confidence procedure C is a valid 75% frequentist confidence procedure and, furthermore, satisfies any number of frequentist

optimality properties. It is clearly misleading, however, to present
$C(x)$ and state "75% confidence" *after* seeing the data x, since if
$|x_1 - x_2| = 2$, one is absolutely certain that $\theta \in C(x)$, whereas if
$|x_1 - x_2| = 0$, one is (more or less) equally uncertain as to whether
θ is $x_1 - 1$ or $x_1 + 1$. Conditional on the data x, one should state
either 100% or 50% "confidence," depending on the value of $|x_1 - x_2|$.

Example 2a: Suppose X is 1, 2, or 3 and θ is 1 or 2, with $P_\theta(x)$ given
in the following table:

		X	
	1	2	3
P_0	.009	.001	.99
P_1	.001	.989	.01

The test, which accepts P_0 when $x = 3$ and accepts P_1 otherwise, is a
most powerful test with *both* error probabilities equal to .01. Hence,
it would be valid to make the frequentist statement, upon observing
$x = 1$: "My test has rejected P_0, and the error probability is .01."
Again this seems misleading, because the likelihood ratio is actually
9 to 1 in favor of P_0, which is being rejected.

Example 2b: One could object, in Example 2a, that the .01 level test
is inappropriate and that one should use the .001 level test, which
rejects only when $x = 2$. Consider, however, the following slightly
changed version:

		X	
	1	2	3
P_0	.005	.005	.99
P_1	.0051	.9849	.01

Again the test that rejects P_0 when $x = 1$ or 2 and accepts otherwise,
has error probabilities equal to .01, and now it indeed seems sensible

to take the indicated actions.   (Suppose an action *must* be taken.)
It still seems unreasonable, however, to report an error probability
of .01 upon rejecting P_0 when $x = 1$, because the data provide very
little evidence in favor of P_1.

Example 3a: Suppose X is $\eta(\theta, 1)$ and that it is desired to test

$\quad$ $H_0: \theta \leqslant -2$ versus $H_a: \theta \geqslant 2$.

Consider the test: reject H_0 if $x \geqslant 0$.   Clearly, for $\theta < 0$,

$$P_\theta(\text{Type I error}) = P_{-\theta}(\text{Type II error}) \leqslant P_{-2}(X \geqslant 0) = .0228.$$

If $x = 0$ is observed, however, it seems misleading to state that "H_0
is rejected, and the error probability is at most .0228."

Example 3b: Suppose X is $\eta(\theta, 1)$ and that it is desired to test

$\quad$ $H_0: \theta \leqslant 0$ versus $H_a: \theta > 0$.

Consider the test: reject H_0 if $x \geqslant 0$.   Clearly,

$$P_{\theta=0}(\text{Type I error}) = \tfrac{1}{2}.$$

If $x = 10$ is observed, we are virtually certain that H_a is true, and
yet (via formal frequentist theory) all we can say is that we reject
with error probability of $\tfrac{1}{2}$.

2.   THE FREQUENTIST APPROACH TO CONDITIONING

Before discussing how frequentists can deal with (at least many)
conditioning problems, it is necessary to define what a valid
frequentist approach is.

2.1 Frequentist Rationale

Until recently, I thought a frequentist was someone with the following approach:

(i) Select a procedure $\delta(x)$ for use.

(ii) Define a criterion (or loss) $L(\theta, \delta)$ that one would like to
know or that measures performance.

(iii) Report $(\delta, R_\delta(\theta))$, where

$$R_\delta(\theta) = E_\theta L(\theta, \delta(X)).$$

Example 4: For confidence set problems, $\delta(x) = C(x) \subset \Theta$ defines a confidence procedure,

$$L(\theta, C(x)) = 1 - I_{C(x)}(\theta)$$

$[I_A(\theta)$ denoting the usual indicator function] is what one would like to know and is the usual measure of performance, and

$$R_C(\theta) = E_\theta L(\theta, C(X)) = 1 - P_\theta(C(X) \text{ contains } \theta). \qquad (2.1)$$

Example 5: In testing, let δ denote a test and L be zero-one loss, so that

$$R_\delta(\theta) = P_\theta(\text{incorrect decision}).$$

Of course, decision-theoretic examples are plentiful.

The usual justification for the frequentist viewpoint outlined above is that if, for a given θ_0, one were to repeatedly use δ_0 on (independent) $X_i \sim P_{\theta_0}$, then (with probability one under mild conditions)

$$\lim_{n \to \infty} \frac{1}{n} \sum_{i=1}^{n} L(\theta_0, \delta(X_i)) = R_\delta(\theta_0). \qquad (2.2)$$

Thus, by reporting $R_\delta(\theta)$, one is stating the long-run performance of δ for that θ.

This leads, however, to two immediate questions:

(i) One does not know θ; so how is $R_\delta(\theta)$ to be used?

(ii) In reality, one will be using δ with different problems having different θ_i; so what is the value of knowing (2.2)?

These questions are perplexing, and in an effort to understand the frequentist answers to them, I returned to Neyman's original papers proposing the frequentist method. [See *A Selection of Early Statistical Papers of J. Neyman*; also relevant is Neyman (1957).] To my surprise, a different and more appealing frequentist viewpoint emerged from these papers, a viewpoint based on use of δ in different problems with different θ_i. In the notation of this paper, this viewpoint can be expressed as follows.

First, consider an infinite sequence of problems in which $X_i \sim P_{\theta_i}$ is observed. Let $\boldsymbol{\theta} = (\theta_1, \theta_2, \ldots)$. Suppose certain sub-sequences $\boldsymbol{\theta}_\omega$, $\omega = (\omega(1), \omega(2) \ldots) \in \{1, 2, \ldots\}^\infty$, are of interest.

Definition 1: A quantity $\bar{R}_\delta^\omega$ will be called a valid frequentist measure of the performance of δ on $\boldsymbol{\theta}_\omega$ if, with probability one,

$$\lim_{n \to \infty} \sup \frac{1}{n} \sum_{i=1}^{n} L(\theta_{\omega(i)}, \delta(X_{\omega(i)})) \leqslant \bar{R}_\delta^\omega. \qquad (2.3)$$

The idea here is that one will report $(\delta, \bar{R}_\delta^\omega)$ and can be assured that, in repeated use of δ for $\theta_i \in \boldsymbol{\theta}_\omega$, the average long-run performance will be at least $\bar{R}_\delta^\omega$.

Often, as in typical estimation or confidence set problems, one will be interested only in $\omega = (1, 2, \ldots)$, that is, one will want to find a quantity R_δ such that, with probability one for all $\boldsymbol{\theta}$,

$$\lim_{n \to \infty} \sup \frac{1}{n} \sum_{i=1}^{n} L(\theta_i, \delta(X_i)) \leqslant \bar{R}_\delta. \qquad (2.4)$$

Indeed, it can be argued that this is the only situation in which a truly meaningful frequentist statement is being made, in that, without knowledge of the θ_i (or ω) that are likely to be encountered, a reporting of $\bar{R}_\delta^\omega$ does not completely convey the performance that is to be expected.

The typical situations in which subsequences are of interest are testing problems, where $\boldsymbol{\theta}_{\omega_0}$ and $\boldsymbol{\theta}_{\omega_1}$ may refer, say, to sequences of null and alternative hypotheses, respectively. Then, for zero-one loss, $\bar{R}^{\omega_0}$ and $\bar{R}^{\omega_1}$ would be bounds on the probabilities of Type I and Type II error, respectively. Of course, if $\boldsymbol{\theta}_\omega$ refers to a subsequence for which all θ_i equal a common value θ, then we are back in the situation described at the beginning of the section, in that setting $\bar{R}_\delta^\omega = R_\delta(\theta)$ will suffice. The point, however, is that only subsequences that are of separate interest and that can be expected to occur should be considered, since the reporting of $\bar{R}_\delta^\omega$ is an attempt to model the actual real-world performance of δ on a variety of problems. Thus, if C is a confidence procedure, one would like to say that, in repeated actual use, C will fail to contain θ_i no more than $100 \times \bar{R}_C\%$ of the time; whereas if δ is a test, no more than $100 \times \bar{R}_\delta^{\omega_0}\%$ of true null hypotheses will be rejected. One can, of course, imagine the thought experiment consisting of repeated use of δ for the same θ, but Neyman explicitly created the frequentist theory precisely to eliminate the dependence of statistics on prior beliefs or supposed structure about the θ that would occur. It is repeatedly stressed in his papers that the measure $\bar{R}_\delta^\omega$ will be valid for any θ_ω, and that this is the breakthrough provided by frequentist theory.

There are a number of reasons why frequentist theory came to be perceived as simply reporting $(\delta, R_\delta(\theta))$. In the first place, for testing problems in which Θ consists only of two points, the null hypothesis θ_0 and the alternative hypothesis θ_1, then the ω_0 and ω_1 referred to above could be considered sequences of identical parameters, with $\bar{R}_\delta^{\omega_0} = R_\delta(\theta_0)$ and $\bar{R}_\delta^{\omega_1} = R_\delta(\theta_1)$ corresponding to the actual probabilities of Type I and Type II error. A natural generalization to problems with more complicated Θ would be to consider $R_\delta(\theta)$ in general, forgetting the original motivation. A second reason for the introduction of $R_\delta(\theta)$ was that $\bar{R}_\delta^\omega$ is often most easily obtained by finding an upper bound on $R_\delta(\theta)$, for the type of subsequence $\{\theta_{\omega_{(i)}}\}$ of interest. It is easiest to provide $R_\delta(\theta)$ and let the user of δ infer the appropriate $\bar{R}_\delta^\omega$. A third natural reason for consideration of $R_\delta(\theta)$ is for comparison of two procedures δ_1 and δ_2. If $R_{\delta_1}(\theta) < R_{\delta_2}(\theta)$, it will almost invariably follow that δ_1 has better performance than δ_2 in actual long-run use. Similarly, for questions of experimental design, $R_\delta(\theta)$ is a basic quantity of interest. It is indeed not surprising that "report $(\delta, R_\delta(\theta))$" came to be perceived as the frequentist viewpoint. Ultimately, however, it is (2.3), or even better (2.4), which seems to provide the justification for the frequentist viewpoint, and so it is Definition 1 that we shall take as basic.

One final historical matter: It is probably not completely clear who first espoused the frequentist viewpoint, although there is no doubt that it was Neyman who provided the first clear general formalizations. The resolution of this historical matter is

complicated by many red herrings, such as significance tests, which
are hundreds of years old and can be given a frequentist interpretation. Indeed, a significance test of the null hypothesis that X
has distribution P_0, which rejects when a statistic $T(x) > t_0$, can
be considered a formal frequentist procedure with

$$\bar{R}_\delta = P_0(T(X) > t_0).$$

Until Neyman, however, the interpretation of such a procedure seemed
to be, upon rejecting, that "either P_0 is false *or* a very unlikely
event has been observed." [See Fisher (1926), for example.] The
frequentist interpretation in terms of long-run behavior did not
become prevalent until after Neyman. (This is not to say that Neyman
supported significance testing of a null hypothesis; indeed, he often
argued that alternatives must be considered.)

　　To many, the definition of frequentism being given here may be
felt to be too strict. For instance, the quoting of a P-value [in
the above setting, $P_0(T(X) > T(x))$] may be felt to be a frequentist
procedure by some, since it involves an averaging over the sample
space. The reporting of P-values can be given no long-run frequency
interpretation in the sense discussed above, however, and cannot
even be given such an interpretation in the more general setup of
the next section. A P-value actually lies closer to conditionsl
(Bayesian) measures than to frequentist measures [see Berger and
Wolpert (1984) for references].

　　This relates to a point that should be mentioned, namely, that
the "practicing" frequentist statistician behaves quite differently
from the "formal" frequentist defined above, recognizing the rarity
of being able to completely state $(\delta, \bar{R}_\delta^\omega)$ [or $(\delta, R_\delta(\theta))$] before
experimentation, and thus admitting the need for "adhocery." Pre-
sumably, however, such a frequentist attempts to stay as close as
possible to the frequentist ideal, so that discussion of the motivation for this ideal is certainly not out of order.

2.2 Conditional Frequentist Approaches

Because of examples such as Example 1, there have long existed conditional frequentist approaches to statistics. The usual idea is to condition on some event or statistic, such as an ancillary statistic [cf. Fisher (1956)] and then to do a frequentist calculation.

Example 1 (continued): Defining the ancillary statistic $T(X) = |X_1 - X_2|$, an easy calculation shows that

$$P_\theta(C(X) \text{ contains } \theta | T(X) = 2) = 1$$

and

$$P_\theta(C(X) \text{ contains } \theta | T(X) = 0) = \tfrac{1}{2},$$

corresponding to intuition.

The most comprehensive development of conditional frequentist theory is that in Kiefer (1975, 1976, 1977), Brownie and Kiefer (1977), and Brown (1978). Two, more or less distinct, approaches are discussed in these papers, namely, *conditional confidence* and *estimated confidence*. These two approaches are reviewed in the next two subsections although, since the setting will be that of general L, the approaches will be termed *conditional risk* and *estimated risk*.

2.2.1 Conditional Risk. This approach is essentially a formalization of conditioning on an ancillary statistic or "relevant" subset. One considers a partition $\{C^b, \ b \in B\}$ of $\mathfrak{X}$, calculates

$$R_\delta^b(\theta) = E_\theta[L(\theta, \ \delta(X)) | C^b], \tag{2.5}$$

and reports, when $x \in C^b$ is observed, the triple $(\delta(x), \ C^b, \ R_\delta^b(\theta))$. The long-run frequentist justification for doing this is, of course, that (when C^b has positive probability)

$$\lim_{n \to \infty} \frac{\sum_{i=1}^{n} L(\theta, \ \delta(X_i)) I_{C^b}(X_i)}{\sum_{i=1}^{n} I_{C^b}(X_i)} = R_\delta^b(\theta) \tag{2.6}$$

with probability one for each θ, so that $R_\delta^b(\theta)$ does measure the average performance of δ in those problems in which X_i falls in C^b (and θ is the same). Furthermore, if one lets $H(x)$ denote that b

for which $x \in C^b$, it is clear that

$$R_\delta(\theta) = E_\theta R_\delta^{H(X)}(\theta), \tag{2.7}$$

so that one also has unconditional frequentist validity.

Example 1 (continued): Let $C^1 = \{x : |x_1 - x_2| = 2\}$, and $C^0(x) = \{x : |x_1 - x_2| = 0\}$. Then

$$R_C^1(\theta) = 1 - P_\theta(C(X) \text{ contains } \theta | C^1) = 0$$

and

$$R_C^0(\theta) = 1 - P_\theta(C(X) \text{ contains } \theta | C^0) = \frac{1}{2}.$$

Example 3a (continued): Let $C^b = \{b, -b\}$ (just the two points) for each $b > 0$. Then

$$R_\delta^b(\theta) = E_\theta[L(\theta, \delta) | C^b] = \frac{f(-(\text{sgn } \theta)b | \theta)}{f(-b | \theta) + f(b | \theta)}$$

$$= \frac{1}{1 + \exp\{2b|\theta|\}}. \tag{2.8}$$

Thus, when x near zero is observed, one reports the relevant decision along with $R_\delta^{|x|}(\theta)$, which will be close to $\frac{1}{2}$ as intuition would suggest. And when x is far from zero, the conditional risk $R_\delta^{|x|}(\theta)$ will be small. Observe also that

$$\sup_\theta R_\delta^b(\theta) = 1/[1 + \exp(4b)]. \tag{2.9}$$

Although the conditional risk approach has a number of attractive features, as seen in the above examples, it also has several deficiencies. First, there is still present the problem that different θ_i are not present in (2.6). In other words, a usable justification, as in Definition 1, cannot be given. This can be corrected if $R_\delta^b(\theta)$ has a usable upper bound $\tilde{R}_\delta^b$, for then, with probability one for all $\boldsymbol{\theta}$ (under reasonable conditions),

$$\lim_{n \to \infty} \sup \frac{\sum_{i=1}^{n} L(\theta_i, \delta(X_i)) I_{C^b}(X_i)}{\sum_{i=1}^{n} I_{C^b}(X_i)} \leqslant \tilde{R}_\delta^b, \tag{2.10}$$

where the X_i arise from different problems with different θ_i. In Example 1, R_δ^b is constant; so (2.10) automatically obtains. In

Example 3a, a usable bound is given in (2.9); so again one has the validity of (2.10).

The obtaining of a useful bound for $R_\delta^b(\theta)$ is deemed to be of primary importance in Brown (1978); and even in Kiefer (1975) and Brownie and Kiefer (1977), the desirability of choosing $\{C^b\}$ so as to achieve constant R_δ^b, or at least a useful upper bound, is stressed. Unfortunately, a useful bound cannot always be achieved. In Example 3b, for instance, it can be shown that, for any partition $\{C^b\}$, $\sup_\theta R_\delta^b(\theta) = \frac{1}{2}$, a useless upper bound.

A second difficulty with conditional confidence is that the choice of the partition $\{C^b\}$ is quite arbitrary, and the justification, even (2.10), depends on this choice. We are not so much referring to the practical difficulty of choosing $\{C^b\}$ (which can be considerable, outside of obvious situations such as Example 1) as to the unappealing arbitrariness in the evaluation of the accuracy of δ that is introduced. In all conditional approaches there will be a certain degree of arbitrariness, but none so bad as the allowance of arbitrary $\{C^b\}$, especially because the choice of $\{C^b\}$ will rarely have any outside justification. More about this will be said in Section 2.3.

2.2.2 Estimated Risk. The estimated risk approach replaces the reporting of $(\delta, R_\delta(\theta))$ by the reporting of $(\delta, \hat{R}_\delta(x))$, where $\hat{R}_\delta(x)$ is, in some sense, an estimate of $R_\delta(\theta)$. If, in fact, $\hat{R}_\delta$ is an unbiased estimator of $R_\delta(\theta)$ for all θ, then with probability one (under reasonable conditions),

$$\lim_{n \to \infty} \frac{1}{n} \sum_{i=1}^{n} [L(\theta_i, \delta(X_i)) - \hat{R}_\delta(X_i)] = 0 \qquad (2.11)$$

for any sequence $\boldsymbol{\theta} = (\theta_1, \theta_2, \ldots)$. Thus the average performance of δ, in actual repeated use, will be the average of the reported $\hat{R}_\delta(X_i)$. Often an unbiased estimator of $R_\delta(\theta)$ may not exist or may, in some sense, be undesirable. In such cases, however, one can usually find a desirable function $\hat{R}_\delta(x)$ such that

$$E_\theta \hat{R}_\delta(X) \geqslant R_\delta(\theta) \text{ for all } \theta, \qquad (2.12)$$

in which case (2.11) could be replaced (under reasonable conditions) by

$$\limsup_{n \to \infty} \frac{1}{n} \sum_{i=1}^{n} [L(\theta_i, \delta(X_i)) - \hat{R}_\delta(X_i)] \leqslant 0, \tag{2.13}$$

establishing frequentist validity of the report $(\delta, \hat{R}_\delta)$ in the same optimum sense as in Definition 1. This is the point to be stressed: The justification for being a frequentist applies equally well to the (δ, R_δ) report as to the $(\delta, \bar{R}_\delta)$ report, and applies much better than to the $(\delta, R_\delta(\theta))$ report, and yet allows much greater latitude for aligning the report with conditional common sense.

Example 1 (continued): Simply choose

$$\hat{R}_\delta(x) = \begin{cases} 0 & \text{if } |x_1 - x_2| = 2 \\ \tfrac{1}{2} & \text{if } |x_1 - x_2| = 0 \end{cases}, \tag{2.14}$$

and one has an unbiased estimate of risk satisfying the "ideal" (2.11), providing good conditional reports.

Example 3a (continued): Although an unbiased estimate of risk cannot be obtained, the choice

$$\hat{R}_\delta(x) = 1/[1 + \exp(4|x|)] \tag{2.15}$$

satisfies (2.12), and hence (2.13). Thus $(\delta, \hat{R}_\delta)$ provides a (conservatively) valid frequentist report and seems much more attractive intuitively than the unconditional error probability bound of .0228.

The above two examples are situations in which $\hat{R}_\delta$ could be chosen to be an appropriate conditional risk or bound on such. The estimated risk approach provides additional flexibility, however, in that it can deal with situations in which the conditional risk approach does not provide useful bounds.

Example 3b (continued): It was noted that, for any $\{C^b\}$, $\sup_\theta R_\delta^b(\theta) = \tfrac{1}{2}$. It is possible, however, to find $\hat{R}_\delta(X)$ such that

$$E_\theta \hat{R}_\delta(X) \geqslant R_\delta(\theta) = \phi(-|\theta|) \text{ for all } \theta, \tag{2.16}$$

where ϕ is the standard normal c.d.f., and such that $\hat{R}_\delta(x)$ decreases to zero as $|x| \to \infty$. Use of such $(\delta, \hat{R}_\delta)$ allows one to report, when $x = 10$ is observed, that the null hypothesis is conclusively rejected. Of some concern here is that any such $\hat{R}_\delta$ must be greater than ½ with positive probability for (2.16) to hold when $\theta = 0$. It would be unappealing to report "I reject H_0 and have estimated error probability of .6." A casual observer would find such a statement most peculiar. Of course, this will tend to arise only when x provides very inconclusive evidence and hence may not be of much practical concern.

It should be mentioned that the idea of estimated risk definitely precedes Kiefer's work on the subject, although earlier work did not explicitly recognize that use of estimated risk was completely valid from a frequentist perspective. Mention of estimated power can be found in Lehmann (1959); and Sandved (1968) found unbiased estimators of risk for a number of problems. There are also a number of current areas of research in which unbiased estimates of risk are commonly employed, such as Stein estimation.

Example 6: Suppose $X \sim \eta_p(\theta, I)$ and that $L(\theta, \delta) = |\theta - \delta|^2$. For $p \geqslant 3$, Stein (1981) shows that

$$\hat{R}_\delta(x) = p - (p - 2)^2 / |x|^2$$

is an unbiased estimator of the risk of

$$\delta(x) = (1 - (p - 2)/|x|^2)x,$$

and hence

$$R_\delta(\theta) = E_\theta \hat{R}_\delta(X) < p = R_{\delta^0}(\theta),$$

where $\delta^0(x) = x$ is the usual estimator. One can thus report $(\delta(x), \hat{R}_\delta(x))$ with total frequentist justification. [Of course, in this case such a report will be fairly silly for $|x|^2 < (p - 2)^2/p$, because $\hat{R}_\delta$ will then be negative.]

Note that the estimated risk approach can be combined with the conditional risk approach to obtain estimated conditional risk [see Kiefer (1977)]. Or one could obtain estimated versions of $\overline{R}_\delta^\omega$ (see

Definition 1), valid for subsequences ω of interest. Again, this would increase flexibility but would result in more ambiguous frequentist justification. Note on the other hand, that reports of $(\delta, \hat{R}_\delta)$ may require different interpretation than, say, $\bar{R}_\delta^\omega$, as the following example shows.

Example 2b (continued): A possible choice of $\hat{R}_\delta$ would be

$$\hat{R}_\delta(1) = .5, \ \hat{R}_\delta(2) = .00749, \ \hat{R}_\delta(3) = .00754. \tag{2.17}$$

It can easily be checked that this is an unbiased estimator of risk, and hence the report $(\delta, \hat{R}_r)$ is a valid frequentist report. This report also has the attractive frequentist property that, when $x = 1$ is observed, the estimated risk reported is .5, corresponding to the intuitive (likelihood ratio) assessment. On the other hand, this estimated risk does not have any interpretation as Type I or Type II error, which may be disturbing to some.

2.3 The Choice of a Conditional Frequentist Measure

The huge variety of possible conditional measures makes choice among them quite difficult. This is especially true for conditional risks $R_\delta^b(\theta)$ and for estimated risks $\hat{R}_\delta^\omega(x)$ where dependence on subsequences ω is allowed. One natural idea is to recognize that a purpose of any version of risk is to communicate some feeling as to the actual loss $L(\theta, \delta)$ and hence to introduce a *communication loss* $L^*(\hat{R}_\delta(x), L(\theta, \delta(x)))$ and corresponding *communication risk*

$$R^*(\hat{R}_\delta, \delta, \theta) = E_\theta L^*(\hat{R}_\delta(X), L(\theta, \delta(X))). \tag{2.18}$$

Or, of course, one could define these same quantities with $R_\delta^b(\theta)$ or $\hat{R}_\delta^\omega(x)$ in place of $\hat{R}_\delta(x)$.

Example 4 (continued): In this confidence set situation, a natural choice for L^* is

$$L^*(\hat{R}_\delta(x), \ 1 - I_{C(x)}(\theta)) = (\hat{R}_\delta(x) - [1 - I_{C(x)}(\theta)])^2. \tag{2.19}$$

Part of the reason that this would be a sensible measure of how well $\hat{R}_\delta(x)$ communicates whether or not θ is in $C(x)$ is that it is a proper scoring rule; any proper scoring rule [cf. Lindley (1982)] would

probably serve as well. Thus, in the situation of Example 1, the
unconditional frequentist report $R_\delta(\theta) \equiv \frac{1}{4}$ has

$$R^*\left(\frac{1}{4}, C, \theta\right) = E_\theta\left(\frac{1}{4} - [1 - I_{C(X)}(\theta)]\right)^2 = \frac{3}{16},$$

whereas the estimated risk report $\hat{R}_\delta$ given in (2.14) has

$$R^*(\hat{R}_\delta, C, \theta) = E_\theta(\hat{R}_\delta(X) - [1 - I_{C(X)}(\theta)])^2 = \frac{1}{8}.$$

Thus $\hat{R}_\delta$ is uniformly better, as intuition would demand.

An analysis of Example 3a similarly shows that $\hat{R}_\delta$, given in
(2.15), is better than the best unconditional (Definition 1) state-
ment of $\bar{R}_\delta = .0228$. It is not always true that a sensible $\hat{R}_\delta$ will
uniformly dominate an unconditional $\bar{R}_\delta$, however, as Example 3b indicates. Here $\bar{R}_\delta = \frac{1}{2}$ can be shown to be better (under squared error
L^*) for θ near zero than any other $\hat{R}_\delta$ satisfying (2.16), although
a decreasing $\hat{R}_\delta(x)$ seems much more satisfying intuitively.

In the above examples, only situations in which the conditional
or estimated risk was independent of θ (or ω) were considered.
Indeed, it is, perhaps, only in such situations that the use of L^*
makes any sense. To see this, note that there is nothing in the
formalism of conditional risk to prevent one from selecting each
singleton $\{x\}$ as a conditioning set, so that one obtains

$$R_\delta^b(\theta) = E_\theta[L(\theta, \delta(X)) | X = x] = L(\theta, \delta(x)). \qquad (2.20)$$

Clearly, this will be optimal from the viewpoint of L^* and is a valid
conditional frequentist conclusion. It tends not to be operation-
ally very useful, however.

Attempting to define meaningful criteria, under which to evaluate conditional measures other than the simple $\hat{R}_\delta(x)$, leads to
something of a morass. A wide variety of evaluation methods and
admissibility criteria are proposed in Kiefer (1975, 1976, 1977),
Brownie and Kiefer (1977), and Brown (1978). In some sense, the
criteria of Brown (1978) are the most appealing in that they attempt
to relate the evaluation of the conditional measure to its likely
use, as opposed to trying to determine intuitively appealing properties of conditional measures. Our own (somewhat naive or more

realistic—take your pick) view of this issue is that one would
really like to communicate the posterior expected loss for the
problem, and only the evaluation criteria that recognize this will
tend to produce reasonable conditional measures. Of course, this
probably also applies to the (L^*, R^*) method for evaluating the
simple $\hat{R}_\delta(x)$.

A few final comments are in order concerning the evaluation of
$\hat{R}_\delta$. First, it should be emphasized that this is a decision problem
with decision space consisting only of the $\hat{R}_\delta$ that satisfy the
property (2.12). If one removes the restriction (2.12), inadmissibil-
ity can result for otherwise admissible $\hat{R}_\delta$. Consider the following
example.

Example 7: Suppose $X \sim n_p(\theta, I)$, where $\theta = (\theta_1, \ldots, \theta_p)$ is unknown,
and consider the classical confidence procedure

$$C^0(x) = \{\theta : |\theta - x|^2 \leqslant \chi_p^2(1 - \alpha)\},$$

where $\chi_p^2(1 - \alpha)$ is the $(1-\alpha)^{\text{th}}$ fractile of the chi-squared distribution with p degrees of freedom. In the framework of Example 4,
this can be considered a frequentist procedure with $\bar{R}_\delta = \alpha$. But with
respect to the loss (2.19), Robinson (1979a, 1979b) proves this to be
inadmissible for $p = 5$. However, the uniformly better procedure is
of the form

$$\hat{R}_{C^0}(x) = \alpha - k(x),$$

where $k(x) > 0$, so that $E_\theta \hat{R}_{C^0}(X) < \alpha$, violating the frequentist
validity requirement.

We make no judgment (in this section) as to whether or not a
violation of the frequentist validity requirement, in situations
such as the above example, is desirable. It is (here) being taken
as given that frequentist validity is required, and the exploration
concerns the leeway still allowed in the choice of $\hat{R}_\delta$. For this
reason we will also not put L^* (or R^*) on the same footing as L
(or R), even though it is very tempting to do so. In the confidence
procedure setting, for instance, one could consider an overall loss
such as

$$L(\theta, \, C(x), \, \hat{R}_C(x)) = k_1[1 - I_{C(x)}(\theta)] + k_2[\hat{R}_C(x) - (1 - I_{C(x)}(\theta))]^2,$$

which recognizes not only the importance of having $C(x)$ contain θ but
also the importance of properly communicating whether or not it does.
Decision theory with such loss functions would be quite interesting,
but it is probably best to keep L and L^* separate. Also, it is
probably unwise to attempt to follow the road much farther, that is,
to actually report (or estimate) R^* as if it were a meaningful quantity (as opposed to a device for selection of $\hat{R}_\delta$). This is mainly a
feeling based on the seeming inevitability of Bayesian analysis as a
mechanism for sensible communication of whether or not θ is in $C(x)$
[cf. Lindley (1982)]. To a non-Bayesian, however, there may be some
appeal to reporting $(\delta, \, \hat{R}_\delta, \, R^*)$, and perhaps there is merit in so
doing. This is especially plausible when the communication aspect
can be considered as part of the real problem.

One final comment about the situation of Example 7 is in order.
A recently much studied problem has been the replacement of the usual
confidence spheres $\{C^0(x)\}$ by spheres $\{C^*(x)\}$ of the same size but
centered at the Stein-type estimator

$$\delta^{J-S}(x) = (1 - c/|x|^2)^+ x.$$

For appropriate c (depending on p), it can be shown [cf. Hwang and
Casella (1982)] that

$$R_{C^*}(\theta) = 1 - P_\theta(C^*(X) \text{ contains } \theta) < \alpha$$

for all θ. But $\bar{R}_{C^*} = \sup_\theta R_{C^*}(\theta) = \alpha$, so that one cannot improve on
C^0 in terms of an unconditional frequentist conclusion, independent
of θ. It is clearly possible, however (though probably difficult),
to find $\hat{R}_{C^*} < \alpha$ such that

$$R_{C^*}(\theta) \leq E_\theta \hat{R}_{C^*}(X) < \alpha,$$

allowing an estimated risk report of $(C^*, \, \hat{R}_{C^*})$ which offers a clear
gain.

3. CONFLICTS BETWEEN FREQUENTIST AND CONDITIONAL ANALYSIS

3.1 Introduction

The examples in Section 1 made it clear that unconditional frequentist theory could not be suitable as a general philosophy of statistics. In Section 2 it was seen, however, that conditional frequentist theory was just as valid from the frequentist viewpoint and seemed to offer much greater scope for correspondence with conditional common sense. The obvious issue, therefore, is whether or not the conditional frequentist theory is itself rich enough to provide a suitable general philosophy. There are, disturbingly, still simple examples indicating concern as to this suitability. For instance, consider the following modification of Example 2b.

Example 2c: Suppose the probability structure is

	X		
	1	2	3
P_0	.05	.15	.8
P_1	.051	.849	.1

Now the risks (error probabilities) of the test δ, which rejects when $x = 1$ or 2, are $R_\delta(0) = .2$ and $R_\delta(1) = .1$. Conditional risk theory will never work well for such a situation, because one of the sets C^b, of the partition, must contain only one point, resulting in a conditional risk of zero or one (unless an extremely artificial randomization is introduced). And since $R_\delta(0)$ and $R_\delta(1)$ differ substantially, the estimated risk theory does not easily apply. Perhaps some version of the estimated risk theory that allows dependence on ω (or θ) would give sensible answers (by which is meant, not only frequentist validity but also an appropriate expression of doubt for $x = 1$), but it is not easy to find such.

Although the difficulty in dealing successfully with simple examples, such as Example 2c, via frequentist theory, is a cause for

concern, the vastness of the conditional frequentist domain makes
unlikely a disproof by counterexample. Serious axiomatic conflicts
with frequentist or conditional frequentist viewpoints exist, however, and it is to a brief discussion of these that we now turn.
Section 3.2 reviews the axiomatics leading to the likelihood principle and the resultant conflict with frequentist ideas. Section
3.3 discusses one of the most important of these conflicts, that
which concerns the role of the stopping rule in statistical analysis.
Indeed, because of the importance and intuitive difficulties concerning this latter issue, a separate argument is given, showing the
incompatibility of frequentist admissibility with the idea that the
stopping rule must be taken into account.

3.2 The Likelihood Principle

We shall forego extensive discussion of the likelihood principle (LP)
here, presenting only a bare-bones outline of its implications and
its axiomatic development [due to Birnbaum (1962)]. A recent
monograph, Berger and Wolpert (1984), extensively discusses the LP,
its history, and its ramifications.

In what follows, x will be considered to be discrete, so that the
likelihood function $f(x|\theta)$, that is, the density considered as a
function of θ for the actual x, is well defined. This restriction
can be removed; indeed, in Berger and Wolpert (1984) a generalization
of the LP called the *relative likelihood principle* is developed,
which yields the same consequences as the LP and yet does not
require the existence of densities (or even parametric models).
Thus, the common criticism that the LP is not valid because it does
not apply to situations in which the model is uncertain, is not
applicable to the appropriate generalization of the LP. [It can
even be argued that, since in reality all x are discrete and finite
and for such x *all* families of distributions are parametric (the
most general possible index θ being simply the vector of probabilities of the elements of x), the simple version of the LP always
applies. Such an argument is given in Basu (1975).] This is not the
place to discuss all the criticisms of the LP that have been raised.

See Berger and Wolpert (1984) and Berger (1984a) for such discussion. We merely wish to make the point that Birnbaum's axiomatic development should be taken seriously and cannot be easily dismissed.

Following Birnbaum's notation, we let E be an experiment consisting of observing $X \sim P_\theta$ and are concerned with the "evidence" or "information" about θ that is obtained (or should be reported) upon observing x. This will be denoted $Ev(E, x)$. (This evidence could be anything at all, including one or several frequentist measures. Note that by listing E we allow Ev to depend on full knowledge of all aspects of the experiment and not on just the observed x.)

The Likelihood Principle: $Ev(E, x)$ should depend on E and x only through the likelihood function $f(x|\theta)$ for the observed x. Two likelihood functions for (the same unknown) θ yield identical evidence about θ if they are proportional (as functions of θ).

Example 2d: Again assume $\mathfrak{X} = \{1, 2, 3\}$ and $\Theta = \{0, 1\}$, and consider experiments E_1 and E_2 which consist of observing X_1 and X_2 with the above $\mathfrak{X}$ and Θ, but with probability densities as follows:

	x_1				x_2		
	1	2	2		1	2	3
$f_1(x_1|0)$	.9	.02727	.07272	$f_2(x_2|0)$	.09	.084545	.82545
$f_1(x_1|1)$	.72	.27272	.00727	$f_2(x_2|1)$	.072	.84545	.082545

If, now, $x_1 = 1$ is observed, the LP states that $Ev(E_1, 1)$ should depend on the experiment only through $(f_1(1|0), f_1(1|1)) = (.9, .72)$. Furthermore, since this is proportional to $(.09, .072) = (f_2(1|0), f_2(1|1))$, it should be true that $Ev(E_2, 1) = Ev(E_1, 1)$. [Another way of stating the LP for testing hypotheses, as here, is that $Ev(E, x)$ should depend on E and x only through the likelihood ratio for the observed x.] It is similarly clear that, according to the LP, $Ev(E_1, 2) = Ev(E_2, 2)$ and $Ev(E_1, 3) = Ev(E_2, 3)$. Hence, no matter which experiment is performed, the same evidentiary conclusion about θ should be reached for the given observation.

The above example clearly indicates the startling nature of the LP. Experiments E_1 and E_2 are very different from a frequentist perspective. For instance, the decision procedure which decides $\theta = 0$ when the observation is 1 or 3, and decides $\theta = 1$ when the observation is 2, is a most powerful test with error probabilities (of Type I and Type II, respectively) .02727 and .72727 for E_1, and .084545 and .154545 for E_2. Thus, the classical frequentist would report drastically different "evidence" from the two experiments. (Conditional frequentist approaches are very unlikely to give similar conclusions: indeed, for E_1 it is very hard to perform any sensible conditional frequentist analysis, because of the three point $\mathfrak{X}$ and widely differing error probabilities.)

This example emphasizes a very important issue. It is clear that experiment E_2 is more likely to provide useful information about θ, as reflected by the overall better error probabilities. The LP, in no sense, contradicts this. Indeed, the LP says nothing about experimental design or any other situation involving an evaluation for not yet observed X. The LP applies only to the information about θ that is available from knowledge of E and the observed x. Even though E_2 has a much better chance of yielding good information, the LP states that the conclusion, once x is at hand, should be the same, regardless of whether x came from E_1 or E_2. The conflict of the LP with frequentist justifications seems inescapable. [See also Birnbaum (1977).]

A committed frequentist might look at this example and reject the LP out of hand, although some unease will undoubtedly be present because of the equal likelihood ratios in the experiments. Very troubling, however, is the fact that the LP is a direct consequence of two other "obvious" principles, the Sufficiency Principle and the Weak Conditionality Principle.

The Sufficiency Principle: If T is a sufficient statistic for θ in an experiment E, and $T(x_1) = T(x_2)$, then $\mathrm{Ev}(E, x_1) = \mathrm{Ev}(E, x_2)$.

Tne Weak Conditionality Principle: Let E_1 consist of observing X_1 with density $f_1(x_1|\theta)$ and E_2 consist of observing X_2 with density

$f_2(x_2 | \theta)$. (Here θ is the same quantity in each experiment.) Consider the mixed experiment E consisting of observing $J = 1$ or 2 with probability $\frac{1}{2}$ each (independent of everything—say, the result of a fair coin flip), and then performing experiment E_J. The random quantity observed from E is thus (J, X_J). Then it should be true that

$$\mathrm{Ev}(E, (j, x_j)) = \mathrm{Ev}(E_j, x_j),$$

that is, the evidence obtained about θ is simply the evidence from the experiment actually performed.

Almost everyone accepts the Sufficiency Principle, and the Weak Conditionality Principle [essentially due to Cox (1958)] seems very natural, being the weakest form of conditioning imaginable. (Some frequentists might reject the Weak Conditionality Principle by essentially rejecting the idea that the goal is to communicate evidence about θ. If the goal is simply to determine repeated performance of a procedure in use, then the repeated performance for E will likely differ from the repeated performance for one of the E_j. It seems unlikely that such a view could ever gain wide acceptance, however; insisting on reporting 75% coverage in Example 1, for instance, is hardly tenable.) Birnbaum's surprising result was that this weakest form of conditioning (together with sufficiency) implies that complete conditioning, down to $f(x | \theta)$, should be done.

Theorem [Birnbaum (1962)]: The Sufficiency Principle and the Weak Conditionality Principle together imply the LP.

All the generalizations of the LP that were referred to earlier also follow from sufficiency and weak conditionality, and so a frequentist is left with the uncomfortable choice of rejecting sufficiency or weak conditionality. It is a conflict which clearly deserves careful thought.

This is not to say that all frequentist procedures violate the LP. In fact, it is well known that a large portion of standard frequentist procedures can be interpreted as Bayesian procedures with "noninformative" priors and hence are consistent with the LP.

[Bayesian procedures always depend on E only through $f(x|\theta)$.] The
frequentist concept of evidence, based on some type of average over
X, is clearly a concept in conflict with the LP, however, and
Example 2d shows how dramatic the conflict can be.

3.3 The Stopping Rule Principle

One of the most important practical applications of the LP is the
Stopping Rule Principle (SRP), developed (at various levels) in
Barnard (1947), Birnbaum (1962), and Pratt (1965). Suppose E is a
sequential experiment, with possible observations X_1, X_2, ... having
probability distribution P_θ (determined by the finite dimensional
distributions, of course), with θ unknown. For convenience, only
nonrandomized stopping rules τ are considered. Such a stopping rule
can, most conveniently, be represented by a sequence of sets
$\{(A_n, B_n)\}$, where

> if $\mathbf{x}^n = (x_1, \ldots, x_n) \in A_n$, stop sampling;
>
> if $\mathbf{x}^n \in B_n$, continue sampling.

$$(3.1)$$

(Without loss of generality, it can be assumed that, if $\mathbf{x}^n \in A_n$, then
$\mathbf{x}^j \in B_j$ for all $j < n$.) Let N denote the stopping time (i.e., the n
for which $\mathbf{x}^n \in A_n$). Only proper stopping rules [i.e., those for
which $P_\theta(N < \infty) = 1$] will be considered.

Stopping Rule Principle: For a sequential experiment E with observed
data $\mathbf{x}^n$, $\mathrm{Ev}(E, \mathbf{x}^n)$ should not depend on the stopping rule τ.

In words, the SRP simply states that the reason for stopping
sampling should be irrelevant to evidentiary conclusions about θ
(providing this reason, as above, does not depend on θ in any fashion
except indirectly through the x_i). One can constantly monitor incom-
ing data and stop at any time that the data look good enough (or bad
enough or whatever), and this optional stopping should play no role
for a valid measure of evidence.

The practical implications of the SRP are enormous, because
optional stopping is a huge problem in many areas of practical
statistics, such as clinical trials. Scrupulous experimenters and

scientists would delight in the freedom to stop an experiment at any point felt to be appropriate, without having to worry about an effect of such optional stopping. And the scientific community, as a whole, would no longer be prey to misleading (frequentist) conclusions arising from situations in which optional stopping was employed but not reported.

Of course, the hitch in all this is that frequentist measures are very dependent on the stopping rule and cannot be used in conjunction with the SRP. Indeed, frequentist intuition will generally react to the SRP with outrage, it being "obvious" that "stopping when the data look good" will bias the results (and, of course, it will in a frequentist sense). Since, however, the SRP can be shown to be a trivial consequence of the LP—or of the relative likelihood principle of Berger and Wolpert (1984), if complete generality is desired—a major conflict in intuition again surfaces; rejecting the SRP corresponds to rejecting either sufficiency or weak conditionality.

In an attempt to resolve the issue (in favor of the SRP) it is interesting to observe that the frequentist intuition that the stopping rule matters is itself inconsistent with the frequentist concept of admissibility. Consider the following example.

Example 8: Suppose $\Theta = R^1$ and that it is desired to estimate θ under squared error loss $L(\theta, \delta) = (\theta - \delta)^2$. Imagine, now, two possible stopping rules, τ_1 and τ_2, determined by $\{(A_n^1, B_n^1)\}$ and $\{(A_n^2, B_n^2)\}$, respectively, and suppose that, for some n_0, there exists a set $A \subset A_{n_0}^1 \cap A_{n_0}^2$ such that $P_\theta(A) > 0$ and on which the estimators δ_1 and δ_2 that would be used under τ_1 and τ_2, respectively, are different. (If no such A exists, then the stopping rules are not really having any effect on the decision.)

To see that this conflicts with admissibility, or more basically with long-run frequentist optimality, imagine that one will be faced with a series of such experiments, in half of which τ_1 will be used and in half of which τ_2 will be used. Then it is a simple matter to show that one could do better (for a sequence of θ_i such that

$P_{\theta_i}(A) > \varepsilon > 0)$ by using the estimator δ_j if τ_j is used and $N \neq n_0$ or $\mathbf{x}^{n_0} \notin A$, while using

$$\frac{1}{2}\delta_1(\mathbf{x}^{n_0}) + \frac{1}{2}\delta_2(\mathbf{x}^{n_0}) \text{ if } N = n_0 \text{ and } \mathbf{x}^{n_0} \in A. \tag{3.2}$$

A formal statement of this would involve, for instance, the consideration of the mixed experiment E, consisting of observing $J = 1$ or 2 with probability $\frac{1}{2}$ each and then doing the sequential experiment with stopping rule τ_J. This E is a well-defined sequential experiment with observation $(J, \mathbf{X}^N)$ (N being the implied stopping time for E), and it is trivial to show that a sufficient statistic for θ (in the experiment E) is

$$T((j, \mathbf{x}^n)) = \begin{cases} \mathbf{x}^n \text{ if } n = n_0, \ \mathbf{x}^n \in A \\ (j, \mathbf{x}^n) \text{ otherwise} \end{cases}.$$

If, now, the stopping rule does matter, then one would presumably use δ_J to estimate θ; but by construction, this is not a function of the sufficient statistic alone. Hence, since the loss is strictly convex, Rao-Blackwellization of the estimator, via (3.2), would result in an estimator with strictly better frequentist risk. Thus, admissibility (or long-run frequentist validity) implies that the stopping rule should be ignored in making the decision (at least for the $\mathbf{x}^n$ that can be observed under either stopping rule).

4.   CONCLUSIONS

Some general conclusions seem possible from the preceding discussions. The first is that the issue of conditioning is serious and deserves careful consideration by all statisticians. It is a tribute to Kiefer's unswerving pursuit of scientific truth that he recognized this issue and sought to resolve it, even though the issue is an uncomfortable one for frequentists.

The second conclusion that can be reached (at least tenuously) is that, even though conditional frequentist approaches can go a very long way toward achieving compatibility with the conditional

view, complete reconciliation appears to be impossible [see also
Birnbaum (1977)]. One thus has an uncomfortable choice: Either
abandon frequentist justification as an absolute must, or resign
oneself to the possibility that from time to time one will be forced
to state a conclusion that is at variance with conditional common
sense.

Neyman and Kiefer made the second choice. I have made the
first choice, essentially because of a refusal to be put in the position of having to give a conclusion for a real statistical problem
that I know is (conditionally) a silly conclusion. This is not to
say that the frequentist view cannot be of great usefulness, but
that, as a philosophical foundation for statistics, I find it
unsuitable.

This raises the issue of what *should* serve as a philosophical
foundation of statistics, and the obvious answer is Bayesian analysis, which seems to be the only approach capable of guaranteeing
sensible conditional answers [cf. Berger (1983, 1984a, 1984b) and
Berger and Wolpert (1984)]. The practical issue of (often extreme)
uncertainty in prior knowledge then raises its head, however, along
with such issues as the need for scientific objectivity and the
practical difficulties in complicated situations of obtaining any
answer at all. The value of frequentist calculations can be
considerable for these and other reasons, as discussed in the above-
mentioned articles (which also have earlier references). Indeed, it
could be argued that, as a practical matter, the frequentist viewpoint will tend to give better answers than the Bayesian viewpoint,
even if the latter is philosophically correct. Having now spent a
number of years attacking problems from both perspectives, I feel
nearly certain that this is not the case, but a discussion of these
matters would clearly take us too far afield.

It should also be observed that the entire discussion in this
paper has been directed toward the statistical conclusion that
will be made once the data is at hand. The problem of designing
good experiments, dependent on knowing the expected performance of
procedures likely to be used, clearly involves a strong frequentist

component (although it can be argued that more attention should be paid to Bayesian matters, here, also). Indeed, the frequentist viewpoint partly arose as an effort to unify these two aspects of statistics [see Pearson (1962)]. There is no questioning the immense contributions to statistics that have been made by Neyman and Kiefer by adopting this frequentist viewpoint. It is, perhaps, time to admit, however, that a grand unification of the two aspects of statistics, under the frequentist banner, is impossible.

ACKNOWLEDGMENTS

Of great help in my struggle to understand these issues were Larry Brown, Leon Gleser, Bruce Hill, Ker-Chau Li, Herman Rubin, and, of course, Jack Kiefer.

REFERENCES

Barnard, G. A. (1947). A review of "Sequential Analysis" by Abraham Wald. *J. Amer. Statist. Assoc. 42*, 658-669.

Basu, D. (1975). Statistical information and likelihood (with Discussion). *Sankhyā, Ser. A 37*, 1-71.

Berger, J. (1983). The robust Bayesian viewpoint. In *Robustness in Bayesian Statistics* (J. Kadane, ed.). Amsterdam: North-Holland.

Berger, J. (1984a). In defense of the likelihood principle: Axiomatics and coherency. In *Bayesian Statistics II* (J. M. Bernardo, M. H. DeGroot, D. Lindley, and A. Smith, eds.). Amsterdam: North-Holland.

Berger, J. (1984b). Bayesian salesmanship. In *Bayesian Inference and Decision Techniques with Applications: Essays in Honor of Bruno deFinetti* (P. K. Goel and A. Zellner, eds.). Amsterdam: North-Holland.

Berger, J., and Wolpert, R. (1984). *The Likelihood Principle: A Review and Generalizations.* Institute of Mathematical Statistics Monograph Series, Hayward, California.

Birnbaum, A. (1962). On the foundations of statistical inference (with Discussion). *J. Amer. Statist. Assoc. 57*, 269-306.

Birnbaum, A. (1977). The Neyman-Pearson theory as decision theory and as inference theory: With a criticism of the Lindley-Savage agrument for Bayesian theory. *Synthese 36*, 19-49.

Brown, L. D. (1978). A contribution to Kiefer's theory of conditional confidence procedures. *Ann. Statist. 6*, 59-71.

Brownie, C., and Kiefer, J. (1977). The ideas of conditional confidence in the simplest setting. *Commun. Statist.—Theor. Meth. A 6(8)*, 691-751.

Cox, D. R. (1958). Some problems connected with statistical inference. *Ann. Math. Statist. 29*, 357-372.

Fisher, R. A. (1926). The arrangement of field experiments. *J. Ministry Agric. Great Brit. 33*, 503-513.

Fisher, R. A. (1956). *Statistical Methods and Scientific Inference.* Edinburgh: Oliver and Boyd.

Hwang, J. T., and Casella, G. (1982). Minimax confidence sets for the mean of a multivariate normal distribution. *Ann. Statist. 10*, 868-881.

Kiefer, J. (1975). Conditional confidence approach in multi-decision problems. In *Multivariate Analysis IV* (P. R. Krishnaiah, ed.). New York: Academic Press.

Kiefer, J. (1976). Admissibility of conditional confidence procedures. *Ann. Math. Statist. 4*, 836-865.

Kiefer, J. (1977). Conditional confidence statements and confidence estimators (with Discussion). *J. Amer. Statist. Assoc. 72*, 789-827.

Lehmann, E. L. (1959). *Testing Statistical Hypotheses.* New York: Wiley.

Lindley, D. V. (1982). Scoring rules and the inevitability of probability. *Int. Statist. Rev. 50*, 1-6.

Neyman, J. (1957). 'Inductive behavior' as a basic concept of philosophy of science. *Rev. Intl. Statist. Inst. 25*, 7-22.

Neyman, J. (1967). *A Selection of Early Statistical Papers of J. Neyman.* Berkeley: University of California Press.

Pearson, E. S. (1962). Some thoughts on statistical inference. *Ann. Math. Statist. 33*, 394-403.

Pratt, J. W. (1965). Bayesian interpretation of standard inference statements (with Discussion). *J. Roy. Statist. Soc. B 27*, 169-203.

Robinson, G. K. (1979a). Conditional properties of statistical procedures. *Ann. Statist. 7*, 742-755.

Robinson, G. K. (1979b). Conditional properties of statistical procedures for location and scale parameters. *Ann. Statist. 7*, 756-771.

Sandved, E. (1968). Ancillary statistics and estimation of the loss in estimation problems. *Ann. Math. Statist. 39*, 1755–1758.

Stein, C. (1981). Estimation of the mean of a multivariate normal distribution. *Ann. Statist. 9*, 1135–1151.

DOES AN ESTIMATOR'S DISTRIBUTION SUFFICE?

COLIN R. BLYTH Queen's University
PRAMOD K. PATHAK University of New Mexico

SUMMARY: Statistical estimators are usually judged solely by their separate, marginal probability distributions: For X with possible distributions P_θ, $\theta \in \Omega$ indexed by the real parameter θ, two estimators $T_1 = T_1(X)$ and $T_2 = T_2(X)$ of θ are usually considered identical if their distributions are identical. But Example 1 shows it is possible to have $P\{|T_2 - \theta| < |T_1 - \theta|\}$ arbitrarily close to 1 uniformly in θ even though T_1 and T_2 have identical distributions and even though T_1 is better than T_2 in the strong sense that $P\{|T_1 - \theta| \leqslant a\} > P\{|T_2 - \theta| \leqslant a\}$ for every $a > 0$. In Section 2, for $X = X_1$, X_2 independent each with the same continuous distribution symmetric about θ, for $T_1 = X_1$ and $T_2 = \bar{X} = \alpha X_1 + (1 - \alpha)X_2$ with $0 < \alpha < 1$, it is shown that $P\{|\bar{X} - \theta| < |X_1 - \theta|\} > 1/2$; but examples show that $P\{|X_1 - \theta| \leqslant a\} > P\{|\bar{X} - \theta| \leqslant a\}$ for all $a > 0$ is possible.

1. INTRODUCTION AND EXAMPLE

In Wald's decision approach to statistics, estimators are looked at only separately or marginally; two estimators with identical

This work was supported by NSERC, the Natural Sciences and Engineering Research Council of Canada.

distributions have random losses with identical distributions, and of course identical risk functions, no matter what the loss function is. More generally, Savage (1954, pp. 225 and 245) writes that if T_1 and T_2 have identical distributions then "there would presumably be nothing to choose between the estimates by any reasonable criterion" and "On any ordinary interpretation of estimation known to me, it can be argued [no argument is given] that no criterion need depend on more than the separate distributions."

On the other hand, as Savage points out, Pitman's criterion, which prefers T_1 to T_2 if

$$P\{|T_1 - \theta| < |T_2 - \theta|\} > \frac{1}{2} \text{ for all } \theta,$$

does depend on the joint distribution of T_1 and T_2 and therefore falls under his description "not reasonable." But a user of Pitman criterion can defend it by arguing that if he knows nothing about loss except that it is increasing in absolute error, then all he learns from each trial (when θ becomes known) is which estimator came closer. This user would choose between T_1 and T_2 by making or imagining a series of real or simulated trials and choosing the estimator that comes closer to θ in a clear majority of the trials for each value of θ.

Pitman-criterion preferences can be nontransitive, and a class of estimators may or may not have a uniformly Pitman-best estimator; but in just the same way, a class of estimators may or may not have a uniformly minimum risk member.

Example 1: This artificial example, adapted from Blyth (1972, p. 367) shows that even though T_1 and T_2 have identical distributions, it is possible to have $P\{|T_2 - \theta| < |T_1 - \theta|\}$ arbitrarily close to 1 uniformly in θ, so that the two estimators are not indistinguishable: Pitman's criterion strongly prefers T_2 to T_1.

If T_1 and T_2 have constant joint density $1/\varepsilon$ on the shaded region in Figure 1, then it is easy to see from the figure that

T_1 and T_2 both have Uniform $\left(\theta - \frac{1}{2}, \ \theta + \frac{1}{2}\right)$ density,

$$P\{|T_2 - \theta| < |T_1 - \theta|\} = 1 - \varepsilon,$$

and that this probability $1 - \varepsilon$ can be made arbitrarily close to 1 by taking $\varepsilon > 0$ close to 0.

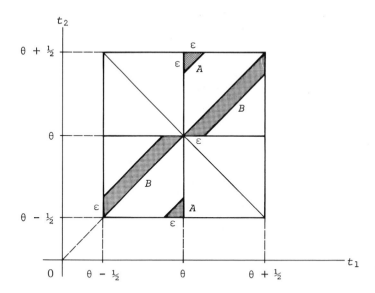

Figure 1. Joint distribution in Example 1.

Modifying this example by a small increase in the constant density on the regions A and a compensating decrease in density on the regions B leaves $P\{|T_2 - \theta| < |T_1 - \theta|\}$ still arbitrarily close to 1 and makes $|T_1 - \theta|$ stochastically smaller than $|T_2 - \theta|$:

$$P\{|T_1 - \theta| \leqslant a\} \geqslant P\{|T_2 - \theta| \leqslant a\} \text{ for all } a > 0,$$

with this inequality strict unless both sides are 1. An example with this inequality strict for all $a > 0$ can be constructed by using $\Phi^{-1}(T_1 - \theta)$, $\Phi^{-1}(T_2 - \theta)$ in place of $T_1 - \theta$, $T_2 - \theta$ with Φ the Standard Normal Cumulative.

Note 1: In this kind of example, the probability that is arbitrarily close to 1 cannot be equal to 1, because if

$$P(Z_1 \leqslant a) \geqslant P(Z_2 \leqslant a) \text{ for all } a, \text{ and } P(Z_2 \leqslant Z_1) = 1,$$

then the first condition gives $EZ_1 \leqslant EZ_2$ and the second condition gives $EZ_2 \leqslant EZ_1$. Therefore $EZ_1 = EZ_2$, and the second condition now gives $P(Z_1 = Z_2) = 1$. If these expectations do not exist, the same proof works, using the expectations of $U(Z_1)$ and $U(Z_2)$ with $U(Z)$ strictly increasing and bounded.

Note 2: In this kind of example, $T_1(X)$ and $T_2(X)$ must be dependent, because if Z_1 and Z_2 with cumulative probability functions F_1 and F_2 are independent, then $F_1(x) \geqslant F_2(x)$ for all x implies

$$P(Z_1 \leqslant Z_2) = \int_{-\infty}^{\infty} F_1 dF_2 \geqslant \int_{-\infty}^{\infty} F_2 dF_2 = \left[\frac{F_2^2}{2}\right]_{-\infty}^{\infty} = \frac{1}{2}.$$

2. COMPARING ONE OBSERVATION WITH AN AVERAGE OF TWO

Theorem 1: If X_1, X_2 are independent, each with the same continuous distribution symmetric about θ, then $T_2 = \bar{X} = \alpha X_1 + (1 - \alpha)X_2$ with $0 < \alpha < 1$ is a better estimator of θ than is $T_1 = X_1$, uniformly in θ, by Pitman's criterion.

Proof: Setting $Z_1 = X_1 - \theta$ and $Z_2 = X_2 - \theta$ in the following lemma gives

$$\frac{1}{2} < P\{|\bar{X} - \theta| < |X_1 - \theta|\} < \frac{3}{4},$$

with this probability strictly increasing in α for $0 < \alpha < 1$, approaching $1/2$ as $\alpha \to 0$ and approaching $3/4$ as $\alpha \to 1$.

Lemma: If Z_1, Z_2 are independent and identically distributed, with the distribution continuous and symmetric about 0, then

$$\frac{1}{2} < P\{|\alpha Z_1 + (1 - \alpha)Z_2| < |Z_1|\} < \frac{3}{4} \text{ whenever } 0 < \alpha < 1.$$

This probability is strictly increasing in α, approaches $1/2$ as $\alpha \to 0$, and approaches $3/4$ as $\alpha \to 1$.

Proof: Symmetry, independence, and equidistribution give

$$P(Z_1 > 0,\ Z_2 > 0) = P(Z_1 > 0) \cdot P(Z_2 > 0) = \frac{1}{2} \cdot \frac{1}{2} = \frac{1}{4},$$

$$P(Z_1 > 0,\ Z_2 > 0,\ Z_1 < Z_2) = P(Z_1 > 0,\ Z_2 > 0,\ Z_1 > Z_2) \text{ so each is } \frac{1}{8}.$$

In this way it is easy to see that the axes and the 45° lines divide the plane into eight symmetric regions S_1, S_2, ..., S_8 of probability $1/8$ each as shown in Figure 2, omitting the boundary lines which have a total probability of 0.

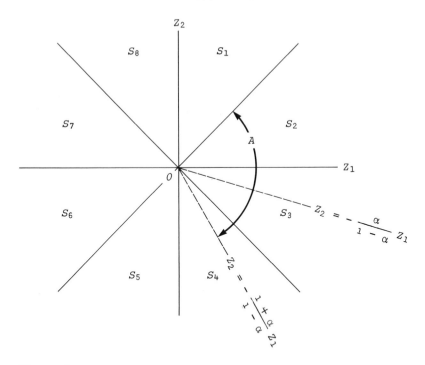

Figure 2. Partitioning of Z_1, Z_2 space in the lemma.

The region $A = \{|\alpha z_1 + (1 - \alpha)z_2| < |z_1|\}$ is easily shown to consist of S_2 and S_3 and the part of S_4 lying above the line $z_2 = -z_1(1 + \alpha)/(1 - \alpha)$, plus the reflections about the origin of these regions. [Consider the inequality of A separately in $(S_1 + S_2)$ and in the four regions into which S_3 and S_4 are divided by the lines $z_2 = -z_1\alpha/(1 - \alpha)$ and $z_2 = -z_1(1 + \alpha)/(1 - \alpha)$]. As α increases from 0 to 1, the position of the line $z_2 = -z_1(1 + \alpha)/(1 - \alpha)$ in S_4 moves clockwise from the boundary with S_3 to the boundary with S_5, and the conclusion of the lemma follows.

Example 2: (In this example X_1 and $\bar{X}$ have identical distributions, yet $\bar{X}$ is better than X_1 in the Pitman sense.)

For X_1, X_2, ... independent, each with Cauchy density given by $f_\theta(x) = [\pi\{1 + (x - \theta)^2\}]^{-1}$, the two estimators $T_1 = X_1$ and $T_2 = \bar{X} = \alpha X_1 + (1 - \alpha)X_2$ with $0 \leqslant \alpha \leqslant 1$ have identical distributions, but the theorem shows that

$$P\{|\bar{X} - \theta| < |X_1 - \theta|\} > \frac{1}{2} \text{ for } 0 < \alpha < 1,$$

with this probability close to 3/4 for α close to 1. Considered separately, the two estimators of θ are indistinguishable; considered jointly, they are not.

In this same example, the two estimators

$$\bar{X}_n = \frac{X_1 + \cdots + X_n}{n},$$

$$\bar{X}_{n+1} = \frac{X_1 + \cdots + X_n + X_{n+1}}{n + 1} = \frac{n}{n + 1} \bar{X}_n + \frac{1}{n + 1} X_{n+1}$$

have identical distributions, the same as X_1, but the lemma applied to $Z_1 = \bar{X}_n - \theta$ and $Z_2 = X_{n+1} - \theta$ shows that

$$P\{|\bar{X}_{n+1} - \theta| < |\bar{X}_n - \theta|\} > \frac{1}{2},$$

with this probability close to 3/4 unless n is small.

Example 3: (In this example $|X_1 - \theta|$ is stochastically smaller than $|\bar{X} - \theta|$, yet $\bar{X}$ is better than X_1 in the Pitman sense.)

For X_1, X_2 independent, with $X_1 - \theta$ and $X_2 - \theta$ each having the same symmetric stable distribution whose characteristic function is given by $\phi(t) = \exp(-|t|^\gamma)$ with $0 < \gamma < 1$, the characteristic function of $\bar{X} - \theta = \alpha(X_1 - \theta) + (1 - \alpha)(X_2 - \theta)$ with $0 < \alpha < 1$ is given by

$$E \exp\{it[\alpha(X_1 - \theta) + (1 - \alpha)(X_2 - \theta)]\} = \exp\{-|t|^\gamma + \alpha^\gamma + (1 - \alpha)^\gamma]\},$$

so that $\bar{X} - \theta$ has the same distribution as $[\alpha^\gamma + (1 - \alpha)^\gamma]^{1/\gamma}(X_1 - \theta)$, where the constant $[\alpha^\gamma + (1 - \alpha)^\gamma]^{1/\gamma}$ exceeds 1 because $0 < \alpha < 1$ and $0 < \gamma < 1$. Therefore,

$$P\{|\bar{X} - \theta| \leqslant a\} = P\{[\alpha^\gamma + (1 - \alpha)^\gamma]^{1/\gamma} |X_1 - \theta| \leqslant a\}$$
$$< P\{|X_1 - \theta| \leqslant a\} \text{ for all } a > 0,$$

while $P\{|\bar{X} - \theta| < |X_1 - \theta|\} > 1/2$ just as in Example 2.

Our thanks to a referee for the following references: Brown and Tukey (1946) show for the symmetric stable distributions with $0 < \gamma < 1$ that the absolute error of $\bar{X}_n = (X_1 + \cdots + X_n)/n$ is stochastically increasing in n, and give similar results for all distributions with heavy enough tails. Yet as we saw in Example 2, for symmetric distributions, increasing n always improves $\bar{X}_n$ in the Pitman sense.

Stigler (1980), for X_1, X_2 independent each with the same continuous distribution symmetric about θ, is concerned with examples showing that using a sample average need not be uniformly advantageous, that is, with the possibility that

$$P\{|X_1 - \theta| \leqslant a\} > P\{|\bar{X} - \theta| \leqslant a\} \text{ for SOME } a > 0. \tag{1}$$

We are concerned with examples showing that using a sample average can be uniformly disadvantageous, that is, with the possibility of the much stronger condition

$$P\{|X_1 - \theta| \leqslant a\} > P\{|\bar{X} - \theta| \leqslant a\} \text{ for ALL } a > 0, \tag{2}$$

and with the conflict between this and the fact that using a sample average is always Pitman-advantageous.

Stigler gives an Edgeworth example showing that the possibility of (1) does not depend on heavy tails: moments up to any order can exist. On the contrary, the possibility of (2) does depend on heavy tails: only absolute moments of order < 1 can exist. For if ϕ is convex then, for $0 < \alpha < 1$,

$$\phi\{\alpha(X_1 - \theta) + (1 - \alpha)(X_2 - \theta)\} \leqslant \alpha\phi(X_1 - \theta) + (1 - \alpha)\phi(X_2 - \theta),$$

giving

$$E\phi(\bar{X} - \theta) \leqslant E\phi(X_1 - \theta)$$

whenever $E\phi(X_1 - \theta)$ exists. Existence of $E\phi(X_1 - \theta)$ for even one nonconstant convex ϕ symmetric about 0 rules out the possibility of (2); so existence of $E|X_1 - \theta|^k$ for even one $k \geqslant 1$ rules out (2): If (2) holds, then EX_1 cannot exist.

REFERENCES

Blyth, C. R. (1972). Some probability paradoxes in choice from among random alternatives. *Jour. Amer. Statist. Assoc. 67*, 366-373.

Brown, G. W., and Tukey, J. W. (1946). Some distributions of sample means. *Ann. Math. Statist. 17*, 1-12.

Pitman, E. J. G. (1937).   The "closest" estimates of statistical parameters.   *Proc. Cambridge Phil. Soc. 33*, 212-222.

Savage, L. J. (1954).   *The Foundations of Statistics.*   New York: John Wiley & Sons.

Stigler, S. M. (1980).   An Edgeworth curiosum.   *Ann. Statist. 8*, 931-934.

BAYESIAN STATISTICS AS HONEST WORK

PERSI DIACONIS Stanford University

ABSTRACT: A review of recent work on exchangeability is offered in the spirit of tolerance and respect for competing views that characterized the work of Jack Kiefer and Jerzy Neyman. The theme is this: If you do a piece of honest work, people will find uses for it no matter what paradigm you work in. This theme is illustrated by tracing the fruits of de Finetti's efforts to understand Hume's problem of induction. His resolution involves exchangeability and his well-known theorem that every exchangeable sequence is a mixture of coin tossing. Subsequent generalizations and an application to the study of human vision are sketched.

NEYMAN, KIEFER, AND BAYES

The organizers of this conference have been kind enough to invite a presentation surveying recent progress in Bayesian statistics. Neither Neyman nor Kiefer was a Bayesian. I found them ever interested in discussing foundational issues. Let me illustrate.

From *Proceedings of the Berkeley Conference in Honor of Jerzy Neyman and Jack Kiefer*, Volume I, Lucien M. Le Cam and Richard A. Olshen, eds., copyright © 1985 by Wadsworth, Inc. All rights reserved.

Neyman and Bayes

When he began working in statistics, Neyman used the Bayesian formulation; almost everyone did. Neyman (1977) gives a charming account of
his change of view. He was teaching a course in statistics and trying
to isolate the properties of a prior distribution that were critical
in various applications. Often Bayes' procedures depend only on the
first few moments of the prior, or in the case of asymptotic properties, on the support of the prior. Neyman continued to wrestle with
how little could be assumed, until Churchill Eisenhart suggested that
he forget about the prior and work on things that did not depend on
prior assumptions. Neyman says "This remark proved inspiring."

I first met Neyman when I began teaching statistics at Stanford
in 1974. It took him several years to recognize me as more than a
passing face. One evening, at a dinner after one of the Berkeley-
Stanford Colloquia, I chanced to be seated next to Neyman. He turned
to me and said "So, what do you do?"

How on earth does one answer such a question? I wanted to have a
serious conversation with Neyman about something I knew; so after a
few seconds of calculation, I managed to say, "Bayesian statistics."
His face fell, and he answered with a clearly disappointed "Oh." We
managed to find something else to talk about.

A few days later, I received a note from Neyman and a copy of a
paper he had written containing a vigorous defense of his foundational
position. It is titled "Frequentist Probability and Frequentist Statistics." Clearly Neyman had not in any sense given up trying to communicate. The paper is in a special foundations issue of the philosophy
journal *Synthese*; my first paper on de Finetti's theorem appears in the
same volume. I was proud to have a paper in the same journal as Neyman;
this gave me courage to read his paper in a critical light. Believing
that I detected much concealed subjectivity, I wrote back. He subsequently sent me other papers clarifying his position and never again
failed to recognize me: "Well, how is Mister Bayesian?"

Over the years, I have continued to read Neyman's papers. For me
his great achievement is the intellectual clarity he brought to our

field. There is never any vague talk about the mysteries of inference.
He says what he means in clear, unequivocal mathematical language.
"Talk" is held to a minimum, and "honest work" is delineated. I think
that the best of Bayesian statistics, the work of de Finetti and
Savage, for example, has this intellectual clarity.

Kiefer and Bayes

Jack Kiefer always worked at the forefront of the paradigm that Neyman
founded. He was an unambivalent frequentist. Yet he never stopped
wrestling with the problems and paradoxes of inference. He has a long
thoughtful paper in that same volume of *Synthese*. His papers on conditional inference represent years of work dealing with one of the most
difficult challenges to the Neyman-Pearson-Wald theory. I talked to
Jack extensively during the conclusion of his preparation of these
papers. His erudition and care in studying the Bayesian position
serves as my model of what it means to be an academic.

I would like to end this recollection of Kiefer and Bayes by proposing an inferential problem. Jack would often reward me with
limericks. For example, after I talked on statistical problems in ESP
at his seminar, he wrote the following:

> A young statistician named Persi
> Proclaimed very loudly a curse, he
>   said "DAMN," and with glee
>   "I'll DESTROY ESP,
> And believe me, I'll show them no mercy!"

My inference problem concerns a limerick that Jack handed me
after I gave a talk on Bayesian statistics at Berkeley. He wrote,

> Our speaker was P. Diaconis,
> On Him was the Bayesian Onus . . .

The problem is: How did it end? I've forgotten, but the limerick is
recorded on a slip of paper, safely put where I'd never lose it. I'm
sure I'll find it someday. But for now, I invite guesses on its ending, based on the evidence (the first limerick) and on the rest of
this paper!

BAYESIAN STATISTICS AS HONEST WORK

The controversies at the foundations of statistics seem different from other controversies in science. The Bayes/non-Bayes controversy has been actively fought for several hundred years with no suggestion of resolution in view. We have no potential crucial experiment that can decide between the subjectivists and objectivists. Indeed, we have theorems proving that "all roads lead to Rome."

The Neyman-Pearson lemma can be regarded as the first complete class theorem. In one version, it states that the only admissible tests for a simple hypothesis versus a simple alternative with zero-one loss are Bayes' for some prior. Abraham Wald (1950), Lucien Le Cam (1955), Charles Stein (1955), and Larry Brown (1976) have developed sweeping generalizations of this result, which states, roughly, that to choose a procedure in any reasonable problem amounts to choosing a prior distribution. This means that computations performed by the Bayesian school are of potential use for frequentists.

The interrelation goes deeper than that. Consider the estimation of a vector of normal means with squared-error loss. *Complete class theorems say you have to choose a formal Bayes' rule.* Common sense suggests that if we have to choose a prior, we may as well choose one that reflects whatever we know, or can guess, about the mean. Work of Jim Berger (1980), and of Bradley Efron and Carl Morris (1973), that emphasizes the practical interplay between admissibility and prior knowledge, has played a large role in the popularity of Stein-like shrinkage estimators.

The above paragraphs point out that frequentists can make use of computations done by Bayesians. In work with David Freedman, I have argued that Bayesians can make use of frequentist properties of procedures, such as consistency and robustness, through a device we call the *what-if method*. The idea is simple: In trying to write a prior distribution, subjective Bayesians have to go through various "elicitation procedures." When considering a specific prior π, consider some possible data x, compute the posterior $\pi^{|x}$ and Bayes' rule $\delta_\pi(x)$, and see if they are satisfactory. The mental exercise goes something like this: What if the data come out to be x. This prior π would lead me to believe $\pi^{|x}$ and take action $\delta_\pi(x)$. Is that really satisfactory? If not, π is probably not a reasonable quantification of what is known.

The what-if method is a variation of what Good (1950) has called *the device of imaginary results*. To some of us, this seems like a perfect characterization of objectivist statistics: A procedure is evaluated through its behavior on data that might have occurred.

The device of imaginary results might seem like the most useless kind of philosophizing to some people. In its defense, let me say that it can suggest concrete computations. Freedman and I have carried out such computations in the problem of estimating the center of symmetry in a location problem with unspecified symmetric errors. These computations are reviewed in Diaconis and Freedman (1983a, 1983b, 1984). It turns out that a straightforward use of the Dirichlet prior in this problem leads to an inconsistent Bayes' rule. This (frequentist) inconsistency in a problem for which a large number of standard procedures result in the right answer suggests (yet again) that the Dirichlet is a strange prior. The inconsistency suggested that certain widely used frequentist procedures—M-estimators with redescending Ψ functions—must also be inconsistent. This turns out to be true for some M-estimators, such as Tukey's biweight, that lack careful scaling.

My point is that philosophical questions can lead to honest work—to taking a fresh look at data or doing a nontrivial piece of mathematics. When one has done an honest piece of work, one can depend that others, perhaps working in very different paradigms, will put that work to use. Here, I am making an inductive inference.

A REVIEW OF RECENT WORK ON EXCHANGEABILITY

I believe that de Finetti's original work on exchangeability was motivated by thinking about a problem in philosophy that I shall call Hume's problem.

Problem: When is it reasonable to believe that the future will be like the past? How can we learn from experience?

De Finetti identifies reasonable belief with probability. Of course, it is not always reasonable to believe that the future will be like the past; past successes may force future failures. Still, in many circumstances, we have no awareness of a hidden mechanism, and our opinions are symmetric or exchangeable. If e_1, e_2, ..., e_n is a given pattern of future successes or failures (take e_1 = 0 or 1), then symmetry becomes $P\{e_1 \ldots e_n\} = P\{e_{\pi(1)} e_{\pi(2)} \ldots e_{\pi(n)}\}$ for all permutations π; de Finetti proved that any such probability could be represented as a mixture of binomial distributions.

Theorem: If P is an exchangeable probability on coin tossing space, then there exists a unique probability μ on $[0, 1]$ such that

$$P\{e_1 \ldots e_n\} = \int p^k (1 - p)^{n-k} \mu(dp) \quad \text{with } k = \Sigma e_i \tag{1}$$

After we observe k successes in the first n trials, the mixing measure for predictions about the future is proportional to

$$p^k (1 - p)^{n-k} \mu(dp)$$

Calculus shows that as n increases, this measure becomes sharply peaked at k/n. Thus, predictions about the future will be approximately consistent with independent identically distributed trials with parameter k/n. We believe that the future will be like the past.

For simplicity, de Finetti's theorem has been stated for two-valued variables. The result holds generally. Though the theorem was proved to answer a philosophical question, the result has applications throughout statistics and probability: Diaconis and Freedman (1984) survey uses and extensions in Bayesian statistics; Kingman (1978) surveys uses in genetics, and Aldous (1983) surveys uses in probability. Koch and Spizzichino (1982) present several further surveys. For de Finetti's views, I recommend de Finetti (1964) and Chapter 9 of de Finetti (1972).

Variations of de Finetti's theorem are available to characterize mixtures of many of the usual parametric models of statistics. Here is an example from Smith (1981): An observed sequence is a location-scale mixture of normal distributions if it is invariant under all rotations fixing the line through the vector of all ones. Freedman (1962) and Diaconis and Freedman (1979) give theorems characterizing mixtures of Markov chains. General theorems have been developed by Scandanavian statisticians under the leadership of P. Martin-lof and S. Lauritzen. A convenient reference is to Lauritzen (1982). When these theorems are sufficiently general, they merge with theorems of modern statistical mechanics; see Dynkin (1978), Georgii (1979), or Ruelle (1978). The interplay between physics and statistics is just starting to be sorted out. A full-length survey of these and other developments is in Diaconis and Freedman (1983c).

The problem of deriving an appropriate version of de Finetti's theorem for two-dimensional arrays arose in the context of applied Bayesian statistics. We wanted to know how to characterize the usual normal models in a two-way analysis of variance. The simplest version of the problem involved random variables X_{ij} with values of zero or one. Such an array is called *row exchangeable* if the law of X_{ij} is the same as the law of $X_{\pi(i)j}$ for any permutation π. Column exchangeability is defined similarly. The set of all row-column exchangeable probabilities is a convex set, and an appropriate version of de Finetti's theorem would be a neat description of the extreme points.

Observe that de Finetti's theorem for exchangeable, two-valued variables can be rephrased in the language of convex sets: The set of exchangeable variables is a convex simplex with coin-tossing measures as the extreme points. The integral representation (1) follows from the fact that every point in a compact convex subset of a metric space can be represented as a mixture of extreme points.

The problem for two-dimensional arrays has been beautifully solved by David Aldous (1981). In his solution, a typical extreme point is indexed by a function

$$\phi : [0, 1]^2 \rightarrow [0, 1]$$

To produce X_{ij} for a specified ϕ, generate independent uniform variables $U_1, U_2, \ldots, V_1, V_2, \ldots$. Fill in X_{ij} as the result of flipping a $\phi(U_i, V_j)$ coin. As ϕ varies, it passes through all extreme points, and thus every row-column exchangeable process is a mixture of such ϕ-processes.

David Freedman and I were hard at work on this problem while I was visiting Bell Labs in 1977. We could show that ϕ-processes were extreme points but didn't know if there were other extreme points. In the midst of this work, I heard about a problem in the psychology of perception. It turned out that ϕ-processes provided insight and counterexamples to 20-year-old conjectures in this field.

The perception research was being carried out by Edgar Gilbert and Bella Julesz at Bell Labs. They were trying to learn exactly when and why the eye sees patterns such as background-foreground. As part of this research they generated random patterns from different distributions, put them side-by-side, and asked if people could see the difference. Many examples are presented in Julesz (1975).

Clearly, two patterns will be visually distinguishable if they have different densities (one dark and the other light). Patterns are also visually distinguishable if they have the same density but different correlations (one pattern smooth and the other clustered). Julesz has been claiming for many years that all the eye can see in these experiments is captured by first- and second-order statistics.

Zero-one processes X_{ij} can be used to generate black and white patterns such as a checkerboard. The conjecture translates to the following: If X_{ij} and Y_{ij} are row-column exchangeable, and

$$E\{X_{ij}\} = E\{Y_{ij}\}, \ E\{X_{ij}X_{k\ell}\} = E\{Y_{ij}Y_{k\ell}\},$$

then the patterns formed by X_{ij} and Y_{ij} will be visually indistinguishable.

In our work on this problem, Freedman and I took Y_{ij} as fair coin tossing. We showed that simple properties of ϕ gave processes with the same low-order moments as coin tossing.

Theorem: a. A ϕ-process has $E\{X_{ij}\} = \frac{1}{2}$ iff $\iint\phi(x, y)\,dx\,dy = \frac{1}{2}$

  b. A ϕ-process has $E\{X_{ij}X_{k\ell}\} = \frac{1}{4}$ for all $(i, j) \neq (k, \ell)$

  iff $\int\phi(x, y)\,dy = \frac{1}{2} = \int\phi(x, y)\,dx$ a.e.

c. A φ-process satisfying condition b has the same third-order probabilities as fair coin tossing: The chance of any three X_{ij} being 1 is 1/8.

d. A φ-process with the same fourth-order probabilities as fair coin tossing *is* fair coin tossing.

The theorem implies that counterexamples had to be found by searching for φ-functions satisfying condition b of the theorem. An example is shown in the figure.

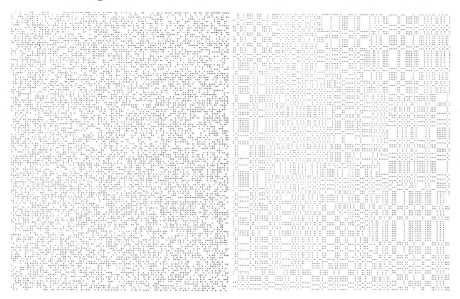

A counterexample to the Julesz conjecture

The two pictures are visually distinguishable. The underlying φ-function can be represented by giving its values at each point of the unit square:

$$\phi(x, y) : y \quad \begin{array}{|c|c|} \hline 0 & 1 \\ \hline 1 & 0 \\ \hline \end{array}$$
$$x$$

This clearly satisfies condition b (and thus condition c), so the patterns have the same first-, second-, and third-order statistics as

fair coin tossing. An alternative description of the process is this:
Fill out the first row of the matrix by fair coin tossing. To fill
out successive rows, flip a coin once for each row: If heads, copy the
first row; if tails, copy the opposite of the first row mod 2. A
third description, given earlier than ours, is in Gilbert (1980).

My paper with Freedman shows dozens of other pictures, which run
from being easy to distinguish through difficult to distinguish. The
pictures allow research to begin on why the conjecture fails.

The point of the example for this discussion is the interplay
between problems in philosophy (de Finetti's theorem), applied
statistics (two-way analysis of variance), mathematics (convex sets
in infinite dimensions), and psychology. Many other areas of Bayesian
statistics can be shown in a similar light: honest work that yields
results of interest in other areas of statistics and science. Perhaps
because of this, the controversy between Bayesians and frequentists
seems to have lost its ability to polarize us. This should give us
time to focus on the new controversy between those who think the computer has taken over and those who proceed only theoretically. I am
sorry that we don't have Kiefer and Neyman on board to help resolve it.

ADDENDA

For those who have read this far, here is the limerick with which the
speaker following me (Erich Lehmann) opened his talk:

> Our speaker was P. Diaconis,
> On him was the Bayesian Onus,
> Though his logic's confusing,
> he's always amusing;
> This limerick here is his bonus.

Jack's limerick was,

*Our speaker was P. Diaconis —
On him was the Bayesian onus.
"He gets highest praise",
Says the Reverend Bayes.
[So we're giving him herewith a bonus.]*

REFERENCES

Aldous, D. (1981). Representations for partially exchangeable arrays. *Jour. Multiv. Analysis*, 581-598.

Aldous, D. (1983). Exchangeability and related topics. Ecole d'e te de St.-Flour 1983. To appear, Springer-Verlag, Lecture Notes in Mathematics.

Berger, J. (1980). *Statistical Decision Theory*. Springer-Verlag, New York.

Brown, L. (1976). *Notes on Statistical Decision Theory* (mimeographed notes, Ithaca, New York).

Diaconis, P. (1977). Finite forms of de Finetti's theorem on exchangeability. *Synthese 36*, 271-281.

Diaconis, P., and Freedman, D. (1980). De Finetti's theorem for Markov chains. *Ann. Prob. 8*, 115-130.

Diaconis, P., and Freedman, D. (1981). On the statistics of vision: the Julesz conjecture. *Jour. Math. Psychol. 24*, 112-138.

Diaconis, P., and Freedman, D. (1983a). Frequency properties of Bayes' rules. In G. Box, T. Leonard, C. F. Wu (eds.), *Scientific Inference, Data Analysis, and Robustness*, Academic Press, New York, pp. 105-116.

Diaconis, P., and Freedman, D. (1983b). On inconsistent Bayes estimates of location. To appear, *Ann. Statist.*

Diaconis, P., and Freedman, D. (1983c). Partial exchangeability and sufficiency. In *Proceedings of the Indian Statistical Institute Golden Jubilee International Conference on Statistics: Applications and New Directions*, Calcutta, pp. 205-236.

Diaconis, P., and Freedman, D. (1984). On the consistency of Bayes' rules. To appear, *Ann. Statist.*

de Finetti, B. (1964). Foresight: its logical laws, its subject sources. In H. Kyburg and H. Smokler (eds.), *Studies in Subjective Probability*, Wiley, New York.

de Finetti, B. (1972). *Probability, Induction and Statistics.* Wiley, New York.

Efron, B., and Morris, C. (1973). Stein's estimation rule and its competitors: an empirical Bayes approach. *J. Amer. Statist. Assoc. 68*, 117–130.

Freedman, D. (1962). Invariants under mixing which generalize de Finetti's theorem. *Ann. Math. Statist. 33*, 916–923.

Freedman, D., and Diaconis, P. (1981). On inconsistent *M*-estimates. *Ann. Statist. 10*, 454–461.

Gilbert, E. (1980). Random coloring of a lattice of squares in the plane. *SIAM Jour. Alg. and Discrete Methods 1*, 152–159.

Good, I. J. (1950). *Probability and the Weighing of Evidence,* Griffen, London.

Julesz, B. (1975). Experiments in the visual perception of texture. *Sci. Amer. 232*, 34–43.

Kiefer, J. (1977a). The foundations of statistics—are there any? *Synthese 36*, 161–176.

Kiefer, J. (1977b). Conditional confidence statements and confidence estimators (with discussion). *Jour. Amer. Statist. Assoc. 72*, 789–827.

Kingman, J. F. C. (1978). Uses of exchangeability. *Ann. Prob. 6*, 183–197.

Koch, G., and Spizzichino, F. (1982). *Exchangeability in Probability and Statistics*, North-Holland, Amsterdam.

Lauritzen, S. (1982). *Statistical Models as Extremal Families*, Aalborg University Press, Aalborg.

Le Cam, L. (1955). An extension of Wald's theory of statistical decision functions. *Ann. Math. Statist. 26*, 69–81.

Neyman, J. (1976). The emergence of mathematical statistics: a historical sketch with particular reference to the United States. In D. B. Owen (ed.), *On the History of Statistics and Probability*, Marcel Dekker, New York, pp. 147–193.

Neyman, J. (1977). Frequentist probability and frequentist statistics. *Synthese 36*, 97–129.

Smith, A. M. F. (1981). On random sequences with centered spherical symmetry. *Jour. Roy. Statist. Soc., Ser. B 43*, 208–209.

Stein, C. (1955). A necessary and sufficient condition for admissibility. *Ann. Math. Statist. 26*, 518–522.

Wald, A. (1950). *Statistical Decision Functions.* Wiley, New York.

DATA ANALYSIS: IN SEARCH OF AN IDENTITY

PETER J. HUBER Harvard University

1. REVOLUTION OR EVOLUTION?

Half a century ago, Neyman started what may be called the measure theory revolution in statistics.

Last month (May 1983), there was a day-long session at the AAAS Meeting in Detroit that was devoted to a more recent revolution: "The revolution in the analysis of scientific data: Concomitant of the computer revolution."

During the Detroit session, I began to wonder whether the above events really were scientific revolutions in the sense of Thomas Kuhn (1982), or mere steps in an evolutionary process.

Evolution, if we take the word in a literal sense (*to evolve: unwind*), progresses along a widening spiral. After some time it returns, although in a different track, to an earlier stage of the

Prepared with the partial support of ONR Contract N00014-79-C-0512 and NSF Grant MSC-8200914.

From *Proceedings of the Berkeley Conference in Honor of Jerzy Neyman and Jack Kiefer*, Volume I, Lucien M. Le Cam and Richard A. Olshen, eds., copyright © 1985 by Wadsworth, Inc. All rights reserved.

development and takes a fresh look at business left unfinished during
the last turn.

The Student-Fisher-Neyman-Egon Pearson-Wald phase of statistics
can be considered a reaction to the preceding period. It stressed
those features in which its predecessor had been deficient and paid
special attention to small sample statistics, to mathematical rigor,
to efficiency and other optimality properties, and coincidentally, to
asymptotics (because few finite sample problems allow closed form
solutions).

During the past decade we seem to have entered a new phase.
People usually associate this phase with the computer, and there is
no doubt that the computer is an important driving force behind it.
But we would be mistaken to treat the move as the mere reaction of
statistics to high tech. I believe what we witness now is another
go-around at an earlier phase of statistics—namely, at the descriptive statistics characteristic of the 19th century.

Between 1920 and 1960, the emphasis on small samples and effi-
cicney inevitably had led to a neglect of situations in which bias
is more important than variability. If you were a mathematical
statistician, you would instinctively shun problems that were unsuitable for a treatment in probabilistic terms. In particular,
descriptive statistics was not dignified enough to be practiced in
print, in a textbook. The eclipse of statistical graphics in the
1940s and 1950s may go back to the same cause.

Tukey (1962) probably was the first to recognize the problem (as
usual); but he was too early, and his paper had little repercussion.
In this article and in his later book on EDA (1977), he stressed the
point about bias being more important than variability, and he
deliberately concentrated on pencil-and-paper (plus pocket calculator)
methods, thereby indicating that he did not regard the computer as
the prime mover. Of course, this creates an unresolved conflict:
Pencil-and-paper methods are matched to small data sets, but in small
data sets, random effects are rarely negligible against the systematic
ones.

2. DATA ANALYSIS AND THE WORKSTATION ENVIRONMENT

Most data analysis is done by applied scientists, not statisticians. Among professional statisticians, and in particular among academic ones, it is still a neglected stepchild. Practicing data analysts come from very diverse fields; correspondingly, they speak very different tongues, and this obscures the basic unity of the subject matter.

Although the specific methods of analysis differ widely, there are many common features: The data sets are large and the analysis is informal and often involves ingenious improvisation.

The data sets, of course, become large and larger because of the increasing computerization of data collection and storage. Their size makes analysis by hand harder, but it favors informality: Either the effect is obvious without a statistical test, or it is spurious (or unimportant) despite its being statistically significant. But without some imaginative preprocessing, the effect may not become visible to the naked eye, and one cannot expect to hit a successful approach without some trial and error. It is easy to document the basic unity in a positive sense through stories of successful data analyses (mostly from the physical and biological sciences, e.g., plate tectonics, pulsars in astronomy, chronobiology), but also negatively, through examples (mostly from the social sciences) in which we have seen analyses go wrong because the investigators neglected to check the qualitative validity of the models informally.

A spectacular example involving the three features mentioned above is the discovery of the *mascons* (mass concentrations) under the lunar maria by Muller and Sjogren. They hypothesized that some seemingly negligible discrepancies between the observed and the calculated motion of the lunar orbiters were caused by irregularities in mass distribution below the surface of the moon. They represented this distribution by a sum of spherical harmonics, fitted the coefficients by least squares, and finally plotted isodensity contours onto a topographical map of the lunar surface. The discovery was made literally when the map emerged from the plotter (Muller and

Sjogren 1968, and personal communication). The persuasive test of the correctness of the hypothesis was not statistical; it consisted in the observation that these concentrations were systematically matched to visible surface features of the moon.

We believe that the basic similarities are sufficiently strong that the computer hardware reaching the market now—powerful, and increasingly affordable, single-user workstations with good and fast graphics—is likely to catalyze a unification of data analysis through a shared hardware and software environment. The most powerful push for unification will come from the economics of software development, because it is so slow and expensive.

Stated in simple terms, the common core requirements for a basic data analysis system are: The data analyst should be able to improvise as easily as with pencil and paper, even if he has quite large data sets, and he should be able to branch out and follow a clue when it pops up. One rarely has the luck of Muller and Sjogren, who hit success on the first try. This mandates the interactive approach, in all cases but those in which one needs archival output or has to process truly huge data sets.

If one is to deal with realistic data sets, the computer must have a large user address space, and to make best use of our eyes to discover patterns, we need high resolution, instantaneous graphics (otherwise, the man-machine feedback loop does not work properly). Hence, we need great local processing power.

If we add up all the hardware requirements implied by these desiderata, we end up at a single-user workstation with a performance level that is just now beginning to be reached by top class micro-computers, which is attached to a fast network with mass storage and high quality I/O (for graphical hardcopy).

Before discussing the software requirements, I need to describe the data analytic paradigm that is slowly emerging.

3. THE DATA ANALYTIC PARADIGM

For a year now, our research group has had a working prototype of a
data analysis system that corresponds to the above description. It
runs on a VAX-11/780 under VMS and uses an Evans and Sutherland Multi
Picture System for graphics. We are currently porting it to an
Apollo DN600 workstation. Our system is deliberately open-ended and
designed to invite improvisation. Some well-defined patterns of
usage begin to emerge. The principal activities of our users can be
grouped into the following five classes:

> Inspection
> Modification
> Comparison
> Interpretation
> Modeling

Inevitably, when you put a new data set on the system, the first
step is *inspection*. On our system, this usually amounts to inspecting
a lot of two- and three-dimensional scatterplots. This inspection
serves to increase your familiarity with the data set and allows you
to learn the layout of the items in data space. One locates and
identifies interesting and typical points—in particular, isolated
points sitting in sparsely populated regions (they should be checked
for validity). Then one looks for clusters. Do the visible clusters
correspond to values of categorical variables in the data? Vice
versa, do the categorical variables aggregate the points into (possibly overlapping) clusters? Should we introduce new categorical
variables to describe the clusters?

As a next step, we shall usually want to *modify* the representation
of the data. Perhaps, we shall perform a nonlinear transformation to
adapt the representation better to the dynamic range of the graphics
device and our eyes. Perhaps, we shall highlight some landmarks in
the data or draw some connecting lines (to show a chronological ordering or a minimal spanning tree). Perhaps, we shall fit some smooth
curve or surface to the data and concentrate either on the smooth part
or the residuals. The ability to create and to modify the graphical

representation of data interactively and to watch it change in real
time is important: We see more when we interact than when we merely
watch, and it is much easier and faster to find informative views
this way.

Structure is discovered by *comparing* two or more things. This
was very clearly expressed by the father of statistical graphics,
William Playfair, in 1821 (p. 54):

> It is surprising how little use is derived from a knowledge
> of facts, when no comparison is drawn between them. Facts
> separately appear to be like diamonds in a rude state, they
> neither serve for ornament nor use.

Playfair wrote these words near the end of his graphical analysis
of the development of wheat prices and wages between 1565 and 1821.
His analysis nicely exemplifies some of the other steps too: His
first graph shows the inflationary rise of both prices and wages
and the wild fluctuation of prices; he comments on these features and
modifies the presentation, averaging the prices over 25-year intervals and plotting them in his final graph in terms also of wages,
to facilitate comparison and interpretation.

Once discovered, structure is *interpreted* by thinking in *models*.
Thus, we shall compare the data not only to other similar data sets
and to different views of the same data, but also to various models;
and we ordinarily will iterate the process by inspecting features of
the model and of the residuals from the model separately.

The first three steps—inspection, modification, and comparison—are amenable to a unified treatment and therefore should be
considered as part of statistics. The last two—interpretation and
modeling—belong to the specific area of expertise of the applied
scientist. However, the five steps are inseparably intertwined, and a
good data analysis system must provide an easy transition between the
common statistics part and the application-specific modeling.

The tools needed for this predominantly visual and impressionis-
tic type of analysis differ somewhat from the usual mainstays of the
statistical packages. The data handling and manipulation tools are
more important than the statistical algorithms, and among the latter,

smoothers (that is, nonparametric curve and surface fitters) are more important than, say, analysis of variance tools; principal components and multidimensional scaling methods totally eclipse hierarchical clustering.

In our experience, interactive data analysis on a workstation is done in an intensely conversational style: The scientist interested in the data and the expert for the data analysis system sit side by side in front of the screen and discuss what they see (it is not always easy to distinguish between features of the data and artifacts of the analysis) and they mutually suggest the next actions to be tried.

Interactive data analysis is hard work: One accomplishes much in a short time, but the load on the human memory is tremendous (because one tends to create so many open ends of investigation). A session of 1 to 1.5 h is about the maximum we could stand before we began to blunder. A typical analysis thus may amount to half a dozen sessions spread over a week, and it may produce a few hundred kilobytes of script files and several megabytes of monitor files (cf. next section).

4. THE SOFTWARE ENVIRONMENT

We find that our own data analysis sytem is most often used for relatively simple inspection and modification tasks. Moral: *Keep simple things simple.* On the other hand, it is abused with increasing frequency for the ad hoc improvisation of very sophisticated tasks that the system designer never could have anticipated. Moral: *The system should be universal* (in the sense of a universal Turing machine, which can compute everything machine computable, albeit inefficiently).

The system designer is confronted with a basic conflict:

The system must be flexible and facilitate improvisation. Graphics must provide instant feedback.

The first requirement mandates an interpretive approach, involving considerable system overhead. The second requirement asks for hardwired, inflexible programming to achieve the required speed. Note that we can steer a picture toward some goal (e.g., toward showing a plane edge-on) without consciously thinking about what we are doing, only if the response time is comparable to our own reaction time.

The solution we chose can be illustrated by the following diagram.

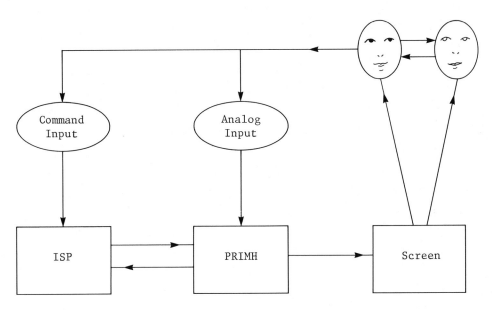

Figure 1. Structure of our data analysis system

The graphics subsystem PRIMH involves the human in a tight feedback loop, whereas the interactive statistical processor ISP forms a much looser loop. It is possible to leave PRIMH to execute an ISP macro, with automatic return to PRIMH.

In detail, the facilities offered are the following:

PRIMH: Graphics

Choose a coordinate triple to be viewed.

Scale, rotate, translate, and produce a 3-D effect by motion parallax.

Identify items (by showing selected labels).

Highlight a subset (by blinking).

Connect any points by lines.

Show stereo pairs.

Show marginal histograms.

Masking/slicing.

Hardcopy.

ISP: Data Analysis

An (almost) full-scale interactive programming language.

Basic statistical primitives, such as regress, boxplot, Fast Fourier transform.

Macros (= procedures) with arguments, local variables, and loops.

Aids to the human memory, such as help files, editable and executable script files (list of the commands typed in by the user), and monitor files (copy of the complete screen output).

Should provide easy access (and does in part) to such outside facilities as data bases, script and macro libraries, editors, text formatters, and other packages.

The principal credits go to my co-workers Hans-Mathis Thoma and to David Donoho, the main developers of PRIMH and ISP, respectively.

The main reason why we chose a version of ISP as our interactive data handler was the fact that Donoho, as an undergraduate, had been the principal designer of the Princeton ISP. Before Donoho joined our group, Thoma and I had considered pressing MINITAB into service, but with serious misgivings about its inherent limitations. The Bell Labs S was out of the question because it was running under UNIX, whereas we were using a VMS VAX; moreover, S would have been too slow for many of our purposes. APL would not interface well. Other packages were too dependent on a particular operating system or programming language, or they were simply too big. We learned to appreciate the advantages of a small and lean system (e.g., building an executable ISP version on the Apollo, starting from the RATFOR/FORTRAN source takes about 30 min).

5. RELATION TO THE NEYMAN-PEARSON THEORY

The recommendation of orthodox mathematical statistics is proverbial:
"Don't look at the data before you test." Conversely, you should not
look afterward either, because you might regret the decision you have
made.

On the other hand, every responsible applied statistician will
want to peek at the data first to check it for irregular features
and to see whether it conforms to the model assumptions on which the
tests or estimates are based. But care is neeeded with the analysis
of features discovered when peeking.

My personal experience with moderately sized data sets is as
follows. If I discover a feature by looking at the data and then
test the hypothesis that the feature does not exist, I always seem
to get P-values in the range 0.01 to 0.1.

My hypothetical explanation is: If $P > 0.1$, I would not see the
feature and therefore would not test. So my test really is conditional on $P < 0.1$. Since, under the null hypothesis, P-values are distributed uniformly in the interval $(0, 1)$, this implies that the actual
level of my test is ten times the nominal level.

On the other hand, if P is small (< 0.01), then I seem to accept
uncritically the feature as real without bothering to test.

Of course, the above numbers 0.01 and 0.1 are crude—perhaps
the former should be 0.001 and the latter 0.05 or 0.2—and of course
they depend on the individual statistician and on the data sets (for
very large sets they certainly are much different). But clearly,
testing the existence of a feature on the basis of the same data in
which it has been discovered is ludicrous. At best, one can gain
some quanitification of the "strength" of the feature seen. Awareness of the size of the estimated standard deviations and of the
associated confidence intervals may prevent you from overinterpreting
the data.

Rather than condemning either visual analysis or mathematical
statistics, as some protagonists of the two opposing viewpoints are
inclined to do, one should treat them as complementary tools.

6. RELATION TO COMPUTER SCIENCE

The recent flurry of activity in expert (or knowledge-based) systems for statistics gives the impression that this is the point at which the common growth edges of statistics and computer science meet.

Such an expert system is supposed to incorporate some of the knowledge of an expert statistician and to guide an inexperienced user (who may have forgotten most of what he learned in an introductory statistics course some ten years ago). This way it should promote the use of "good" statistics and prevent the well-known abuses.

I believe that, unfortunately, the prototype expert systems (of which I have seen some fragments) are doomed to fail, for two reasons. The first is superficial and therefore correctable. In all the sample dialogues I have seen so far, the system talks down to the user in an unbearable fashion. The user may know little about statistics, but he should be assumed to be a mature and intelligent scientist. Such users may need crutches, but not straightjackets, and they prefer to be in the driver's seat. Dialogues of this sort will make a system unacceptable to all but the proverbial, hapless graduate student in ???ology, who needs statistical topping to make the paper acceptable.

The second reason goes deeper: The activities of the statistical consultant involve much more than straight statistical expertise. The problem, in Neyman's words, is, "for a satisfactory performance of a statistician's duty it is . . . necessary that he or she fully understands the circumstances of experiments, whatever their nature, to which statistical methods are applied" [Reid (1982, p. 183)].

Without these precautions, even experienced statisticians and their clients are likely to fall into one of the common traps: that they use the same words but mean different things; or that the client has prejudiciously and mistakenly settled on one of the few statistical methods he happens to know and then truncates his account, in the manner of Procrustes, to fit that method.

It is not easy for the human consultant to acquire enough background insight to circumnavigate these traps, and for the machine it

may be impossible. Here we run into one of the big unsolved problems
of artificial intelligence: knowledge acquisition [cf. Duda and
Shortliffe (1983)].

Thus, I believe that the present attempts at building expert
systems for statistics are barking up the wrong tree. Nevertheless,
I think they are in the right forest. More precisely, I believe that
expert systems have a future in statistics, provided they are written
by an expert for an expert (or for somebody on the way to becoming one).
That is, we should assume that the user knows what he wants to do but
needs assistance in doing it. Instead of trying to emulate a high
priest, pontificating on the straight and narrow path of doing statistics properly, the system should try to emulate the role of the
intelligent lab assistant, who quietly performs many chores, who knows
much better than his boss how to operate the lab equipment, who keeps
the experimental records in order and is able to retrieve them again.
Once a system exceeds a certain size, even the expert needs intelligent help to assist his or her memory; and the problem gets worse if
the data base and the available tools can be changed and added to
interactively. Instead of trying to make the system an expert stati-
tician, we might better heed the classical admonition inscribed on
the Apollo temple in Delphi—"Know thyself"—and make the system have
active, expert knowledge of itself, so that it can assist the user in
making the translation from the tasks to the tools and from the concepts to the commands.

7. OUTLOOK

I believe that data analysis is at a crucial juncture now. Statisticians have a unique opportunity to exert a unifying influence on its
practice and to add healthy impulses on their own. It would be a
pity if they missed it.

But statisticians will have to change some of their ingrained
ways of thinking. The present-day intellectual role of mathematics
will have to be complemented by an equally strong role of computer
science.

Data analysis will be a source of some new, mathematically challenging problems—for example, in connection with dimension-reducing techniques such as projection pursuit, whose sampling properties are practically unknown, and with various approaches to multivariate smoothing. But this should not obscure the fact that the really new challenges of data analysis lie elsewhere and that they go much beyond the typical statistician's view of computational statistics (statistical algorithms).

These challenges range from the man-machine-graphics interface (involving also the psychology of perception and interaction) to interactive statistical languages and AI approaches to intelligent on-line help.

Interactive statistical data analysis systems are, on one hand, severely constrained by the reality of the data. On the other hand, they must allow for essentially unlimited, unstructured improvisation, and the system must assist the user to impose a structure on what might otherwise become an unmanageable mess. Third, graphics introduces a component of real-time computing. All this adds up to one of the most demanding problems in interactive computing one can imagine.

ACKNOWLEDGMENTS

Data analysis research is a collaborative effort; while the views expressed here are my own, they were shaped through the efforts of my co-workers David Donoho, Ernesto Ramos, and Mathis Thoma, and I have made free use of their ideas.

Note: At the meeting, a videotape was shown, illustrating the inspection, modification, and comparison steps, using the following three data sets:

1. The 3-dimensional distribution of galaxies, based on the Smithsonian Redshift Catalog.

2. Consumer Reports cars data for the model years 1970-1982.

3. Epicenters of earthquakes near Fiji.

REFERENCES

Duda, R. O., and E. H. Shortliffe (1983).   Expert systems research.
    Science 220, 261-268.

Kuhn, T. S. (1962).   *The Structure of Scientific Revolutions*.
    Chicago: University of Chicago Press.

Muller, P. M., and W. L. Sjogren (1968).   Mascons—lunar mass
    concentrations.   *Science 161*, 680.

Playfair, W. (1821).   *A Letter on Our Agricultural Distresses, Their
    Causes and Remedies*.   London.

Reid, C. (1982).   *Neyman from Life*.   New York: Springer.

Tukey, J. W. (1962).   The future of data analysis.   *Ann. Math.
    Statist. 33*, 1-67.

Tukey, J. W. (1977).   *Exploratory Data Analysis*.   Reading, Mass.:
    Addison-Wesley.

IMPROVING CRUCIAL RANDOMIZED EXPERIMENTS–
ESPECIALLY IN WEATHER MODIFICATION–
BY DOUBLE RANDOMIZATION AND RANK COMBINATION

JOHN W. TUKEY Princeton University and Bell Laboratories

ABSTRACT: Crucial experiments involve few treatments and deserve highly
stringent analyses that allow effectively for various sources of possible bias, alternative forms of functional dependence, and alternative
stochastic models. Thus they deserve carefully randomized experiments
and flexibly chosen summaries, for which empirical crucial values are
usually needed. Empirical comparison with *all* randomizations is not
likely to be feasible; comparison with a sample reduces stringency.
Thus choice of a limited "assignment set" of randomizations, followed by
a choice of one assignment from the chosen set at random, is indicated.
If all randomizations fall into such assignment sets, then the assignment set can be chosen by one randomization, and the particular
assignment within the set by a second. When the experiment involves
two treatments and many blocks, we frequently find a few blocks more
important than the rest. Balance over these few blocks is a useful
characteristic of the assignment set, and it is guaranteed by the use
of orthogonal arrays as assignment sets. Desire for stringency in the
face of alternative stochastic models often leads to analyses based on
complicated summary statistics, making calculation of theoretical randomization distributions unlikely. Desire for stringency in the face
of alternative functional models often leads to a desire to use analyses based upon two or more summary statistics. Double randomization
easily accommodates complicated summary statistics. Exhaustion of a

Prepared in part in connection with research at Princeton University
sponsored by the Army Research Office, Durham.

From *Proceedings of the Berkeley Conference in Honor of Jerzy Neyman
and Jack Kiefer*, Volume I, Lucien M. Le Cam and Richard A. Olshen,
eds., copyright © 1985 by Wadsworth, Inc. All rights reserved.

complete or limited set of randomizations makes use of two or more summary statistics in parallel simple and effective (k-ads of individual ranks can themselves be ranked). Simplified numerical examples are given. Attention is also given to less than complete balance and a corresponding notion of average strength.

PART A: PRINCIPLES

1. Crucial Experiments

The main characteristics that make crucial experiments recognized as such are

Need for large efforts and large expenditures, and
Pressure for high standards of quality of inference.

As a consequence,

Crucial experiments involve few treatments, often only two
We want our overall analysis to have as high stringency as we can attain
We must be sensitive to all conceivable sources of bias
We must examine the logic of our inferences as carefully as we can.

Indeed we shall frequently find that:

Many replicates (or many blocks) will be needed to get the necessary precision
Many variations of "the response" may seem to deserve analysis.

Though our discussion will be somewhat more general, this paper focuses on crucial experiments. Weather-modification experiments and large clinical trials are perhaps the most obvious examples.

We shall begin pseudohistorically.

2. The Classical Experimental Test

The pattern of the classical experimental test, of the sort leading from a stochastic model to a P-value, is shown and commented upon in exhibit 1.

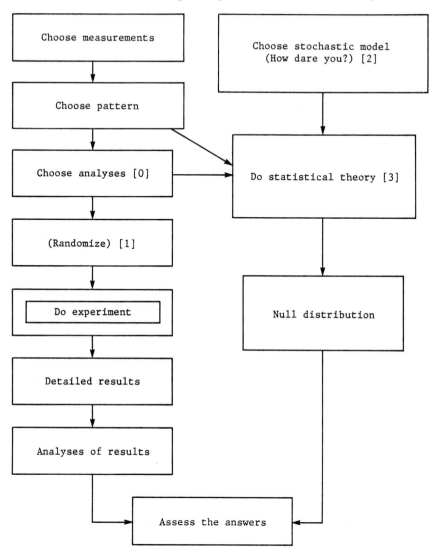

[0] This must be done before the later steps, if the logic is to be seamless.

[1] You're a fool if you don't!

[2] DANGER, DANGER, DANGER!

[3] Often hard, hence too likely to influence choice of analysis, which should be chosen (a) to relate well to the likely functional (nonstochastic) behavior of the treatment(s) and (b) to provide results of high stringency.

Exhibit 1. The classical experimental test.

Except for the comments, there is little unusual. Notice that we use *pattern* to cover everything except how the treatments are assigned, often within blocks. sometimes more generally. Together, *pattern* and *assignment* specify the experiment. Similarly, the "blind details" are all that we learned from the experiment with treatment identification replaced by dummy labels, different for each block.

Those who adhere to the principle of conditionality, and have already been wise enough to randomize the assignment of treatments, have to face the facts that we DO know the blind details and that the treatments might have fallen in any of many ways. They may find it hard to avoid a need to make their inferences conditionally upon the blind details.

Going conditional could strengthen, or weaken, what we perceive to be the stringency of our test. There is a real question, however, as to whether we ought to define the stringency of an experimental test in advance or only after we have the blind details. If the latter, then this definition will also need to be conditional and the actual stringency will not be changed by going conditional.

3. The Classical Conditional Experimental Test

So let us be conditional. To do this, we should separate the experimental results into two pieces:

The blind details, with all the numbers, but with dummy labels whenever treatment names should appear

The *actual assignment* which translates dummy labels into treatment names, almost certainly differently in different blocks.

We often condition on the blind details.

The schematic structure is now as in exhibit 2.

We still have a stochastic model, so we are still in greater or lesser danger, though now, perhaps, the randomization, if done, might protect us. To go further, we need to admit that we needed the randomization.

But if we put our faith in our randomization, at least so far as validify (e.g., that "5%" is no more than something very close to 5%)

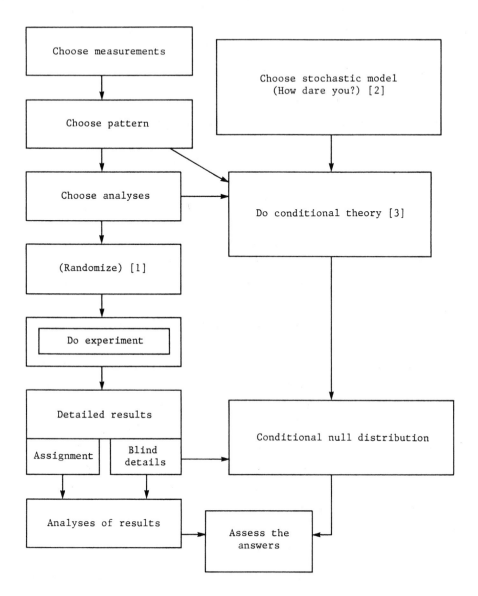

[1] You're a fool if you don't (see [4]).

[2] DANGER possible (see [4]).

[3] Often seems harder.

[4] Your only real protection against DANGER is randomization—*if* you do it.

Exhibit 2. The classical CONDITIONAL experimental test.

goes, we no longer need a stochastic model for validity's sake. Do
we need it for anything? Can we make good use of it?

4. The Completely Randomized Experiment

We now turn to exhibit 3, where both

A stochastic model, for the inevitable confusing details of bahavior
 and

A functional model, for how the treatment really works when it is not
 null,

combine to give us guidance about which analyses we might choose to get
as much stringency as we can.

  We are likely to need to consider alternative functional models.
We are almost certain to need to consider alternative stochastic
models. (And of the two, getting at least one functional model nearly
right is likely to be more important.)

  It is in picking analyses that do well against a diversity of
functional and stochastic behavior that the skills of the experimenter
and the experiment designer combine to make the difference. Danger
has gone away because of randomization, but there is still an important
place for skill and knowledge.

  We cannot be entirely happy with either exhausting or sampling all
possible assignments. (No matter how fast the computer, there will be
jobs that we can't afford to do too many times. For example, just
because we can afford to analyze 1000 assignments, it does NOT follow
that we can afford to do this for 1,000,000. And sampling can waste
stringency.

  The difficulties with getting low moments by theory are two-fold:

To do it we may have to oversimplify our analyses, which we should not
 accept.

Though the resulting facsimile of the randomization null distribution
 may well be close enough, it may not be easy to prove that this
 is so (and in a critical experiment, we must avoid potential
 cracks in our logic).

  Unrestricted randomization has done much, but we want still more,
something better. Can restricted randomization—in which we select,

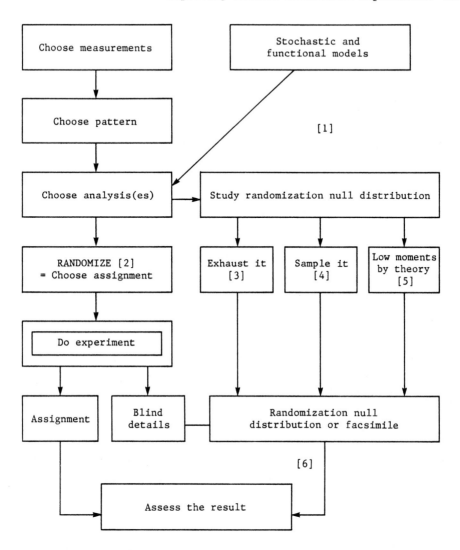

[1] DANGER has gone away.

[2] Now a MUST.

[3] May involve too much effort; 2^{20} or 2^{30} alternatives!

[4] Likely to be sloppy at [6].

[5] Generally sound, but precision may be limited—and hard to establish.

[6] Additional uncertainty associated with either sampling or use of only low moments.

Exhibit 3. The completely randomized experiment.

not randomly from *all* assignments, but randomly (with equal probability) from a carefully selected set of assignments—help us to take the final steps?

5. The Limited Randomization Experimental Test

Look then at exhibit 4, neglecting the dashed boxes and arrows, in which we plan to use an assignment set that is

Large enough to make 5%, 2.5%, ..., 0.5% correspond to enough assignments so that granularity will not really matter, but also

Small enough to make computing the answers once for each assignment a bearable effort.

(This seems likely to have us use sets of from 500 to a few thousand assignments.)

We have now

Sent DANGER (to validity) away,

Avoided extra sloppiness in finding *P*-values,

Left an important place for skill and knowledge (a) in choosing the alternative functional models, (b) in choosing the alternative stochastic models, and (c) in using these choices to guide the choice of analyses.

Some problems still remain—problems that could be lived with, but problems still worth dealing with, namely,

How do we make it hard to question our initial choice of the assignment set?

How do we learn to live with choosing more than one answer in the beginning?

How does our combination of answers affect the number of answers we shall be wise to choose in the beginning?

We shall address these questions in turn.

6. The Doubly Randomized Experimental Test

Look again at exhibit 4, this time including the dashed boxes and arrows.

If we could conveniently subdivide "all assignments" into disjoint "assignment sets" and reveal the resulting family of assignment

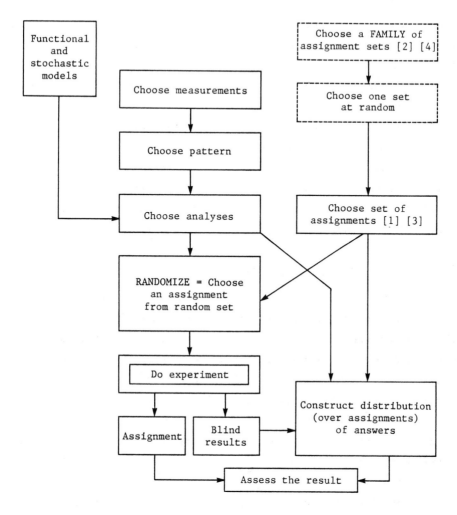

[1] NO appreciable LOSS of stringency because of limited number of assignments in set, if ≥ 512, say.

[2] Especially IF analyses were selected to do well on average (over many sets of blind results), we may lose some stringency IF the chosen assignment SET does not resemble the set of ALL assignments in whatever ways are important for the blind results obtained.

[3] Flack possible about "did you choose the assignment set you said you did?".

[4] Convenient to have disjoint assignment sets that together include all assignments compatible with the chosen pattern.

Exhibit 4. The limited-randomization experiment. (Neglect dashed portion until Section 6.)

sets in advance of doing an experiment, we should avoid any argument about which one we used. Revealing the family does nothing to reveal the assignment actually used to anyone supposed to be blind to it. Yet when the assignment actually used is ultimately revealed, since it belongs to one and only one assignment set, the choice of that set is equally well documented.

If we want to choose our analyses on the basis of stringency of overall performance, we want to make the few thousand assignments of each assignment set look like the millions, billions, or trillions of all assignments—in whatever way is most important. Meeting this requirement is more likely to differ from one subject matter area to another than is meeting any of the other requirements we have discussed.

7. The Dog with Many Heavy Tails

If one block in our experiment is like the tail that wagged the dog, we are in trouble. If the outcome of one of the blocks dominates, then the random assignment of treatments in that block determined the experiment's outcome. (Since we are likely to have only two treatments, our outcome depends on a single coin flip: hardly a good basis for a statistical inference, except at $P = 50\%$!)

In a weather-modification experiment this could happen, and our planning of analyses might well be such as to protect us from such a contingency. (Explaining why we are putting in such protection may require considerable expository skill.)

A more likely situation in weather modification is one with two treatments (one is usually a control) in which, out of perhaps 30 to 60 blocks, something like 4 to 6 show very large responses (perhaps torrential rains). If total quantity (say, of rainfall) is what should concern us, we may well not wish to scale down such large values (as we otherwise might do, for example, by summing the square roots or the cube roots of rainfall in each storm).

In such a situation, the 4 to 6 "heavy" blocks are the dog's many heavy tails. If we knew in advance which blocks would be heavy, we would have an easier task. But we do not. Instead, we want to choose the assignment sets so that, whatever set of j blocks nature picks to be heavy, the assignment sets will each contain 2 choices of assignment in those j blocks equally often. (They will surely occur equally often among all assignments.)

If we write our assignment set as a matrix of + or − with perhaps 30 rows (corresponding to blocks) and perhaps 1024 columns (corresponding to assignments), this requirement of balance for any set of j rows is just the definition of an orthogonal array of strength j.

The use of orthogonal arrays in this way was suggested by Brillinger et al. (1978, pp. G1-G4). Through the courtesy of my colleague Neil (N. J. A.) Sloane, we were able to call attention to the known orthogonal arrays in exhibit 5.

In each of seven instances cited in exhibit 5, the orthogonal arrays are co-sets of a corresponding code [see MacWilliams and Sloane (1977)], so that the corresponding families of assignment sets are made up of disjoint sets and cover all possible randomizations just once. So these orthogonal arrays give us exactly what we should like for our assignment sets.

Exhibit 5. Some Useful Orthogonal Arrays*

Size of assignment set	Strength = number of blocks always balanced		
	4	5	6
5112	⩽ 23	—	—
1024	⩽ 32	⩽ 24	—
2048	⩽ 63	⩽ 33	⩽ 23
4096	—	⩽ 63	—

*Entries are maximum total number of blocks.

8. Two or More Analyses—Possibly Quite Similar But Possibly Not

Our next question is dealing with a multiplicity of analyses. If we feel we must control the simultaneous significance, or confidence, level of our procedure, the simplest and quickest thing to do is to call on the Bonferroni inequality, multiplying by K the more extreme

of the K different tail areas (one for each of K different analyses).
This need not be wasteful, but it can be.

Sometimes the full Bonferroni adjustment is necessary; we are
all used to a two-tail P-value—covering A unusually high or A unusually low—that is just twice whichever single-tail P-value is least.
At the other extreme, there are times where there need be no adjustment at all; if the assignments that give the 5% highest values of
Result A are, collectively or individually, the same as those that
give the 5% highest values of Result B, then either A or B among its
highest 5% of all values occurs exactly 5% of the time by chance. At
5%, in this case, no adjustment is required.

Notice that this second special case can arise for results (statistics) that have any of a wide range of correlations (either product-
moment or rank). Knowing how well the extreme 5% matches tells us
little about the other 95%. Equally, knowing a correlation tells us
little about just how well the extremes match. So a correlation-
adjusted Bonferroni is not a viable answer.

The Spirit of Bonferroni. The spirit of the Bonferroni approach is
the understanding that we shall focus on the answer with the smallest
P-value. Bonferroni itself then takes adequate care of the consequences of this focusing by multiplying the smallest P-value by the
number of alternatives considered. What we need to do is to follow
the spirit of Bonferroni: look to the most extreme, but make only
the necessary allowance for our focusing on this extreme.

Rank Combination. Once we say this, what we ought to do (but are
unlikely to be able to do by formula manipulation) in the classical
case of Section 2, is clear. We fix on H different answers. We find
a transient individual P-value, P_h ($1 \leqslant h \leqslant H$) for each. We look at

$$P^\star = \min_h \{P_h\}$$

and then we ask what the null distribution of $P^\star$ is, under our stochastic model. Given this distribution, we can assign a P-value to $P^\star$. If
we could do the formula manipulation, which we probably cannot, all
would be easy.

In the limited randomization case, including the double-random-
ization case, everything is relatively easy, very easy if there are no
ties. So let us start with H kinds of answers for each of the n
assignments in the chosen assignment set. If we stick to one kind of
answer, we can rank the n instances of that answer (with 1 being the
most favorable). Thus we have

$$\left.\begin{array}{r} \text{kind of answer } h \\ \text{assignment } n \end{array}\right\} \text{rank } r_{hn}$$

The spirit of Bonferroni tells us to look at

$$r_n^\star = \min_h \; r_{hn}$$

and then to order the $r_n^\star$, considering the $N/20$ smallest values as
significant at 5% = 1/20. Absent ties, all is simple.

Present ties, as almost always will be the case, we need to be a
little more careful. We illustrate for $H = 3$. Suppose r_{1n}, r_{2n}, r_{3n}
are the ranks for the three kinds of answers corresponding to assignment set n. Consider the pairs $(r_{1n}, 1)$, $(r_{2n}, 2)$, $(r_{3n}, 3)$ and sort
these in lexicographical order $(r_n^\star, h_n^\star) <= (r_n^{\star\star}, h_n^{\star\star}) \leqslant (r_n^{\star\star\star}, h_n^{\star\star\star})$,
where $\leqslant$ means *either* the first entry is smaller *or* the first entries
are tied, *and* the second entry is smaller. (There can be no ties
among such 2-vectors, since $h_n^\star$, $h_n^{\star\star}$, and $h_n^{\star\star\star}$ are all different.)

Now assign a 6-vector

$$(r_n^\star, \; r_n^{\star\star}, \; r_n^{\star\star\star}; \; h_n^\star, \; h_n^{\star\star}, \; h_n^{\star\star\star})$$

to assignment n, and order these 6-vectors lexicographically. There
can be no more ties among these 6-vectors than there were complete
simultaneous ties among the (r_{1n}, r_{2n}, r_{3n}). Thus, if we use this
order to assign a final rank

$$R_n$$

to assignment n, we will face no new ties. (Since we are engaged in
$\leqslant$ significance, the value of R_n, relevant to significance has, in the
presence of tied 6-vectors, to be the largest rank of any 6-vector tied
with the 6-vector corresponding to assignment n.)

Such calculations are not difficult and have no trouble from ties except from those that are inevitable. Their extension to $H > 3$ is almost trivial. Since we can do this for every assignment in our assignment set, knowing only

The assignment set, and
The blind details,

we make a valid randomization test when we ask "how small is R_n for the actual assignment?".

Scalar Equivalents of Vector Ordering. While most computer systems will easily do a lexicographic sort on many fields, so that our 2-vector and 6-vector sorts are easily practiced in vector form, it may be helpful to notice that it is easy to turn such vector sorts into scalar form. If we wish to sort all (a, b), for example, we may always do this by sorting

$$a - \frac{1}{b}.$$

The extension to (a, b, c) is simple, namely,

$$a - \frac{1}{b - \dfrac{1}{c}} = a - \frac{1}{b-} \left| \frac{1}{c} \right.$$

where we have used continued-fraction notation on the right. For our 6-vectors we could sort

$$r_n^\star - \frac{1}{r_n^{\star\star}-} \left| \frac{1}{r_n^{\star\star\star}-} \right| \frac{1}{h_n^\star-} \left| \frac{1}{h_n^{\star\star}-} \right| \frac{1}{h_n^{\star\star\star}-}.$$

Degree of Invariance. The procedure just described, particularly because it uses $h_n^\star$, $h_n^{\star\star}$, and $h_n^{\star\star\star}$, depends on the order in which the kinds, 1, 2, ..., H, of answer were written down. This usually seems a small penalty compared with the possible reduction in ties. This is certainly the case when both the details of the kinds of analyses *and* their order are either published or put in protected neutral storage before the experiment is begun.

When this is not so, we have only

to sort r_{1n}, r_{2n}, r_{3n}, as $r_n^\star \leqslant r_n^{\star\star\star}$,
to sort the vectors $(r_n^\star, r_n^{\star\star}, r_n^{\star\star\star})$ over n to find a (finite reference distribution

to look at the rank corresponding to the actual assignment as a basis for setting a *P*-value.

9. A More Perspicuous Version of Combination, Probably Chiefly of Pedagogic Interest

The essentials of the rank combination just described are

Looking at the most extreme answer,

Comparing different kinds of answers by ranking each in an appropriate distribution, one associated with that kind of answers,

Reranking the more extreme answers (in a form in which we have broken ties as far as we can).

It may be clearer to some if we separate the distributions in which the ranking takes place from the assignment set we are actually using.

In doing this, we do not wish to make the combining distribution very different from that corresponding to the assignment set. "Separate but similar" should be our watchword. This is easy to do by choosing as the comparison distribution (for this kind of answer and these blind details) one or more *other* assignment sets drawn from the same family.

Since the value of answer h, for assignment n of the assignment set, need not be any of the values of that answer for assignments of another assignment set, our modified ranking will usually involve interpolation, thus reducing the number of ties somewhat further.

10. A Different Style of Combination

If we think that our answers, y_{1n}, y_{2n}, ..., y_{hn}, ..., y_{Hn} will be good—particularly that they are likely to give stringent tests for most values of h (all, indeed, except for an unknown few)—the spirit of Bonferroni is inappropriate. Instead of

$$R_n^\star = \min_h\{r_{hn}\}$$

we should probably do better to take

$$R_n^\star = \operatorname*{median}_h\{r_{hn}\}$$
or
$$R_n^\star = \operatorname*{lower\ hinge}_h\{r_{hn}\}.$$

No other change in procedure will be required.

11. "Weighted" Combination = Combination by Stages

There may be times when, of eleven kinds of analysis, we may want to take one very seriously and the other ten somewhat less so. One easy way to do this is to begin by combining the ten and then going on, combining

The one taken seriously, and
The combination of the ten,

as just two analyses, to be treated equally. This will put roughly half our attention on the one taken seriously, leaving perhaps 5% for each of the others.

More than two stages of combination can be used, as can a variety of ways of putting them together.

12. And Still We Need Both!

At this point, we need to note important large-scale diversity, because the results of even the most carefully and narrowly posed critical experiment should be used for other purposes than narrowly answering the question to which the experiment was directed. We do need to extract all the insight and all the uncertain but helpful guidance from it that we can. After we are through with our preplanned analyses, which are all that can contribute to the most trustworthy answer, we need to look for clues, suggestions, and possibilities with the tools of exploratory data analysis. It is only by using the two antithetical approaches, and by keeping their results clearly separate in our minds, that we can obtain full value from such experiments.

PART B: EXAMPLES

13. Preliminary Calculations for the Microexamples

We shall now give some examples designed to show how the "breath goes in here and the music comes out there." To keep everything down to an overviewable size, we shall use double randomization in a situation in which it should never be used, because exhaustion (randomizing *and* analyzing over all assignments) is both easy and more effective. But our examples do illustrate rather clearly just what goes on, both in double randomization and in response combination.

This section also illustrates how much of the analysis can be done "blind" without knowledge of which assignment was randomly selected and actually used. This knowledge can then be used to turn the analysis results into P-values.

Patterns. We envisage an experiment with eight blocks—block 1, block 2, ..., block 8—each producing one treatment response and one control response. We are going to use orthogonal arrays of strength two so that there will be a convenient number, 16, or assignments in each, even though this will mean that our assessed probabilities come in unrealistically large lumps (of 6.25% each). Also, for convenience, we assume a one-sided interest by the experimenter, who wants only to know if the treatment causes a positive change in response.

Exhibits 6, 7, and 8 show three of the $2^8/16 = 16$ possible choices of an assignment set from the chosen family of arrays of strength two.

Exhibit 6. An Assignment Set of Strength Two of (CAPS) 16 Assignments of ± to 8 Blocks. (For Unrealistic Microexample.)

	CAPS randomization set, first 8 assignments							
Block #	A	B	C	D	E	F	G	H
1	+	+	+	+	+	+	+	+
2	+	+	+	+	−	−	−	−
3	+	+	−	−	+	+	−	−
4	+	−	+	−	+	−	+	−
5	+	+	−	−	−	−	+	+
6	+	−	+	−	−	+	−	+
7	+	−	−	+	+	−	−	+
8	+	−	−	+	−	+	+	−

(continued)

Exhibit 6.   Continued

	CAPS randomization set, second 8 assignments (the opposites of the first 8)							
Block #	J	K	M	N	P	Q	R	S
1	-	-	-	-	-	-	-	-
2	-	-	-	-	+	+	+	+
3	-	-	+	+	-	-	+	+
4	-	+	-	+	-	+	-	+
5	-	-	+	+	+	+	-	-
6	-	+	-	+	+	-	+	-
7	-	+	+	-	-	+	+	-
8	-	+	+	-	+	-	-	+

Exhibit 7.   A Second Assignment Set of Strength Two (LOCASE) of 16 Assignments of ± to 8 Blocks.   (Also for Unrealistic Microexamples.)

	First 8 assignments							
Block #	a	b	c	d	e	f	g	h
1	+	+	+	+	+	+	+	+
2	-	-	-	-	+	+	+	+
3	+	+	-	-	+	+	-	-
4	-	+	-	+	-	+	-	+
5	+	+	-	-	-	-	+	+
6	-	+	-	+	+	-	+	-
7	+	-	-	+	+	-	-	+
8	-	+	+	-	+	-	-	+

	Second 8 assignments (the opposites of the first 8)							
Block #	j	k	m	n	p	q	r	s
1	-	-	-	-	-	-	-	-
2	+	+	+	+	-	-	-	-
3	-	-	+	+	-	-	+	+
4	+	-	+	-	+	-	+	-
5	-	-	+	+	+	+	-	-
6	+	-	+	-	-	+	-	+
7	-	+	+	-	-	+	+	-
8	+	-	-	+	-	+	+	-

Exhibit 8. A Third Assignment Set of Strength Two (GREEK) of 16 Assignments of ± to 8 Blocks. (Also for Unrealistic Microexamples.)

	First 8 assignments							
Block #	α	β	γ	δ	ε	η	ζ	ω
1	−	−	−	−	−	−	−	−
2	+	+	+	+	−	−	−	−
3	−	−	+	+	−	−	+	+
4	−	+	−	+	−	+	−	+
5	−	−	+	+	+	+	−	−
6	+	−	+	−	−	+	−	+
7	+	−	−	+	+	−	−	+
8	+	−	−	+	−	+	+	−

	Second 8 assignments (the opposites of the first 8)							
Block #	λ	μ	ν	φ	ψ	ρ	σ	τ
1	+	+	+	+	+	+	+	+
2	−	−	−	−	+	+	+	+
3	+	+	−	−	+	+	−	−
4	+	−	+	−	+	−	+	−
5	+	+	−	−	−	−	+	+
6	−	+	−	+	+	−	+	−
7	−	+	+	−	−	+	+	−
8	−	+	+	−	+	−	−	+

14. Microexamples: Responses and Initial Calculations

We assume that the responses at the individual plots are known as the integers in the second and third columns of exhibit 9, that the two responses chosen are

(Result 1) *total* of ("treated")^3 − ("control")^3

(Result 2) *median* of ("treated" − "control")$^{1/3}$

and that we know that only the assignment sets CAPS, LOCASE, and GREEK might be involved. Here "treated" and "control" are the responses in any one block, and the total or "median" is over the eight blocks. The chosen results seem different enough in structure to make their combination seem likely to be illuminating.

Exhibit 9.   Responses and Initial Calculations

Individual yields and results

Block #	L	and R	Results 1* (+)	(−)	Results 2[†] (+)	(−)
1	1	2	−7	7	−1	1
2	3	6	−189	189	−1.442	1.442
3	3	4	−37	37	−1	1
4	2	3	−19	19	−1	1
5	10	6	784	−784	1.587	−1.587
6	8	12	−1216	1216	−1.587	1.587
7	14	6	2528	−2528	2	−2
8	2	1	7	−7	1	−1

Assignment results

Ass't[‡]	LOCASE Res1	Res2	Ass't[‡]	CAPS Res1	Res2	Ass't[‡]	GREEKS Res1	Res2
a(j)	4685	1.221	A(J)	1851	−1	$\alpha(\lambda)$	409	1
b(k)	−2827	−1	B(K)	−749	−1	$\beta(\mu)$	−2267	−1
c(m)	−1851	1	C(M)	−4713	−1.221	$\gamma(\nu)$	−3167	−1
d(n)	721	−1	D(N)	2827	1	$\delta(\phi)$	4297	1
e(p)	321	−1	E(P)	3079	−1	$\epsilon(\psi)$	4773	1.221
f(q)	−2355	−1	F(Q)	−4357	−1	$\eta(\rho)$	−2739	1
g(r)	−3107	−1	G(R)	735	1	$\zeta(\alpha)$	−1911	1
h(s)	4357	1	H(S)	2327	1	$\omega(\tau)$	661	−1

*Result 1 = (left)^3 − (right)^3 for a block, or the negative of this, and the *total* of these for an assignment.

[†]Result 2 = (left − right)$^{1/3}$ for a block, or the negative of this, and the *median* of these for an assignment.

[‡]Results for assignment in ( ) are the negatives of those shown.

The calculations in the top panel of exhibit 9 proceed as follows:

Responses on the *L* and *R* plots of each block

Result 1 for each block (a) if the assignment is +, which makes "treated" fall on *L*, (b) if the assignment is −, which makes "treated" fall on *R*, giving the same absolute value but the opposite sign for the net response at this particular block.

Result 2 similarly

In the lower panel, we give the results for each of the 16 assignments in each of the assignment sets that might be concerned. We actually give only eight because, as noted in exhibits 6, 7, and 8, the latter eight assignments in each assignment set are exactly the opposites (+ for − and − for +) of the eight that make up the first half. Because of the odd symmetry of the results we are studying, this ensures that the results, individual *and* combined blocks, for the second eight assignments will be the negatives of those in the first eight.

We can now notice how thoroughly the *sets* of 16 results for one assignment set balance those of another. Exhibit 10 has the detail.

Exhibit 10. The (Separately) Ordered Sets of 16 Results for the Three Assignment Sets and the Two Results

	Result 1			Result 2	
CAPS	LOCASE	GREEK	CAPS	LOCASE	GREEK
4713	4685	4773	1.221	1.221	1.221
4357	4357	4297	1.	1.	1.
3079	3107	3167	1.	1.	1.
2827	2827	2739	1.	1.	1.
2327	2355	2267	1.	1.	1.
1851	1851	1911	1.	1.	1.
749	721	661	1.	1.	1.
321	321	409	1.	1.	1.
−321	−321	−409	−1.	−1.	−1.
−749	−721	−661	−1.	−1.	−1.
−1851	−1851	−1911	−1.	−1.	−1.
−2327	−2355	−2267	−1.	−1.	−1.
−2827	−2827	−2739	−1.	−1.	−1.
−3079	−3107	−3167	−1,	−1.	−1.
−4357	−4357	−4297	−1.	−1.	−1.
−4713	−4685	−4773	−1.221	−1.221	−1.221

Note how closely, particularly for Result 1, the columns of 16 results for the 16 assignments for the different assignment sets resemble one another. (This couldn't happen for most random samples of the 256 possible assignments.)

15. Microexamples: Combination Results

Exhibit 11 begins with the assignments of LOCASE, giving for each its result of each kind and the rank of that result among its 16 fellows of the same kind.

Exhibit 11.   The (Auto) Combined LOCASE Results

Ass't	Results and (median) ranks				Minimum individual rank (tie breaker)	Inclusive rank	P-value
	Result 1		Result 2				
a	4685	1	1.221	1	1(1)	1	6.25%
h	4357	2	1.	5	2(5)	2	12.5%
r	3107	3	1.	5	3(5)	3	18.75%
k	2827	4	1.	5	4(5)	4	25.%
q	2355	5	1.	5	5(5)	5	31.25%
m	1851	6	-1.	12	6(12)	9	56.25%
d	721	7	-1.	12	7(12)	10	62.5%
e	321	8	-1.	12	8(12)	11	68.75%
p	-321	9	1.	5	5(9)	6	37.5%
n	-721	10	1.	5	5(10)	7	43.75%
c	-1851	11	1.	5	5(11)	8	50.%

The column head "Minimum individual rank" shows the lesser of the two ranks (for the two results) with ties broken by the other rank.   Thus

$$(4, 5) < (5, 5) < (5, 6) < (5, 7) < (5, 8) < (6, 12).$$

So far all the ranks have been individual ranks.

The remaining columns refer to ranks for the combined order, which we shall call final ranks.   Thus p, with individual ranks of 5 and 9, is the sixth from the top in terms of minimum individual ranks and has actual (inclusive) rank 6.   (If it were part of a tie, its inclusive rank would be the highest final rank participating in that tie.) Since we have 16 assignments in our set, we multiply the inclusive rank by $1/16 = 6.25\%$ to get the P-value of 37.5%.

Notice that the first five assignments according to Result 1 are also the first five in the final ranking, in which they appear in the same order.   For this particular assignment set, these two results, and this set of blind details, considering both Result 1 and Result 2 rather than Result 1 above, have not changed any P-values $\leqslant$ 37.5%. (Assignment m, individually ranked 6 for Result 1, is, however, finally ranked $9 \neq 6$.)

For this combination of assignment set, choice of results, and blind details, there was no need to pay a Bonferroni penalty for

P-values less than 37.5%, because Result 2 did nothing to modify the order produced by Result 1 until we went to ranks corresponding to larger *P*-values.

16. Mesoexample: Exhaustive Analysis of the Two Single Results and of Their Combination

It is not hard to enumerate the values of, say, Result 1 when all 256 assignments are possible. The positive half of the distribution for Result 1, given the observations of exhibit 9 but not the assignment, is given in exhibit 12.

Exhibit 12. The 128 Positive Values of Result 1 When All 256 Assignments Are Considered (in Groups of 8 with the Corresponding Values for the Three Assignment Sets of 16)

	All assignments	(CAPS)	(LOCASE)	(GREEK)
1 to 8 (1) 9 to 16	4787,4773,4773,4759,4749,4735,4735,4721 4713,4699,4699,4685,4675,4661,4661,4647	(4713)	(4685)	(4773)
17 to 32 (2) 25 to 32	4409,4395,4395,4381,4371,4357,4357,4343 4335,4321,4321,4307,4397,4383,4383,4269	(4357)	(4357)	(4297)
33 to 40 (3) 41 to 48	3219,3205,3265,3191,3181,3167,3167,3153 3145,3131,3131,3117,3107,3092,3093,3079	(3079)	(3107)	(3167)
49 to 56 (4) 57 to 64	2841,2827,2827,2813,2803,2789,2789,2775 2767,2753,2753,2739,2729,2715,2715,2701	(2827)	(2827)	(2739)
65 to 72 (5) 73 to 80	2355,2341,2341,2327,2317,2303,2303,2289 2281,2267,2267,2253,2243,2229,2229,2215	(2327)	(2355)	(2267)
81 to 88 (6) 89 to 96	1977,1963,1963,1949,1939,1925,1925,1911 1903,1889,1889,1875,1865,1851,1851,1837	(1851)	(1851)	(1911)
97 to 104 (7) 105 to 112	787,773,773,759,749,735,735,721 713,699,699,685,675,661,661,647	(749)	(721)	(661)
113 to 120 (8) 121 to 128	409,395,395,381,371,357,357,343 335,321,321,307,297,283,283,269	(321)	(321)	(409)

Because of the symmetry of the calculation of Result 1, the negative half of this distribution is the mirror image of the positive half.

Because of the heavily grouped nature of the distribution, the distribution of the corresponding 256 values for Result 2 is easier to write.   Exhibit 13 has the details.

Exhibit 13.   The Distribution of Result 2, given the Observations of Exhibit 9

Value	Cases out of 256	Inclusive P-value
1.221	15*	5.86%
1	77	35.94%
.221	1*	36.33%
0	70	63.67%
-.221	1	64.06%
-1	77	94.14%
-1.221	15	100.00$

*Together, they are the 16 cases[†] in which the four blocks scored ±2, ±1.587, ±1.587, ±1.442 are *all* scored $+ > 0$.   The one for .221 is the one for which the other four blocks, scored at ±1, ±1, ±1, ±1, are *all* scored $-1 < 0$.   The other 15 arise when at least one is scored $+1 > 0$.

[†]The four largest individual values of (Result 1) correspond to the four largest individual values of (Result 2) and are each larger than the sum of the absolute values of the other four individual values for (Result 1).   These 16 cases are those with the 16 largest values of (Result 1).

We can now do, quite easily, the all-assignments combination calculation for the most extreme cases.   Exhibit 14 has the details. We see that it is easy to work out the most extreme 20 combined p-values, which are the same as the first 20 individual p-values for Result 1.

Because of (a) a high correlation between Result 1 and Result 2 and (b) a rather heavy grouping of the Result 2 values, we see that for these extreme assignments, although *not* for all 256 possible assignments, the final, combined P-value has reduced to the Result 1 P-value, showing us that Result 2, as formulated, can be of no additional help to us—for *these* blind details—in reaching $p \leqslant 5\%$ (or $p \leqslant 10\%$, or $p \leqslant 20\%$). Thus this method of combination, by contrast with Bonferroni, charges us nothing extra for looking at either of *these two* results.

Exhibit 14. All Assignments; Combination Near the Upper Tail

				P-values		
Label*	Result 1	Result 2	Result 1	Result 2	Min	Combined
4787	4787	1.221	.39%	5.86%	.39%	.39%
4773A	4773	1.221	1.17%	5.86%	1.17%	1.17%
4773B	4773	1.221	1.17%	5.86%	1.17%	1.17%
4759	4759	1.221	1.56%	5.86%	1.56%	1.56%
4749	4749	1.221	1.95%	5.86%	1.95%	1.95%
4735B	4735	1.221	2.73%	5.86%	2.73%	2.73%
4735B	4735	1.221	2.73%	5.86%	2.73%	2.73%
4721	4721	1.221	3.12%	5.86%	3.12%	3.12%
4713	4713	1.221	3.52%	5.86%	3.52%	3.52%
4699A	4699	1.221	4.30%	5.86%	4.30%	4.30%
4699B	4699	1.221	4.30%	5.86%	4.30%	4.30%
4685	4685	1.221	4.69%	5.86%	4.69%	4.69%
4675	4675	1.221	5.08%	5.86%	5.08%	5.08%
4661A	4661	1.221	5.86%	5.86%	5.86%	5.86%
4661A	4661	1.221	5.86%	5.86%	5.86%	5.86%
4647	4647	.221	6.25%	36.33%	6.25%	6.25%
4409	4409	1.006	6.64%	35.91%	6.64%	6.64%
4395A	4395	1.006	7.03%	35.94%	7.03%	7.03%
4395B	4395	1.006	7.42%	35.94% †	7.42%	7.42%

*Based on the Result 1 value.

†The remaining 256 – 19 = 237 (Result 2) values in this column will be $\geqslant 35.94\%$, so the corresponding minimal % will be $\geqslant 7.42 + .39 = 7.81$, the least value for Result 1.

We were able to calculte exhibit 14 knowing only the assignment set (here "all assignments") and the blind details, without regard to what assignment was actually used. The only requirement to make the final combined P-values valid is that the assignment actually used was "drawn at random (with equal probability)" from the assignment set made up of all 256 assignments.

17.   Microexample with a Third Kind of Analysis

To illustrate the diversity of circumstances that can arise, we now
introduce a third kind of analysis in our microexample, in which

$$\text{Result } 3 = \left(\frac{-1}{\text{"treated"}}\right) - \left(\frac{-1}{\text{"control"}}\right).$$

Exhibit 15 has the resulting numbers.

Exhibit 15. Values of Result 3 for Blocks and Certain Assignment Sets

	Individual contributions for Result 3			
Block	L	R	+/−	−/+
1	−1	−.5	−.5	+.5
2	−.333	−.167	−.167	+.167
3	−.333	−.250	−.083	+.083
4	−.500	−.333	−.167	+.167
5	−.1	−.167	.067	−.067
6	−.125	−.083	−.042	.042
7	−.071	−.167	+.095	−.095
8	−.5	−1	+.500	−.500

		Assignment results			
Assignment	Result 3	Assignment	Result 3	Assignment	Result 3
a(j)	−.646	A(J)	−.297	$\alpha(\lambda)$	+.069
b(k)	−.153	B(K)	−1.070	$\beta(\mu)$	−1.204
c(m)	.297	C(M)	−1.474	$\gamma(\nu)$	−1.320
d(n)	−.831	D(N)	.153	$\gamma(\phi)$	−.213
e(r)	−.098	E(P)	−1.013	$\varepsilon(\psi)$	−.377
f(g)	−1.539	F(Q)	+.046	$\eta(\rho)$	.180
y(r)	−.987	G(R)	−.098	$\zeta(\sigma)$	−.036
h(s)	−.046	H(S)	+.463	$\omega(\tau)$	−1.097

If we try to combine all three results over LOCASE, we get exhibit 16.
Notice that the final P-values are now quite different from the Result
1 P-values. This is because the high value of Result 3, for example,
does not occur for the assignment that gives the highest value of
Result 1. As a consequence, we do pay penalties that often come close
to Bonferroni penalties for two results (Result 2 is still ineffective).

Exhibit 16. Combining Three Kinds of Results for LOCASE

Assignment	Individual rank of Result* 1	2	3	Ordered triad	Final rank	Final P-value
a	1	1	13(−.646)	(1,1,13)	1	6.25%
h	2	5	9(−.046)	(2,5,9)	4	15.%
r	3	5	2(.987)	(2,3,5)	3	18.75%
k	4	5	11(.153)	(4,5,11)	6	37.5%
q	5	5	1(1.537)	(1,5,5)	2	12.5%
m	6	12	13(−.297)	(6,12,13)	10	—
d	7	12	14(−.931)	(7,12,14)	11	—
e	8	12	10(−.098)	(8,10,12)	12	—
p	9	5	7(.098)	(5,7,9)	9	—
n	10	5	3(.931)	(3,5,10)	5	31.25%
c	11	5	5(.297)	(3,5,11)	8	—
f	12	12	16(−1.537)	(12,12,16)	15	—
b	13	12	11(−.153)	(11,12,13)	14	—
g	14	12	15(−.987)	(12,14,15)	16	—
s	15	12	8(.046)	(8,12,15)	13	—
j	16	16	4(.646)	(4,16,16)	7	—

*Value shown in ( ) for Result 3.

Notice also that we need never pay a Bonferroni-like penalty for every possible assignment when, as here, there are no final ties. With N assignments in our assignment set, all N final ranks will a appear; so all integer multiples of $1/N$ will occur as final (inclusive) P-values. (With full Bonferroni, only the multiplies of k/N could appear when there are k different results.) If the different results have their 5% most extreme values for different, nonoverlapping subsets of assignments, however, we shall find ourselves paying close to a full Bonferroni penalty at all significance levels up to $5k\%$. (We will not pay this panalty when the P-value is uninterestingly high.)

PART C: AVERAGE STRENGTH

We have asked for orthogonal arrays in terms simply of *strength*, which we now want to call *complete strength*. Complete strength k means that every choice of k blocks receives balanced attention in our N assignments.

But what matters to us is whether the particular set of blocks that matter in a specific set of blind details is balanced. If many sets of $k + 1$ blocks receive balanced treatment, we should count this to the credit of that assignment set. Though we are then not sure to balance the $k + 1$ blocks important in some particular instance, we do have a good change of doing this.

So let us look at one of our assignment sets, say CAPS, and see what happens.

When we turn to exhibit 6 and ask what happens if we multiply two blocks together, i.e., what happens if we multiply the two corresponding entries in each column and regard the result at the new entry in that column? We find that if we write i for 1, a for 2, b for 3, c for 4, ab for 5, ac for 6, bc for 7, and abc for 8, that

Multiplying any two rows always gives another row, and

The product row is just what $a^2 = b^2 = c^2 = i$ with i the identity
    would call for. (Thus $ab \times abc = c(a^2b^2) = c$ and $a \times bc = abc$.)

We begin by noting that each row is balanced.

Consider any two rows, whose formal interaction is another row. Then we can use the balance of the three rows to show that the two given rows have full 2×2 balance. Suppose, for instance, that the two rows are ab and abc. To be sure of 2×2 balance, it suffices to have three instances of one-way balances, first the direct two, namely,

ab by itself is balanced (8+ and 8-)

abc by itself is balanced (8+ and 8-).

If there are x instances of ++ there must be 8-x of -+, an equal number of +-, and consequently x of -. The product will then have 2 +'s and 16-2 -'s and will be balanced when $x = 4$; so it has 8 +'s and 8 -'s and there are exactly four occurrences of each sign pair. But

Since $ab \times abc = c$, and c is balanced,

the three balance conditions ensure balance of the 2×2 table for ab and abc.

Let us go further and look at triples. Of the 56 triples, 49, such as i, c, and ac, are balanced (since $i \times c = c$, $i \times ac = ac$, $c \times ac = a$, $i \times ac \times c = a$, none of which is i) while 7, such as a, b, and ab, are not balanced (because $a \times b = ab$, $a \times ab = b$, $b \times b = a$, and $a \times b \times ab = i$, where i is one of the products). [The 7 unbalanced triples are a, b, ab; a, c, ac; b, c, bc; a, bc, abc; b, ac, abc; c, ab, abc; and ab, ac, bc.]

Going further shows that 19/35 of the 70 quadruples are balanced, as are 1/7 of the 56 quintuples. None of the sextuples or septuples are balanced, neither is the lone octuple.

In a very reasonable sense, the average strength of this assignment set is

$$1 + 1 + \frac{7}{8} + \frac{19}{35} + \frac{1}{7} + 0 + 0 + 0 = 3.56$$

whereas the guaranteed strength is only 2.

If four blocks require special attention, so that we are concerned with a quadruple and its subtuples, it is easy to show, using the symmetry of the unbalanced seven, that at least three of the four triples in any quadruple are balanced.

This shows us that

For one heavy tail, the corresponding 2 is balanced

For two heavy tails, the corresponding 2×2 is balanced

For three heavy tails, the corresponding 2^3 is balanced with probability 7/8

For four heavy tails, at least three of the 2^3 that are subtuples of the 2^4 are balanced, and the whole 2^4 is balanced with probability 19/35.

This is just about what we would like to see for an average strength of 3.5.

Exhibit 17 summarizes matters. Clearly we are better off than if all triples, but no quadruples, were balanced. A 54% chance of improvement is made because we include the fourth (in order of size or effect), which outweighs a 12% chance that we may have to replace the third by the fourth (in order of size or importance) in order to gain balance.

Exhibit 17. What a Specific Orthogonal Array of Strength 2 (CAPS) Provides

---

If we look at the *two* most important blocks:

• They are surely balanced.

If we look at the *three* most important blocks:

• There are seven chances in eight that all three will be balanced.

• If the first, second, and third are not balanced, the first, second, and fourth will be.

If we look at the *four* most important blocks:

• There is a 54% chance that all four will be balanced.

• If not, there is a 34% chance that the three most important will be balanced.

• If neither (a 12% chance), the first, second, and fourth will be balanced.

REFERENCES

Brillinger, D. R., Jones, L. V., and Tukey, J. W. (1978). "The role of statistics in weather resources management." Volume 2 of *The Management of Weather Resources*. Washington, D.C., U.S. Government Printing Office.

MacWilliams, F. J., and Sloane, N. J. A. (1977). *The Theory of Error-Correcting Codes*. New York, North-Holland Publishing Company.

PRELIMINARY ANALYSIS OF GROSSVERSUCH IV
HAIL DAMAGE ON EXPERIMENTAL DAYS

SHYRL M. DAWKINS, JERZY NEYMAN, and ELIZABETH L. SCOTT
University of California, Berkeley

This study records our last research with Jerzy Neyman, part of a
long series of papers on weather modification. Grossversuch IV is
indeed the fourth experiment organized by the Swiss to study the
possibilities and reliability of hail modification. The experiment
has just ended after seven years, a long, expensive, international
cooperative effort (Federer et al. 1978). The seeding methods and
requirements for a seeding opportunity are Soviet, the seeding
rockets were made in Yugoslavia, the target area and the extensive
radar observations are Swiss, part of the hailpad measures are French
and the rest are Italian, and specialized cloud physics observations
were made by several groups but especially by the Swiss and the
United States, who provided a plane equipped to fly into the thunder-
storm clouds during the latter part of the experiment. Finally, the
direction and the organization of the experiment were Swiss, with
extensive international cooperation.

From *Proceedings of the Berkeley Conference in Honor of Jerzy Neyman
and Jack Kiefer*, Volume I, Lucien M. Le Cam and Richard H. Olshen,
eds. Copyright © 1985 by Wadsworth, Inc. All rights reserved.

For the last thirty years, the idea underlying efforts to modify the weather is that additional nuclei introduced into the incipient storm clouds at the right time and place will alter the effects. The physical processes are not completely understood, but it is thought (Sulakvelidze, Kiziriya, and Tsykunov 1974) that the release of artificial ice nuclei into the heart of the cloud at the -5°C isotherm will induce glaciation, because the large super-cooled water droplets in the region will freeze around the artificial nuclei (this process is called seeding) and thus reduce the liquid water supply and decrease the radius of the hailstones that do form. Under this competing embryo idea of hail modification, the resulting number of hailstones may be increased or decreased but the hailstones will tend to be smaller and thus do less damage.

The weather is extremely variable and not well predicted so that very long and expensive randomized experiments are needed. Hail modification experiments involve the most variable and difficult to measure observations. Our role in the analyses of Grossversuch IV is to study the extended area and extended time effects and to contribute to a better understanding of the physical processes involved. Interesting statistical problems arise, and there are lots of complicated data that contribute to the studies even though a complete resolution is not available.

Hailstones tend to arise in thunderstorms. All such experiments in the United States indicate large decreases in rainfall over wide areas often many hours later. At least partly, these decreases are due to the lessening of the build-up of the typical afternoon thunderstorms. Seeding the storm appears to decapitate it. However, the Swiss experiment Gosssversuch III, intended to decrease hail, showed large increases in rain, even far downwind on the other side of the Alps. The suggestion is that we cannot anticipate what will result in Grossversuch IV.

There are several types of observations available in the target area, essentially the canton of Lucerne. These include radar measurements made during the experimental period, hailpad observations over an extensive network, and verified reports of hail damage made to the national cooperative hail insurance company. Over the rest of Switzerland, there are only reports of hail damage supplemented by observations of hail and precipitation made by the Swiss Meteorological Service along with other physical data on each storm situation. No type of data is very satisfactory because (i) the extreme variability, indeed the patchiness, of the hailfall induce variability in the hailpad measurements made on a systematic grid in the target area, and also create variability in the hail insurance claim data since only insured plots—typically plots planted with more sensitive crops—are reported as damaged. No observations are collected on hail that misses the hailpads or, correspondingly, hail that falls outside of insured plots. Further, (ii) the radar cannot distinguish between hail and intense rainfall over a wide range of wavelength that is of importance in hail modification interpretation. In addition, the radar observations are integrated over distance and occur in the atmosphere and thus can be rather different from local measurements of the corresponding hail damage on the ground. It is important to carry out analyses for each type of observations, to interpret the differences that may result, and to try to construct a combined analysis that can further our understanding of hail modification possibilities.

The confirmatory test set up at the start of the experiment (Federer et al. 1978) involves the deviation from the kinetic energy estimated from actual measurements compared to the predicted energy under the given storm conditions fitted to data from the preliminary years 1975 and 1976. The observations used first were from radar, and it was assumed that one can discriminate between hail cells and rain-only cells. The deviation for cell m occurring on day k in year j of the experiment is estimated to be

$$D_{jkm} = \alpha_j + \eta H_{jkm} + \beta[(T_B)_{jk} - T_B] + \Delta\gamma S_{jk}$$
$$+ \Delta\beta S_{jk}[(T_B)_{jk} - T_B] + \varepsilon_{jkm}$$

where

H_{jkm} is 1 if the cell originates in the target, and 0 if outside,
S_{jk} is 1 if this day is seeded, and 0 if not,

T_B is the concomitant variable,

α_j is a possible year effect.

We want to test whether $\Delta\beta = 0$ and whether $\Delta\gamma = 0$. If both of these coefficients are zero, then there is no effect of seeding on the kinetic energy. If it is true that seeding the clouds tends to reduce hail energy, then both $\Delta\beta < 0$ and also $\Delta\gamma < 0$. On the other hand, it may be that seeding tends to increase hail energy, and this will imply the converse effect on the two coefficients. Since there is a two-tail alternative, we expect a two-tail test.

The first confirmatory test employed is the randomization test, that is, the assignment of the days for which $S_{jk} = 1$ was permuted among all possible days and all the computations repeated correspondingly, so as to find the distribution of the linear coefficient $\Delta\gamma$ under the hypothesis of no seeding effect and also that of the interaction coefficient $\Delta\beta$, as well as the joint distribution of the two coefficients. The actual experiment was randomized with probability one-half of seeding on any experimental day. At the hour when it was decided whether the day will be an expermental day or not, it was not known whether seeding will occur or not; such decisions were made independently by a separate group and placed in sealed envelopes. The observations were collected and analyzed blind to the occurrence of seeding as far as possible.

There were some difficulties with the predicting equation for the kinetic energy as estimated from radar, but the preliminary estimate is $\Delta\gamma = +0.93$, which corresponds to an increase of 150% under seeding. As determined from the randomization computations, the confidence interval ellipsoid is $-0.5 \leqslant \Delta\gamma \leqslant +2.16$, which corresponds to a decrease of -40% up to an increase by a factor of 11. The corresponding significance probability is $P = 0.08$, which is not significant at the standard level but does suggest that there is low probability of no effect. The confidence interval does cover 0 but does no cover -90% as had been claimed by the Soviet operators. The permutation results are illustrated in Figure 1. The point corresponding to the actual experiment lies in the upper right

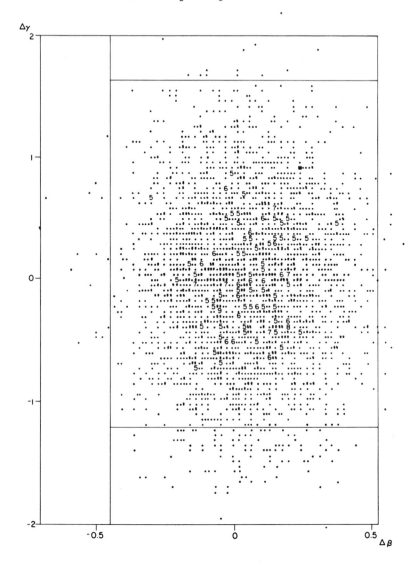

Figure 1. Permutation test for effects of seeding on kinetic energy of hail, as estimated from radar observations. Distribution of 3000 values of $\Delta\gamma$, the linear effect of seeding, and $\Delta\beta$, the interaction effect of seeding, all calculated under the hypothesis of no effects of seeding using 3000 rerandomizations of the label "seeded" among the observed points. The actual point is marked by a square, and appears in the upper right of the distribution, indicating a non-significant increase in both coefficients ascribed to seeding.

portion of the 3000 points produced by rerandomization done under
the hypothesis that seeding has no effect on kinetic energy. The
observed point (marked by a square symbol) has a location corresponding to a somewhat positive effect of seeding, both positive linear
coefficient and positive interaction.

The second confirmatory evaluation of the effect of seeding
according to the Soviet technique used the hailpad observations in
the French test zone (Mezeix and Caillot 1982). In the area shown
in Figure 2, essentially the northeastern part of the target, the
French maintained a dense and regular network of 211 hailpads, each
0.1 m^2, made of polystyrene and elevated above the ground so as to
be sensitive to impacts of hail. The diameters, number, and depths
of impact were measured directly and used then as secondary variables.

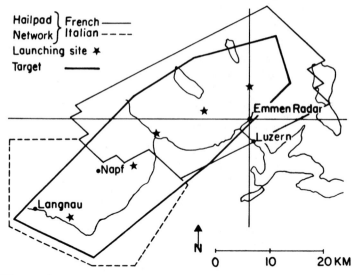

Figure 2. Location of the target area and of the French and Italian
hailpad networks. The five rocket launching sites are also shown,
along with the location of the radar installations.

They were also used to estimate the area of hailfall and the distribution of kinetic energy for each storm cell's hailfall occurring on an experimental day. The distribution on seeded days was
compared to that on not-seeded days as the confirmatory test. The

preliminary results are that there is no significant change in the number of experimental days with zero hailfall, and that there is no significant change in the mean kinetic energy of hailfall on the nonzero experimental days that were seeded versus on those not seeded. There is an indication of a nonsignificant increase in kinetic energy on seeded days. Analyses of the secondary variables produce similar conclusions.

The observations from the Italian network should be included in the analysis before coming to any final conclusions. At present, the 90% confidence results on the hail kinetic energy ratio, seeded to not-seeded, ranges from 0.33 to 3.92, which is wide. The inclusion of the Italian data and the possible stratifications such as seeding coverage, storm type, and the meteorological situation should provide narrower bounds.

The confirmed observations of hail damage from the records of the national hail insurance association, Schweiz Hagel Versicherungs Gesellschaft, were not part of the conformatory evaluation. We decided to use the hail damage data after the experiment had started because only these data appear to allow the possibility of extended area evaluations. We are indebted to Director H. Scharpf for allowing us to use the damage observations and to U. Braun for making the data available and explaining the collection process. We are especially indebted to M. Schüepp who helped us so much in collecting and interpreting the observations.

However, before discussing extended area evaluations, we can examine the hail damage that occurred within the target area so as to provide another assessment of the effect of seeding. One possible measure of the hail damage is the number of communes in the target that received hail damage on an experimental day. Since the communes differ in area and in the proportion of their area that is insured and for what indemnity, a better measure of hail damage may be the proportion of the indemnity coverage that is paid out in an experimental day. As noted above, the expected effect of seeding a hail cell in a cloud may be either to increase or to decrease the probability that hail will fall. Further, given that hail does fall, the damage resulting may be increased or decreased by seeding. We examine each of these possibilities and also their combination using the Neyman $C(\alpha)$ test, which is locally best for these purposes (see Neyman and Scott 1967). In Table 1 are shown the analyses of the possibility of change in the probability that an experimental day will have some hail, that there will be no change in the damage per hail day, and that there will be no change in the damage per experimental day when seeding occurs versus when there is no seeding. The two measures of hail damage, the number of communes with hail damage and the proportion of indemnity coverage that receives hail damage, are analyzed separately.

Table 1 also shows the same Neyman–Scott test applied to the radar observations and then to the hailpad observations so that the conclusions may be compared. The data refer to the period 1977–1981 and provide only preliminary results. There are three columns referring to kinetic energy estimated from three Z–levels for the radar observations; the third level is a weighted combination of the first two, and is believed to provide a better discrimination between observations of hail and of water. The significance probabilities shown are two-tail since the alternative to no effect of seeding may be either increase or decrease. Further study is underway to provide better understanding of the use of radar observations. The next column provides a similar analysis using kinetic energy estimated from hailpad observations, with systematically lower estimates. We are indebted to Waldvogel, Schmid, and Schiesser for carrying out these analyses, using a program prepared by M. A. Wells.

The various kinds of observations and the different measures are not examining the same effect, and none of them is completely satisfactory. Nevertheless, there is rather good agreement in the conclusions. There is no evidence that seeding affects the probability that an experimental day will have hail. On the other hand, the second mechanism, the kinetic energy on a day with hail or, correspondingly, the damage on a hail day measured either in number of communes damaged or in indemnity paid out, tends to be about 50% larger on the experimental days with seeding. Although the increase

Table 1. Comparison of Analyses of Seeding Effect on Hail Modification Using Kinetic Energy Estimated from Radar Observations, from Hailpad Observations, and Using Hail Damage Number of Communes Damaged and Proportion of Indemnity Coverage Receiving Damage

Instrument	Radar$_{Z-level}$			Pad
Kinetic Energy:	E_{56}	E_{61}	E_{GR}	E_{pad}
Percent hail days				
Seeded	77	43	77	50
Not seeded	76	39	76	51
P-value	1.00	0.65	1.00	1.00
Energy/hail day				
Seeded	4.5	3.1	2.2	1.3
Not seeded	2.7	2.1	1.3	0.8
P-value	0.09	0.31	0.13	0.20
Energy/experimental day				
Seeded	3.4	1.3	1.7	0.6
Not seeded	2.0	0.8	1.0	0.4
P-value	0.08	0.24	0.14	0.29

Measure of hail damage (insurance):	% of communes damaged	% indemnity coverage damaged
% days with ins. damage		
Seeded	55	55
Not seeded	63	63
P-value	0.60	0.60
% damaged per hail day		
Seeded	63	0.07
Not seeded	44	0.04
P-value	0.16	0.23
% damaged per exper. day		
Seeded	35	0.04
Not seeded	28	0.03
P-value	0.60	0.48

is not significant at standard levels, there is consistent indication that the significance probability is not large. More complete
studies are needed. The effect of seeding per experimental day
reflects both possible mechanisms created by seeding. The estimated
kinetic energy tends to be about 50% larger on seeded days, and the
damage about 25% larger. The differences in the parallel analyses
of the radar measurements are of the same order of magnitude as the
differences among different instruments.

The extended area computations have been completed only for the
whole of Switzerland, that is, all of the more than 3000 communes
considered together. On those days that are ruled to be experimental days in the target area, there may be hailfalls originating
from storms near the target and also from far away. The total
observations reflect a mixture of effects, if there are any. Not
surprisingly, the percentage of experimental days with hail somewhere is large, 91% without seeding and 95% with seeding in the
target area, a quite nonsignificant increase ($P = 0.64$). The mean
number of communes damaged by hail increases from 41 to 46 per hail
day, and from 38 to 44 per experimental day; both increases and
nonsignificant ($P = 0.70$ and 0.56). The percentage of indemnity
coverage damaged by hail, which presumably is a better measure of
hail damage, increases under seeding from 0.04 to 0.06 per hail day
and from 0.04 to 0.05 per experimental day. The significance probabilities are smaller, 0.34 and 0.27, but not significant. The power
of these analyses is not large even though the time span is six hail
seasons, 1977-1982, partly because the total number of experimental
days is only 95 and also because these days have different storm
types not only in the target area but also in the extended areas.
Further study is in progress.

Hail tends to fall in long swaths. It also tends to accompany
severe storms, moving generally eastward across the country. One
swath will usually strike a series of parcels of land so that it is
not convenient to consider the sum of the number of strikes in small
areas, such as the parcels planted to a single crop by one farmer.
There will be problems of dependence in considering damage in nearby

small areas. The commune has been used here as the unit and seems
to provide a good fit to the Gamma distribution also used by Neyman
and Scott (1967) to develop the locally best test of the effects of
seeding. We are now constructing a more realistic model, using the
hailswath as the experimental variable.

We have plotted the communes with reported hail damage on a
large map for each experimental day, indicating by the shading
employed 12 levels of proportional damage for each commune. We are
indebted to J. Hoben and to Professor H. Häfner for the detailed
construction of these large maps which were drawn at the Computer
Center of the University of California, San Francisco Medical Center.
A reduced copy of the central portion, including the target area, of
a typical map is shown in Figure 3.

We are deeply indebted to Professor M. Schüepp for delineating
the hailswaths on each map, and then estimating the center and time
of each swath and recording any hailfalls reported by the Swiss
Meteorological Service, with their time and location. These help
in assessing the times when hail damage occurs. These times are
shown on Figure 3, marked by a triangle for an officially reported
hailfall, a dot for the times and locations of thunderstorms, with
the time of individual swaths shown at the upper left. The identification number of the swath is arbitrary, moving from west to east.
It is usual for the hailswaths to appear to march across Switzerland
in a discontinuous pattern. Dr. Schüepp has also provided meteorological data, such as storm types throughout the day and wind speed
and direction at a series of levels, for each experimental day.

With the cooperation of Dr. Schüepp, we are starting the study
of the stochastic process of hail damage downwind as possible seeding effects travel downwind. We can study extended time as well as
extended area effects. For each commune within a hailswath, we
propose to study the proportional hail damage separately for each
major crop category, since the value of the crops vary differentially
during the hail season. These studies need new methods of statistical analysis, which we hope will be of interest and will provide new
insight into the mechanisms of hail modification.

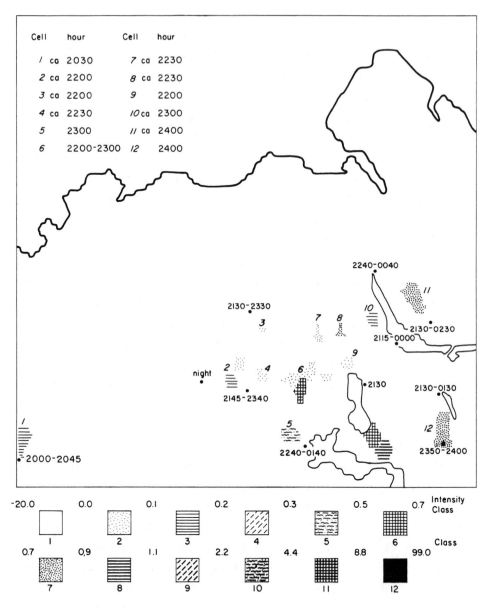

Figure 3. Communes in Central Switzerland with confirmed reports of hail damage, on a typical experimental day. The shading indicates the intensity of damage category. The triangle marks a hailfall near a Swiss Meteorological Station with its officially reported time; the dots give similar information for thunderstorms near a hail area. The time of hailfall on each damaged commune has been estimated by M. Schüepp.

REFERENCES

Federer, B., Waldvogel, A., Schmid, W., Hampel, F., Rosini, E., Vento, D., Admirat, P., and Mezeix, J.-F. (1978). Plan for the Swiss randomized hail suppression expreiment: Design of Grossversuch IV. *Pageoph. 117*, 1-24.

Sulakvelidze, G. K., Kiziriya, B. I., and Tsykunov, V. V. (1974). Progress of hail suppression work in the USSR. In *Weather and Climate Modification* (W. W. Hess, ed.), pp. 410-431. New York: Wiley.

Mezeix, J.-F., and Caillot, P. (1982). A confirmatory evaluation of the Grossversuch IV experiment using hailpad data (French network 1977-1981). *Fourth W.M.O. Scientific Conference on Weather Modification*, Sofia.

Neyman, J., and Scott, E. L. (1967). Note on techniques of evaluation of single rain stimulation experiments. In *Proceedings Fifth Berkeley Symposium on Math. Statistics and Probability 5*, 371-384. Berkeley: University of California Press.

ECONOMETRICS AND THE LAW: A CASE STUDY IN THE PROOF OF ANTITRUST DAMAGES

R. S. DAGGETT Brobeck, Phleger, Harrison
D. A. FREEDMAN University of California, Berkeley

PREFACE

An econometric model was used to prove damages in antitrust litigation.[1] The difficulties in such proof will be explored. For plaintiffs' counsel, an econometric analysis may seem quite attractive.
Here is scientific, objective proof of market power and damages, backed
by the prestige of mathematics, economics, and the computer. However,
defense counsel need not despair. On closer examination, the econometric analysis may turn out to be no more than a series of unsupported
assumptions, even if these are expressed in formidable equations. Once
articulated, the assumptions may conflict with each other or with common
sense. Also, the statistical calculations may turn out to be vulnerable

---

[1]Enrico Farms and Diedrich Ranch v. H. J. Heinz et al. Daggett
represented one defendant, and Freedman gave expert testimony on
behalf of that defendant. This paper was adapted in part from
materials prepared for the Federal Court Practice Program, Northern
District of California, Lecture and Workshop session on Expert Witnesses, February 1982 and in part from Freedman's deposition testimony.

From *Proceedings of the Berkeley Conference in Honor of Jerzy Neyman
and Jack Kiefer*, Volume I, Lucien M. Le Cam and Richard A. Olshen,
eds., copyright © 1985 by Wadsworth, Inc. All rights reserved.

to attack. Furthermore, in many econometric studies the data are of poor quality; this alone may vitiate the analysis.

The validity of models, the reasonableness of their assumptions, their usefulness in forecasting and policy analysis, and their probative value in the courtroom must therefore be assessed very carefully on a case-by-case basis. No general rule can be given except this: Mathematical symbols and computer printouts are not in themselves reliable indicators of scientific merit. All depends on where the equations come from and what the computer programs do.

Econometric models are often used in damage litigation. The present case is used only as a concrete example, to focus the issues and make them clearer. In any endeavor involving fields as different as the law and economics, there is ample room for misunderstanding. The object of this paper is to identify the sort of technical assumptions that are involved in the legal use of econometrics and the sort of questions that can be raised about those assumptions. For other views, see Finkelstein (1980), Fisher (1980), and the opinion of Higginbotham (1980) in Vuyanich v. Republic National Bank 505 F Sup 224. We tried to make this paper accessible to a large group of readers and hope that specialists will not be put off by the explanations that are unnecessary for them.

THE CASE

The litigation was a private antitrust action. The plaintiffs, who grow and sell tomatoes in the Fresno and Stockton regions, alleged that defendants Heinz and five other major canners conspired throughout the 1970s in violation of Section One of the Sherman Act to fix, stabilize, and maintain at depressed levels the prices that the defendant canners paid plaintiff growers for tomatoes. Plaintiffs contended that, as a result of the conspiracy, the prices plaintiffs received from defendants for tomatoes were lower than the prices they would have received in a free market, entitling plaintiffs to an award of treble damages and attorneys' fees. Defendants maintained that no such conspiracy existed, that each defendant canner determined the price it wanted to offer for tomatoes in the exercise of its own independent judgment, based upon estimates made from year to year of market conditions, and that, in fact, tomato prices for the greater portion of the period in suit were fixed and controlled by the growers themselves through the California Tomato Growers Association, which enjoys immunity from antitrust liability under the Capper-Volstead Act.

Plaintiffs proposed to call as an expert witness Richard E. Just, Professor of Agricultural and Resource Economics, University of California, Berkeley. This expert had developed an econometric model of the California tomato processing industry, which he used to prove the exercise of market power by the canners in the purchase of raw tomatoes and to estimate what the free market prices for the tomatoes would have been without collusion by the canners. These prices were the basis for plaintiffs' proof of damages. A preliminary version of the model and its data base are described in a published monograph, Chern and Just (1978). The model itself is described in a journal article, Chern and Just (1980). Additional detail was given in deposition testimony.[2]

The defense engaged David A. Freedman, Professor of Statistics, University of California, Berkeley, to review the model and rebut it if possible. The present article will give some of the institutional background and summarize the model and the review. Some tentative conclusions will be drawn about the usefulness of econometric analysis in litigation.

The case was Enrico Farms and Diedrich Ranch v. H. J. Heinz et al., Nos. CV-75-206-EDP and CV-79-186-EDP (E.D.Cal). This case was called for trial in April 1981 at Fresno, but the Court declared a mistrial upon grounds, among others, that a fair and impartial jury could not be obtained in the Fresno area. A new jury trial was set for Sacramento; before trial, all defendants except one settled. The trial was

---

[2]In a civil case, before trial each party can examine the other's witnesses to discover their anticipated testimony; these proceedings are "depositions."

then rescheduled in Fresno; depositions were taken, after which plaintiffs dismissed the case voluntarily in March 1982.

BACKGROUND

Throughout the San Joaquin Valley in the Eastern District of California, canning tomatoes are harvested over a short season beginning in early July and ending about mid-October in each year. Canners, sometimes called processors, locate, purchase, and pay for tomatoes for a variety of end-uses, e.g., canned tomatoes, catsup, canned tomato juice, tomato soup, spaghetti sauce, and tomato paste. During the harvest season the work goes on 24 hours a day, seven days a week. Tomatoes are picked up in the fields and transported to processing plants by trucks. If the quantity of tomatoes delivered to a particular processor's plant on a given day exceeds the capacity of the plant, tomatoes may be delivered to another canner's plant with open capacity, and accounts are settled later among the processors.

Until 1963, tomatoes in California were picked by hand by migrant workers. In 1963, mechanical harvesting equipment began to be used by commercial growers as the result of development of tomato varieties suitable for harvesting by machine. Other varieties of tomatoes are still grown and picked by hand for sale as fresh tomatoes. It is uncommon in the industry for a grower to raise tomatoes for both the canning and the fresh tomato market. Introduction of the mechanical harvester enabled growers to produce greater quantities of tomatoes per acre at lesser harvesting costs, but purchase of the equipment required growers to borrow money and incur substantial financing and other fixed costs. Canners purchasing tomatoes harvested by machine incurred significant changeover costs that resulted from new requirements for handling tomatoes damaged by machine picking. It is not likely that consumer demand for tomato products was changed by the introduction of mechanical harvesting.

During the 1950s, about 57 canners purchased and processed California tomatoes, and by 1972 the number decreased to 28. Twenty-five

of these were at first named as defendants in this action. Smaller
canners were dismissed before trial, voluntarily by plaintiffs in some
instances and by the Court in others; the remaining defendants were
the largest purchasers of processing tomatoes in the California market.
The named plaintiffs were large growers of financial means and each
had invested in mechanical harvesting equipment. In their complaints,
plaintiffs alleged the litigation was a class action in which plaintiffs represented the claims and interests of all growers of California
tomatoes. The Court denied motions to certify the litigation as a
class action upon grounds, among others, that conflicts of interest
among growers precluded plaintiffs from representing a class.

The purchaser market for California processing tomatoes includes,
in addition to the defendant processers at trial and the smaller
processors that were previously dismissed, several large member-owned
cooperatives. In some instances, cooperatives take delivery of
tomatoes grown by members in return for nonmonetary scrip or points
and distribute funds to members from cash received from the harvest
in proportion to the quantity supplied to the cooperative by the
grower-member. In other instances, cooperatives buy tomatoes from
members and other growers for cash like any other purchaser.

As with other fungible agricultural commodities, tomatoes
available to canners have tended to seek and find a prevailing market
price during each growing year or season. The prevailing pattern of
contract purchase of processing tomatoes by canners from growers is
this: Large growers tend to sell their tomatoes to the same canner
year after year, but not always. To conserve the soil, crops are
rotated from year to year among tomatoes, grains, alfalfa hay, sorghum,
and, in the Fresno area, cotton and safflower. A grower does not grow
tomatoes on the same land for many years; hence each grower may decide
in a particular year to grow tomatoes or some other crop suitable to
the region. After the harvest each year, at about Thanksgiving or
Christmas, canners' field representatives discuss with growers how
many acres of tomato production the canner is likely to want to purchase from the grower under contract and how many acres of tomatoes
the grower is planning or willing to plant for the coming season.

A few months later, usually in the January-March period, each canner
informs each grower of the base price the canner proposes to pay for
tomatoes that are harvested beginning in July. Sometimes, but not
always, the canner will offer a premium price over the base price for
purchase of late tomatoes, harvested in September or October. This
is done because the risk of bad weather makes growers more reluctant
to plant tomatoes for late harvest than for maturity in earlier warm
months.

When agreement on price is reached between the canner and the
growers who have decided to sell tomatoes to the canner, written con-
tracts are signed by the canner and grower, under which the grower
will sell the canner the tomatoes harvested from stated acreage at the
contract price. The grower, having already prepared the soil, then
plants his tomato crop. During some growing seasons, the contract
price has been modified upward, on occasion retroactively. Generally,
prevailing base prices have risen from about $20 a ton in 1951 to over
$50 a ton in 1975, when the premium paid for late tomatoes was about
$5 a ton.

During the 1960s advocates of "fair prices" for growers attempted
to obtain, but did not obtain, a government marketing order that would
have fixed by law the price canners paid growers for tomatoes. More
recently, beginning in the mid-1960s, grower advocates formed the
California Tomato Growers Association (CTGA) to present a united front
to the canning industry in negotiating a price at which members would
contract each year to sell their processing tomatoes. By about 1974
CTGA had assembled enough grower support to influence significantly,
some say, and exercise control over, say others, the annual prevailing
contract price for processing tomatoes in the region.

California Canners League (CCL) is a trade association, formed
many years ago, to which most of the defendant canners belonged.
The purpose of CCL and the nature of its activities were in dispute.
CCL and its members maintained that CCL exists and acts to deal with
industry-wide problems of a technical nature unrelated to competitive
behavior in the market. Plaintiffs' counsel contended that CCL has at
least provided a fund of information concerning anticompetitive

activities in the canning industry and that CCL might be found by the jury to have provided some assistance in, or to have served as a vehicle for, illegal price fixing by canners of prices paid to growers for tomatoes.

MAIN POINTS IN THE DEPOSITION TESTIMONY OF PLAINTIFFS' EXPERT

An econometric model was used to show the existence of "price-leadership oligopsony" in the market for California canning tomatoes, on the basis of statistical tests of hypothesis.[3] The price leader may have been a single firm or a set of firms acting "as if" in collusion to set grower prices. The model, together with additional assumptions and calculations, was then used to estimate the "effective colluding share" of the market. This share turned out to be substantially larger than the market share of the largest single firm. The conclusion: There was a group of firms acting as if they were colluding to fix grower prices to maximize the joint profits of the colluders. Finally, the model was used to estimate what "free-market" prices would have been without this collusion. To establish background for this argument, we present a brief review of two basic concepts in economics.

SUPPLY AND DEMAND

Two basic concepts in economics are *supply* and *demand*.[4] These are terms of art, the meaning of which differs from ordinary usage.

---

[3]An *oligopoly* is a small group of firms controlling the sales of a commodity; an *oligopsony* is a group controlling purchases. A *price leader* sets the price, taking into account the interests of the other firms in the group.

[4]A standard principles-of-economics text is Samuelson (1980). A more advanced reference is Mansfield (1982). A rigorous treatment of competitive markets is in Debreu (1959). A standard text of oligopoly theory is Scherer (1980).

Supply, for example, does not refer to a single quantity of some commodity but to the relationship between the price and the total quantity
that producers will offer to the market for sale at that price.   Such a
relationship is represented by the *supply curve* in Figure 1.   Price is
shown on the horizontal axis, quantity on the vertical; the curve shows
the total quantity that would be offered for sale in the market at each
price.   Notice that the curve slopes up; other things being equal, as
price increases so does the quantity offered for sale.

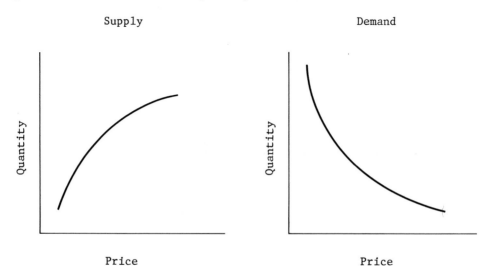

Figure 1.   Supply and demand

The *demand curve* in Figure 1 shows, for each price, the total
quantity of the commodity that would be demanded at that price by
buyers in the market.   This curve slopes down; other things being
equal, demand goes down as price goes up.   As shown in Figure 2, the
market price and the quantity transacted are considered to be determined by the intersection of the supply curve and the demand curve.
At the intersection of the two curves, the market clears: Supply
equals demand.   At a higher price, economists argue, the quantity
sellers offer will exceed the quantity buyers want; this excess supply
should drive the price down.   Similarly, at a lower price the quantity
buyers want will exceed the quantity sellers offer; the excess demand
should drive the price up.

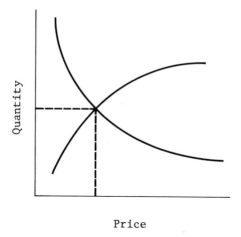

Figure 2. Market clearing

The supply and demand curves express an important behavioral assumption: there is a stable relationship between the quantity of a commodity supplied and its price; likewise for demand and price. This relationship is assumed, as economists say, *ceteris paribus*, other things being equal. That is, there are factors other than the price of the commodity that affect its supply; these are called *the determinants of supply*. For tomatoes, the price of fertilizer, the price of alternative crops, and agricultural labor rates may all be important determinants of supply. Likewise, there are *determinants of demand*, such as consumer income. In Figure 1, the determinants of supply and demand are held constant.

Of course, the curves are hypothetical constructs because they represent quantities and prices in transactions that are not observed; only the market-clearing prices and quantities are observable. Furthermore, the curves in Figure 1 are qualitative rather than quantitative (no numbers appear on the axes); but by imposing additional assumptions, economists can make the curves quantitative and estimate them from actual transactions data.

To proceed, a crucial assumption must be made about the algebraic form of the curves in Figure 1. Economists refer to the choice of algebraic form as a *specification*. They may specify linear forms:

$$\text{Supply} \quad Q = a_0 + a_1 P \tag{1a}$$

$$\text{Demand} \quad Q = b_0 - b_1 P \tag{1b}$$

Or they may specify log linear forms:

$$\text{Supply} \quad \log Q = a_0 + a_1 \log P \tag{2a}$$

$$\text{Demand} \quad \log Q = b_0 - b_1 \log P \tag{2b}$$

Other choices are possible, and economic theory is not sufficently well developed to dictate which is the right choice in any given situation.

In these equations, Q represents quantity and P the price; a_1 and b_1, for example, are parameters, or constants of nature, that govern the market. Usually, the numerical values of the parameters are unknown and must be estimated from data.

Plaintiffs' expert specified the linear form (1) for California canning tomatoes,[5] and we shall focus on that by way of example. It is customary to think of a succession of *market periods*, taken as calendar years in the tomato example. These are represented by the subscript t for time. The primary data are the price P_t that obtained in year t and the quantity Q_t transacted in that year. For the tomatoes, t ran from 1951 through 1975. Rewriting the equations,

$$\text{Supply} \quad Q_t = a_0 + a_1 P_t \tag{3a}$$

$$\text{Demand} \quad Q_t = b_0 - b_1 P_t \tag{3b}$$

The determinants of supply are usually considered to come in through the intercept a_0 of equation (3a). For example, suppose the only important determinant of tomato supply is the agricultural wage rate. Let W_t be this wage rate in year t. An economist may specify that

$$a_0 = \alpha_0 - \alpha_1 W_t \tag{4a}$$

As W changes, the supply curve (3a) shifts up and down parallel to itself; the slope a_1 does not change. Likewise, consumer income may

---

[5]In Chern and Just (1980); in Chern and Just (1978), however, the specification was log linear.

be the important determinant of demand for tomatoes, and in the framework of (3b) an economist may assume that

$$b_0 = \beta_0 + \beta_1 I_t \qquad (4b)$$

Here, I_t denotes income in period t.

Again, economic theory does not dictate specific choices for the major determinants of supply and demand. Even given these choices, equations such as (4a) and (4b) are by no means logically inevitable. The right equations, for example, might turn out to be

$$a_0 = \alpha_0 - \alpha_1 W_t{}^2 \qquad (5a)$$
$$b_0 = \beta_0 + \beta_1 \sqrt{I_t} \qquad (5b)$$

Still other choices are possible.

It is usually considered that many random phenomena will influence supply and demand; for example, the supply of an agricultural commodity can be influenced by weather. Likewise, many variables will have some small, indirect influence on supply and demand; for example, fluctuations in the stock market may have a marginal effect on consumer decisions to spend or save. The impact of random phenomena, and of omitted variables, is usually summarized by adding a "random disturbance term" to the equations which, in effect, shifts the supply and demand curves up and down. These terms may be denoted by ε_t or δ_t. Our linear supply and demand equations now take the form:

$$\text{Supply} \quad Q_t = \alpha_0 - \alpha_1 W_t + a_1 P_t + \varepsilon_t \qquad (6a)$$
$$\text{Demand} \quad Q_t = \beta_0 + \beta_1 I_t - b_1 P_t + \delta_t \qquad (6b)$$

At this point, it is convenient to introduce a distinction between *exogenous* and *endogenous* variables. In equations (6), W and I are exogenous, determined outside the system. However, Q and P are endogenous, determined within the system. Thus, the two simultaneous linear equations can be solved for the two endogenous variables Q and P in terms of W, I, ε, and δ. A pair of equations such as (6a) and (6b) constitute an *econometric model*.

The equations above have six parameters: α_0, α_1, a_1, β_0, β_1, b_1. To estimate these from actual transactions data, statistical assumptions must be made about ε and δ. This completes the *specification* of the model, i.e.,

Selection of the major determinants of supply and demand

Decisions about the algebraic form of the equation

Assumptions about the statistical behavior of the disturbance terms
sufficient to enable the estimation of the parameters from the
data

The statistical assumptions might be that $(\varepsilon_t, \delta_t)$ are independent and identically distributed from year to year but have some fixed, arbitrary covariance matrix within a year. To make this more vivid, imagine a box of tickets. Every ticket shows two numbers: the first an ε and the second a δ. The average of the ε's is zero, and likewise for the δ's. For each year t, draw a ticket at random: the first number is ε_t in (61), the second is δ_t in (6b). The draws are made with replacement, so that the box stays the same from year to year, whatever the values of the exogenous variables I and W. (Technically, the exogenous variables are assumed to be statistically independent of the disturbances.) Then the pair of simultaneous equations (6a, b) can be solved for Q_t and P_t, in terms of the exogenous variables and disturbances.

Estimation of the parameters of a model from data is a complicated topic, but the idea is straightforward. Suppose that over time, the demand curve is static while the supply curve shifts. The market-clearing prices and quantities will then trace the demand curve (see Figure 3). Conversely, if the supply curve is static and the demand curve shifts, the supply curve will be traced. In general, both curves will be shifting simultaneously, and statistical estimation techniques must be used. Since Q and P are obtained by solving (6a, b), both involve ε and δ. This simultaneity is a crucial technical feature of econometric models: since P is correlated with ε and δ, ordinary least squares is inconsistent. To overcome this difficulty, econometricians have developed a technique called *two-stage least squares*; a reference is Theil (1971), or see Appendix C.

Supply Shifts   Demand Shifts

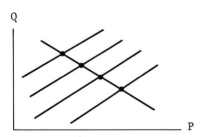
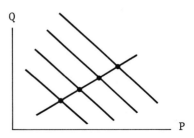

Figure 3.   Tracing the demand and supply curves

Note that the specification of a model involves making a series of assumptions. These may be more or less reasonable. However, in its present state of development, the science of economics does not permit any exact, objective way of finding the major determinants of supply and demand nor the algebraic form of the equation; nor is there usually any way to deduce the requisite statistical assumptions about the disturbance terms from generally accepted theory. Moreover, there are no generally accepted statistical procedures for testing the validity of an econometric model; some of the assumptions can indeed be examined, but only by introducing still other assumptions.

One powerful empirical test of a model is often available: Use it to make predictions about the future and see how accurate these turn out to be. Such tests, which are seldom made, give conflicting results; see [9] or [13] and the references cited there, especially Christ (1975) and Zarnowitz (1979). The ability of a model to predict the future must be distinguished from its ability to reproduce the past. Given the flexibility described above in specifying a model, it is often possible by trial and error to get equations that reproduce the past very well; such equations can be quite unsatisfactory when used to predict the future and may not capture the essential features of the market in question. For a more technical discussion, see [2] and [10].

THE MODEL AND THE PROOF OF MARKET POWER

Plaintiffs' expert used an econometric model to conclude that the
market behavior he examined was consistent with collusion among
canners in buying processing tomatoes and inconsistent with competitive behavior in that market. Summarizing the previous discussion, the econometric model includes as hypothetical constructs a
supply curve and a demand curve for processing tomatoes in the area
studied. The supply curve represents the quantity of tomatoes that
will be sold at a given price; the demand curve represents the quantity of tomatoes that will be purchased at a given price. The supply
curve and demand curve thus represent relationships between quantities and prices. They are hypothetical constructs in the sense that
they represent quantities of tomatoes that would be supplied or
demanded at prices that have not been observed in actual transactions.
The curves are estimated by statistical methods, however, from data
derived from actual transactions.

In a market consisting of numerous tomato growers in competition
with each other, introduction of machine harvesting must cause a shift
in the supply curve in two respects, according to plaintiffs. First,
at any given price the supply of tomatoes will be larger than it was
before machine harvesting. Second, the tomato supply will be less responsive to any given change in price than it was before the advent of
machine harvesting. The explanation is that investment in harvesting
equipment made the growers less free to grow crops other than tomatoes
while the harvesting machinery remained unused. In short, the grower
is "locked in." This shift in the supply curve is illustrated in
Figure 4.

If the canners compete in buying tomatoes, a shift in the supply
curve will not cause a shift in the demand curve, because demand for
canning tomatoes is then a function of consumer demand for the tomato
end products and is not determined by supply. On the other hand, if
tomato prices are determined or tend to be determined by canners acting in collusion, after the introduction of machine harvesting the

perceived demand curve, i.e., the demand curve facing the growers, will shift in two respects. First, at any given price, the demand for tomatoes will be less. Second, tomato demand will be less responsive to changes in prices than before machine harvesting (see Figure 4). Likewise, tomato demand will become less responsive to changes in the determinants of demand, such as income.

Supply Shift   Market Power: Demand shift

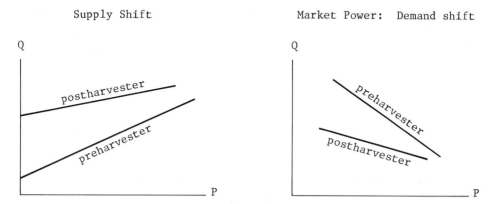

Figure 4. The harvester shifts the supply curve. If the canners exercise market power, the perceived demand curve will shift when the supply curve does.

Perceived demand is a nonstandard construct. In brief, with a linear supply curve for canning tomatoes and a linear demand curve by consumers for tomato end products, and with canners exercising market power, there exists a linear perceived demand curve that intersects the supply curve to give the transacted price and quantity for canning tomatoes. The slope of this curve depends on the slopes of the consumer demand curve and the supply curve; the intercept depends on the foregoing and on the intercept of the consumer demand curve. The intercept of the supply curve does not come in. See Appendix A for details.

From the data in Chern and Just (1978), plaintiffs estimated supply and demand curves for the years before and after introduction of machine harvesting, omitting the transition period 1964-1966 when hand picking and machine harvesting were both used. The supply curve shifted after machine harvesting to show that supply was less responsive to changes in price, consistent with free competition among growers. Furthermore, the demand curve shifted to show that demand was less responsive to changes in price after machine harvesting, which was consistent with the existence of a "colluding share" of the canner market and inconsistent with tomato price behavior in a market of canner-purchasers acting in free competition with each other; likewise for the determinant of demand (income). See Tables 1 and 2 on pages 596-97 of Chern and Just (1980).

WEAKNESSES IN THE ARGUMENT

There are three major reasons for concluding that plaintiffs' statistical argument failed:

The logic of the model is open to serious question.

The statistical calculations are wrong.

Even if one takes the empirical results at face value, they admit several plausible interpretations other than price-leadership oligopsony.

These reasons will now be considered in more detail.

With respect to the logic of the model, many assumptions are made without empirical support; for example,

1. That there was competition among the growers
2. That the supply curve for processing tomatoes was linear in price
3. That the demand curve for tomato end products was linear in price
4. That there were constant unit processing costs
5. That the colluding processors had perfect knowledge of the supply and demand curves and the processing costs

Assumption 1 is important, because if any significant number of the growers exercise market power, there is no supply curve to estimate: A monopoly or oligopoly does not have a supply curve.[6] Assumptions

---

[6]This may seem like a paradoxical assertion, but it is standard doctrine. The reason is that a monopolist sets price to maximize profit, and in the calculation uses the slope of the consumer demand

2-4 are also critical, because they are needed to derive the perceived demand curve, without which construct the model is vacuous. Assumption 5 is needed to do the profit maximization on behalf of the colluding processors.

Plaintiffs' expert admitted in deposition testimony to having made no study supporting these assumptions: Indeed, he made no study of individual processors and no study of demand at the consumer level. No arguments were given in support of the linearity assumptions, except that any smooth function can be approximated by a linear one: A function differentiable at a point can be approximated near that point by a line. In short, for small changes, linear supply and demand curves may be good enough. But the relevance of such observations to the issues at hand is not so clear. Plaintiffs used linear supply and demand curves over a 25-year period, during which California tomato production increased threefold while grower prices doubled and consumer income went up by a factor of four.[7] These are not small changes.

What does it mean to say that the consumer demand curve is linear in price? Just this: A 10¢ increase in the price of a can of tomatoes has the same impact on consumer demand in 1951 as in 1975, whether the can costs a quarter or a dollar, whether the consumer is making $5000 a year or $15,000. A very strong assumption.

Two additional assumptions are needed to develop plaintiffs' theory of price-leadership oligopsony:

6. A segmented market for tomato end products
7. Fungible product within each segment

Assumption 6 is easiest to explain in terms of a hypothetical example. Suppose that there are red-label and blue-label canned tomatoes, virtually identical but differentiated by the label and

---

curve. Hence, there is no stable relationship between the price set and the quantity offered. All depends on the demand curve. See p. 285 of Mansfield (1982).

[7]We are following plaintiffs here in using current rather than constant dollars, an issue to be discussed later.

perhaps by advertising. The competitive canners, such as cooperatives, sell the red-label product. The colluding canners fix the grower price to maximize joint profits; the competitive canners take the grower price as given.

No evidence is presented to support assumption 6, which is a nonstandard way to develop price-leader theory.[8] This unorthodox approach is necessitated by assumption 4 of constant unit costs. It is also worth noting that plaintiffs treated the two markets independently, admitting no cross-elasticity of demand. To use the hypothetical example given above, consumer demand for red-label canned tomatoes is assumed to depend only on the price of the red-label product and not at all on the price of the blue-label product. This, too, is against standard doctrine.

Assumption 7 is unrealistic. Consumers do not drink a glass of chilled tomato paste before dinner; neither do they have a bowl of hot, nourishing tomato juice for lunch. On occasion, plaintiffs suggested that though canned tomatoes, juice, puree, catsup, and paste may not be interchangeable, their quantities can be measured in terms of the weight of raw tomatoes needed for their manufacture. However, the cost of a tomato end product does not depend solely on the weight of its raw tomato content. (The raw tomatoes in a No. 303 can of tomatoes, for example, cost less than the can itself.) In sum, this suggestion of plaintiffs violates the idea that in a given market a commodity has only one price. Specifically, the price of a ton-equivalent of the commodity will depend on how that quantity is distributed among the end products.

We turn now to another set of issues. Plaintiffs began with an "analytical model" of the market for processing tomatoes. This considers only one geographical market for the tomatoes and does not bring in determinants of supply and demand. A linear supply curve is assumed, of the form

---

[8]For the standard theory, see Mansfield (1982, pp. 349–50) or Scherer (1980).

$$Q = a_0 + a_1 P$$

Plaintiffs then introduced an econometric model that considered eight county markets for the tomatoes and does bring in determinants of supply and demand. Thus, the post-harvest supply equation is assumed to be

$$Q_{ct} = a_c + a_1 P_{ct} + a_2 Y_{ct} + \varepsilon_{ct}$$

In county c and year t,

Q_{ct} is the quantity of canning tomatoes sold in thousands of tons.

P_{ct} is the price of canning tomatoes, in dollars per ton.

Y_{ct} is the average yield over the previous three years, in tons per acre.

ε_{ct} is the disturbance term.

a_c is a county-specific intercept.

a_1 and a_2 are parameters, the same for all counties.

Likewise for the demand equations.

The econometric model is at the level of raw tomatoes: the supply considered is by the growers; demand is by the canners. The following aspects of the econometric model must be considered:

1. The supply equation includes a lagged three-period moving average of yields as an *expectations variable* and the corresponding standard deviation as a *risk variable* in the pre-harvester period.

2. The supply equation for the pre-harvester period includes the price of only one alternative crop, viz., sorghum; no alternative crop prices are included for the post-harvester period.

3. The demand equation includes an end-product price index.

4. The analytical model applies to a single, integrated market for raw tomatoes; the econometric model applies to eight county markets.

Points 1 and 2 may be minor, but they are indicative of the arbitrariness in plaintiffs' approach. There is little justification for the choice of the expectations and risk variables. For example, why not use a four-period moving average or the mean absolute deviation? There is also little justification for including sorghum rather than, say, alfalfa or sugar beets. To quote (Chern and Just, 1978, p. 19):

For example, more than 100 crops are grown in San Joaquin County, and it is impossible to single out one or two crops as the most common alternative crops to processing tomatoes.

In his deposition, plaintiffs' expert argued at times that the omitted crop prices are collinear with the sorghum price, so there is no point in putting them into the equation, and at other times, that including such prices gave rise to estimated coefficients with signs known à priori to be incorrect. These two points stand in contradiction to one another, and both seem wrong. First, the correlation between, e.g., the price of sugar beets and the price of sorghum in the pre-harvester period (1951-1963) is only about 0.32, so these two crop prices moved nearly independently of each other. Second, if sugar beets belong in the model, but putting them in gives incorrect signs, plaintiffs' expert concluded he should drop the beets; rethinking the model seems a better alternative.

Point 3 is a bit more serious. The development of the analytical model is aimed in part at eliminating the price P^d for end products from the perceived demand curve for raw tomatoes; see Chern and Just (1980), pages 591-92, or Appendix A. All that belongs in the demand equation are the determinants of demand, i.e., the variables that affect consumer demand, other than price. There is only one such variable in the model: income. Putting product price back into the demand equation represents a theoretical inconsistency and makes quite a difference in the results, as shown in Appendix B.

Point 4 is another major difficulty. The analytical model shows how processor demand for raw tomatoes is driven by consumer demand for tomato end products. This consumer demand is national in scope; indeed, the major processors sell into a national distribution network. Plaintiffs should have indicated how this national consumer demand is to be parcelled out among the eight counties and failed to do so.

To see the tension between the one-market analytical model and the eight-market econometric model, consider this basic consequence of the existence of price-leadership oligopsony in the one-market analytical model for canning tomatoes: After the introduction of the machine harvester, according to Chern and Just (1980, p. 590), "the perceived

demand (the estimated econometric demand) would be shifted downward."
That is, other things being equal, after the machine harvester comes
in, at any given price, the canners will demand fewer tomatoes from the
growers. Now consider the estimated demand curve for canning tomatoes
in Yolo county, according to the eight-county econometric model. (In
1975, Yolo was the largest producer among the eight counties considered
in the econometric model.) In round numbers, the demand equations are

Pre-harvester (1951–1963) $Q = -8 + .7I + 132R - 14P$
Post-harvester (1967–1975) $Q = 696 + .14I + 46R - 3.3P$

where

Q = tomato demand (thousands of tons)
P = grower price (dollars per ton)
R = product price index (dollars)
I = U.S. disposable personal income (billions of dollars)

To implement the idea of "other things being equal," choose in
round numbers values for I and R typical for the whole period 1951–1975
as follows: $I = 500$ and $R = 5$. The equations become

Pre-harvester (1951–1963) $Q = 1002 - 14P$
Post-harvester (1967–1975) $Q = 996 - 3.3P$

Demand in the analytical model

Estimated demand in the
econometric model; Yolo county

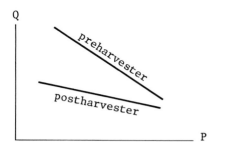
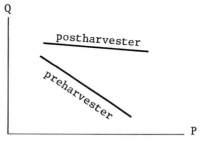

Figure 5. The analytical model versus the econometric model: the
demand for canning tomatoes

Historically, P ranges from about $20 to $60 a ton. Over this
entire range, the post-harvester demand by processors for raw tomatoes

exceeds the pre-harvester demand, at any given price. See Figure 5. In other words, one of the most basic conclusions of the single-market analytical model is contradicted by the findings of the eight-county econometric model. Only two resolutions of the difficulty seem possible: either the models are wrong, or no market power is being exercised by the canners. This completes our discussion of the first main reason for rejecting plaintiffs' argument: The logic of the model is open to serious question.

We turn now to the second reason: The statistical calculations are wrong. Plaintiffs' argument depends on differences that are "statistically significant" between the estimates of the demand curves made separately for the periods before and after machine harvesting, by the t test. For this it is necessary to compute standard errors. The formulas used by plaintiffs to compute standard errors are founded on assumptions shown to be false by a mere statement of them, for example, that weather patterns in neighboring counties are unrelated. (For quantitative estimates of the impact of such assumptions, see Appendix C.) Further, in contexts such as the present one, in which the amount of data is limited relative to the number of parameters being estimated, the parameter estimates may be seriously biased; plaintiffs' statistical procedures do not take this bias into account (see Appendix C). Apart from such considerations, plaintiffs' admission that to avoid implausible results, many variations of the econometric model were attempted before settling on the final version is more than sufficient to invalidate the statistical computations; see [2] and [10].

We turn now to the third main reason for rejecting plaintiffs' statistical argument. Taking the empirical findings at face value, we find that they admit of several plausible interpretations other than price-leadership oligopsony:

1. The plaintiffs did not take inflation into account. If a dollar is worth less in 1975 than in 1951, a dollar increase in the price of a ton of tomatoes will matter less. Adjusting for inflation makes a big difference in the empirical results; see Appendix B.

2. Plaintiffs excluded Fresno and Kern county tomato production from the eight-county econometric model. In 1961 these counties were not significant producers, but by 1975 they contributed 25% of

California processing tomato production (Brandt et al., 1978). A very short version of the statistical argument offered by plaintiffs' expert is as follows: assuming competition, he can figure out how many tomatoes the canners ought to have wanted to buy, post-harvester, in the eight counties; he sees they bought less; he concludes they did not compete as buyers. The alternative: the canners bought the missing tomatoes in Fresno or Kern, or for that matter, anywhere else. Extending this line of thought, a canner who commits himself to buying the bulk of his tomatoes in Fresno may well become less sensitive to variations in price or consumer demand when buying tomatoes in plaintiffs' eight counties; this would explain the change in coefficients reported by plaintiffs. So would an increase in processing costs arising from the introduction of the harvester.

3. Awareness or belief on the part of canners that growers, after 1963, were not only enabled to supply increased quantities of tomatoes to the market by their investment in harvesting machines but were impelled to do so. In short, seeing that the growers were locked into tomato production, the canners bargained harder on price.

Plaintiffs' expert excluded the period 1964-1966 as transitional. To rebut the sort of argument now made, in terms of alternative explanations, he conducted a "sensitivity analysis," moving the excluded period back and forth, and found the most significant differences to be when this excluded period is close to the original one. Plainly, this argument does not meet point 3, which attributes the change in perceived demand to the introduction of the harvester. Nor does it meet point 2, for Fresno and Kern counties became significant shippers of canning tomatoes only at the time of the introduction of the harvester. It does not even cope with point 1 especially well, because the rate of inflation accelerated substantially, by coincidence, during the period of the introduction of the harvester. For example, as measured by the CPI, consumer prices increased only by about 20% from 1951 to 1963; but they increased by about 60% from 1967 to 1975. Likewise, fuel prices and fertilizer prices increased rapidly, even in real terms, from 1973 through 1975, due to the Arab oil embargo.

FREE MARKET PRICES

To prove damages, plaintiffs used the model to estimate the size of the "effective colluding share," and to estimate what the free-market prices would have been without collusion by the canners. Even granting the validity of the model, these subsequent calculations are seriously flawed.[9]

To estimate the model, plaintiffs' expert disaggregated to the eight counties. To estimate free-market prices, he reaggregated back to the state level. However, there is nothing to stop us from using plaintiffs' equations to compute the free-market prices in 1975 for Yolo and Sacramento counties, after the introduction of the harvester, assuming the "effective colluding share" was 80%. The exogenous variables are set in round numbers at typical values for 1975 as follows: $I = 1000$, $R = 10$, $Y = 25$. Large differences across county lines in the model free-market prices are almost inevitable, due to the county-specific intercepts. (For details of our computation, see Appendices A and B.)

The results are shown in Table 1. Comment may be superfluous, but Sacramento County is adjacent to Yolo County. Differences in hypothetical prices of $50 or more a ton across the county line are implausible. The actual difference in 1975 was only 25¢ a ton.

---

[9]The methodology is not explained in Chern and Just (1978, 1980), but it is described in deposition testimony, summarized in Appendix A. Plaintiffs' expert first estimated the effective colluding share and then used the estimate as a datum in computing the prices; his estimate changed from time to time during the deposition, but 80% seems to be a representative figure.

Table 1. The Model's Free-Market Prices in Two Counties in 1975

County	Case 1*	Case 2†
Yolo	$126 per ton	$137 per ton
Sacramento	$74 per ton	$87 per ton

*Case 1: The canners colluded in buying the tomatoes, but com-
peted in selling end products. The table is computed using
plaintiffs' formulas, which include a parameter to differentiate
between the two cases. However, it may be questioned whether
plaintiffs have a coherent theory for Case 1. On plaintiffs'
assumptions, it can be argued that if the canners compete in
sales and fix the price of raw tomatoes, they can only fix it
at what the free-market price would have been.
†Case 2: The canners colluded both in buying and in selling.

APPENDIX A: The Analytical Model

This appendix gives a concise statement of plaintiffs' analytical model
for price-leadership oligopsony, defines perceived demand, and indicates the methodology used in computing free-market prices. We follow
Chern and Just (1980). The econometric model will be considered in
Appendix B.

The supply equation for growers is

$$Q = a_0 + a_1 P_g \tag{A1}$$

where Q is quantity (1000s of tons), and P_g is the grower price
(dollars per ton).

There is a dominant firm, or set of colluding firms, that sells
into one segment of a market and faces the demand equation

$$Q = b_0 - b_1 P_r \tag{A2}$$

where P_r is the price of the end product, assumed fungible and
measured in the same units as supply, viz., raw tomato content. These
firms all have the same constant unit processing cost θ. They set the
grower price P_g.

Next, there is a fringe of competitive firms who take the grower price P_g as given, sell into another segment of the market, facing the demand equation

$$Q' = c_0 - c_1 P_r' \tag{A3}$$

These competitive firms all have constant unit processing costs θ', so $P_r' = P_g + \theta'$. They will sell

$$Q' = (c_0 - c_1 \theta') - c_1 P_g \tag{A4}$$

and are allowed to buy this much from the growers.

The dominant share now faces the supply relation

$$Q = a_0 + a_1 P_g - Q' = \alpha_0 + \alpha_1 P_g \tag{A5}$$

where

$$\alpha_0 = a_0 - c_0 + c_1 \theta' \quad \text{and} \quad \alpha_1 = a_1 + c_1 \tag{A6}$$

Take the case in which the colluding firms exercise market power in selling the end product as well as buying the raw tomatoes. They set P_r and P_g to maximize joint profits,

$$Q(P_r - P_g - \theta) \tag{A7}$$

where Q is defined by (A5). They must also clear the market for canning tomatoes, so

$$\alpha_0 + \alpha_1 P_g = b_0 - b_1 P_r = Q \tag{A8}$$

The optimal P_g can be shown to satisfy the equation

$$a_0 + a_1 P_g = d_0 - d_1 P_g \tag{A9}$$

where

$$d_0 = (c_0 - c_1 \theta') + \frac{\alpha_1}{2\alpha_1 + b_1}(b_0 - b_1 \theta) \tag{A10}$$

$$d_1 = c_1 + \frac{\alpha_1}{2\alpha_1 + b_1} b_1 \tag{A11}$$

The right-hand side of (A9) is the perceived demand curve.

It is assumed that b_1, c_1, θ, and θ' are constant over the period 1951-1975, that $c_0 > c_1 \theta'$ and $b_0 > b_1 \theta$. If a_1, the slope of the supply curve, diminishes due to the introduction of the harvester, then $\alpha_1 = a_1 + c_1$ will diminish too, and so will d_1. Likewise for d_0.

This is plaintiffs' argument relating the shift in post-harvester perceived demand to the exercise of market power by the canners.

We turn now to the model's free-market prices, following the deposition testimony of plaintiffs' expert. He assumes

$$\frac{b_0}{b_0 + c_0} = \frac{b_1}{b_1 + c_1} = \eta \tag{A12}$$

No rationale was given for this strong assumption. The idea is that if the product price were P in both segments of the market, the colluding firms would have the market share

$$\frac{b_0 - b_1 P}{(b_0 - b_1 P) + (c_0 - c_1 P)} = \eta$$

Since the model is set up to allow different product prices in the two markets, it seems peculiar to interpret η as a market share. However, he calls η the *effective colluding share*.

Let $\phi = (1 - \eta)/\eta$; so $c_0 = \phi b_0$ and $c_1 = \phi b_1$. To avoid additional complications, we follow plaintiffs and set $\theta = \theta' = 0$; in essence, this redefines the intercepts b_0 and c_0. Recall (A6). Substitution into (A10) and (A11) gives

$$d_0 = \phi b_0 + \frac{b_0(a_1 + \phi b_1)}{2(a_1 + \phi b_1) + b_1} \tag{A13}$$

$$d_1 = \phi b_1 + \frac{b_1(a_1 + \phi b_1)}{2(a_1 + \phi b_1) + b_1} \tag{A14}$$

Plaintiffs consider (A14) separately for the pre-harvester and post-harvester periods. The two demand price coefficients (d_1-before, d_1-after) and supply price coefficients (a_1-before, a_1-after) are estimated in the econometric model. This gives two equations in two unknowns, viz., ϕ and b_1: it is assumed that b_1 is constant over the entire period. Thus ϕ can be estimated, and then η; also b_1 is estimated. Now (A13) can be used to determine b_0 from d_0, the latter being estimated in the econometric model. In a competitive market with zero processing costs, the free-market price $P_r = P_g = P$ can now be determined by solving (A8):

$$P = \frac{b_0 - \alpha_0}{b_1 + \alpha_1} = \frac{(1 + \phi)b_0 - a_0}{(1 + \phi)b_1 + a_1} \tag{A15}$$

The empirical results will be presented after the econometric model is discussed in Appendix B.

APPENDIX B: The Econometric Model

This appendix states briefly the econometric model in Chern and Just (1980) and shows how the results depend on the specification. The free-market prices will also be computed.

The pre-harvester supply equation in the econometric model is

$$Q_{ct} = a_c + a_1 P_{ct} + a_2 Y_{ct} + a_3 D_{ct} + a_4 F_t + a_5 G_{t-1} + a_6 W_{ct} + \varepsilon_{ct} \qquad \text{(B1)}$$

The post-harvester supply equation is

$$Q_{ct} = a_c + a_1 P_{ct} + a_2 Y_{ct} + \varepsilon_{ct} \qquad \text{(B2)}$$

The demand equation for both periods is

$$Q_{ct} = d_c + d_1 P_{ct} + d_2 I_t + d_3 R_t + \delta_{ct} \qquad \text{(B3)}$$

In county c and year t:

Q_{ct} is the quantity transacted of processing tomatoes in 1000s of tons.

P_{ct} is the grower price in dollars per ton.

Y_{ct} is the average yield over the preceding three years in tons per acre.

D_{ct} is the SD of those three yields, in tons per acre.

W_{ct} is the agricultural wage rate, in dollars per hour.

F_t is the price of fertilizer, in dollars per ton.

G_t is the price of grain sorghum, in dollars per ton.

I_t is the total U.S. personal disposable income, in billions of dollars.

R_t is an index of tomato product prices in dollars.

All money variables are in nominal dollars; a_c and d_c are county-specific intercepts; $a_1, \ldots, a_6$ and d_1, d_2, d_3 are parameters, the same across counties; $G_t, I_t,$ and R_t are the same for all counties.

ε_{ct} and δ_{ct} are stochastic disturbance terms, assumed to have mean 0 and to be independent and identically distributed across c and t, given the variables other than quantity and price. For each c

and t, the pair (ε_{ct}, δ_{ct}) has a fixed arbitrary 2 × 2 covariance matrix.

To explain more vividly the stochastic assumptions made by plaintiff, imagine a box of tickets; each ticket shows two numbers, the first being an ε and the second a δ; the ε's average out to zero, and so do the δ's. Focus on, for example, the pre-harvester period. For each year t in that period and county c, draw a ticket from the box at random, with replacement; the first number on the ticket is ε_{ct} in (B1), and the second is δ_{ct} in (B3). The same box is used for each year and each county, irrespective of the values of the exogenous variables (Y, D, F, A, W, I, and R). Then equations (B1) and (B3) are solved for the two endogenous variables Q_{ct} and P_{ct} in terms of the exogenous variables and the disturbances.

In the present case, the stochastic assumptions seem unrealistic. For example, these assumptions imply that every county shows the same random variation, big counties and small alike, in good years and bad; also, after we adjust for the exogenous variables, the remaining variations of supply and demand in one county are assumed to be uncorrelated with those in any other county.

Coming back to the data-processing, we attempted to verify plaintiffs' estimates, and succeeded for the pre-harvester period. For the post-harvester period, as it turns out, there seems to have been one key-punch error made by plaintiffs in transferring data from Chern and Just (1978) to the computer. In 1973 in San Joaquin, Chern and Just (1978) give the price of grain sorghum as 2.87 dollars per ton; plaintiffs' computer data set (printed out as one of the deposition exhibits) has 1.43. When this error is corrected, the post-harvester demand price coefficient changes from -3.3 to -1.7, showing how sensitive the model must be to small errors in the data.[10]

---

[10]The corrected model is better for plaintiffs. The computer work described in this section was done by Dr. Thomas Permutt in SAS on the IBM 4341 at the University of California, Berkeley. (See Table B1.)

Table B1.   Plaintiffs' Parameter Estimates

	Pre-harvester 1951–1963		Post-harvester 1967–1975	
	Supply	Demand	Supply	Demand
P	11.7	−14.2	3.9	−3.3
Y	9.9	—	15.4	—
D	−6.5	—	—	—
F	−2.1	—	—	—
G_{t-1}	−91	—	—	—
W	−335	—	—	—
I	—	.7	—	.14
R	—	132	—	46
County Intercepts				
San Joaquin	871	217	261	453
Yolo	625	−8	497	696
Solano	316	−319	−148	55
Sutter	295	−331	−147	67
Sacramento	378	−260	−294	−104
Stanislaus	300	−349	−307	−103
Santa Clara	274	−381	−325	−103
San Benito	227	−406	−354	−99

We attempted to respecify the model by dropping the product price index R. In the pre-harvester equations, both price coefficients came in with the wrong sign: Demand increases with price, and supply decreases. Post-harvester, demand and supply both increase with price. We also attempted to respecify the model, keeping R but deflating all monetary variables to 1967 dollars, using the CPI. Pre-harvester, the signs are right. Post-harvester, however, demand increases with price. In short, two attempts to correct misspecifications caused the model to produce inconsistent results. (We did not attempt to adjust for population increase, which must also have a strong impact on demand.)

As part of the data processing, we fit a very simple dynamic model of our own:

$$Q_{ct} = a + bQ_{ct-1} + \zeta_{ct} \tag{B4}$$

The estimates are $a = 14$ and $b = 1$. This simple model has only two parameters; it takes no advantage of data on price, income, yield, etc., and it runs right through the transition from hand harvesting to machine harvesting. But it fits the data just about as well as plaintiffs' model. Our equation explains just over 90% of the variance; their equations explain between 87% and 94%.

We return now to the free-market prices, Case 2 in Table 1: $\eta = .8$ corresponds to $\phi = .25$; the post-harvester version of (A14) entails $b_1 = 5.78$.[11] These quantities are the same for all counties, by assumption. Then (A13) can be used to compute the county-specific b_0. The post-harvester econometric results from Table B1 give us the left side of (A13):

$$d_0 = \text{county demand intercept} + .14I + 46R$$

Now (A15) can be used to obtain the county-specific free-market price P, with a_0 from the post-harvester results of Table B1:

$$a_0 = \text{county supply intercept} + 15.4Y$$

Case 1 is similar and proceeds from the assumption that the dominant firms set the product price equal to marginal cost for the entire coalition, i.e., taking into account price reaction by growers to increased demand. Plaintiffs described this as the case in which the canners in the coalition compete with each other in sales, a description the merit of which is not apparent.

APPENDIX C: Two-Stage Least Squares and the Bootstrap

This appendix presents a brief account of the estimation procedure used by plaintiffs, known as two-stage least squares, and some details on the *bootstrap*, a statistical procedure used to test plaintiffs' calculations.[12] We begin with the more basic *generalized least squares*.

---

[11]Pre-harvester, $b_1 = 26.18$, casting some doubt on the constancy of b_1, over the period 1951-1975.

[12]A standard reference on two-stage least squares is Theil (1971). For a more extended discussion of the bootstrap, see [11] and [12]. This appendix is joint work with Dr. Stephen Peters, Center for Computational Research in Economics and Management Science, MIT.

Consider the regression model

$$Y \atop n \times 1 \;=\; X \atop n \times p \; \beta \atop p \times 1 \;+\; \varepsilon \atop n \times 1 \;, \quad E(\varepsilon) = 0, \; \operatorname{cov} \varepsilon = \Sigma \tag{C1}$$

For historical reasons, X is called the *design matrix*. With Σ known, the generalized least squares (gls) estimate is

$$\hat{\beta}_{\text{gls}} = (X^T \Sigma^{-1} X)^{-1} X^T \Sigma^{-1} Y \tag{C2}$$

As is standard,

$$E(\hat{\beta}_{\text{gls}}) = \beta \tag{C3}$$

$$\operatorname{cov}(\hat{\beta}_{\text{gls}}) = (X^T \Sigma^{-1} X)^{-1} \tag{C4}$$

When Σ is unknown, statisticians often use (C2) and (C4) with Σ replaced by some estimate $\hat{\Sigma}$. Iterative procedures can be used, as follows. Let $\hat{\beta}^{(0)}$ be some initial estimate for β, typically from a preliminary ordinary least squares (ols) fit. There are residuals $\hat{e}^{(0)} = Y - X\hat{\beta}^{(0)}$. Suppose the procedure has been defined through stage k, with residuals

$$\hat{e}^{(k)} = Y - X\hat{\beta}_{\text{gls}}^{(k)}$$

Let $\hat{\Sigma}_k$ be an estimator for Σ, based on $\hat{e}^{(k)}$: an example will be given below assuming a block diagonal structure for Σ. Then,

$$\hat{\beta}_{\text{gls}}^{(k+1)} = (X^T \hat{\Sigma}_k^{-1} X)^{-1} X^T \hat{\Sigma}_k^{-1} Y \tag{C5}$$

This procedure can be continued for a fixed number of steps, or until $\hat{\beta}_{\text{gls}}^{(k)}$ settles down. A convexity argument shows that $\hat{\beta}_{\text{gls}}^{(k)}$ converges to the maximum–likelihood estimate for β, assuming ε is independent of X and multivariate gaussian with mean 0.

The covariance matrix for $\hat{\beta}_{\text{gls}}^{(k+1)}$ is usually estimated from (C4), with $\hat{\Sigma}_k$ put in for Σ:

$$\widehat{\operatorname{cov}}^{(k+1)} = (X^T \hat{\Sigma}_k^{-1} X)^{-1} \tag{C6}$$

This may be legitimate, asymptotically. In finite–sample situations, all depends on whether $\hat{\Sigma}_k$ is a good estimate for Σ or not. If $\hat{\Sigma}_k$ is a poor estimate for Σ, the standard errors estimated from (C6) may prove to be unduly optimistic: an example is given in [11]. Unfortunately, approximate gls estimators are often used when there are too few data to offer any hope of estimating Σ with reasonable accuracy. In such circumstances, the bootstrap is a useful diagnostic, and in many cases it gives a more realistic estimate of the standard errors.

To ease notation, $\hat{\beta}_{gls}^{(k)}$ will be referred to as the (gls, k)-estimator. The paper [11] considers only the (gls, 1)-estimator, which in many situations has full asymptotic efficiency. In some examples, further iteration seems to make the coefficient estimates better, but also exaggerates the optimism of the standard-error estimates. In other examples, iteration actually makes the coefficient estimators worse.

In econometric work, it is usual to constrain β to fall in some linear space Λ; these are the *identifying restrictions*. Typically, many elements of β are constrained to vanish, and some are constrained to equal each other. The constraints are often incorporated by re-expressing the model in terms of linearly independent parameters; this involves linear manipulations of the columns of the design matrix X. A more elegant solution is to make unconstrained estimates, as indicated above, and then to project the unconstrained estimator, $\hat{\beta}_u = \hat{\beta}_{gls}^{(k)}$, say, into the constraint space Λ. However, the projection must take into account the covariance structure of $\hat{\beta}_u$, i.e., the constrained estimator $\hat{\beta}_c$ is the element of Λ minimizing the distance

$$(\hat{\beta}_u - \hat{\beta}_c)^T \hat{W}^{-1} (\hat{\beta}_u - \hat{\beta}_c) \tag{C7}$$

where $\hat{W}$ is the estimated covariance matrix of $\hat{\beta}_u$. Thus, the constrained gls estimator is found by two applications of unconstrained gls.

Changing the meaning of n, p, and X, consider next an econometric model of the form

$$\underset{1 \times q}{Y_i} = \underset{1 \times q}{Y_i} \underset{q \times q}{A} + \underset{1 \times p}{X_i} \underset{p \times p}{B} + \underset{1 \times q}{\varepsilon_i} \tag{C8}$$

In this equation, A and B are coefficient matrices of unknown parameters, to be estimated from the data, subject to identifying restrictions; Y_i is the vector of endogenous variables at data point i; X_i is the vector of exogenous variables; and ε_i is the vector of disturbances. This is a system of q equations in q endogenous variables; there are p exogenous variables. The matrix $I - A$ is assumed invertible. To be more specific about the stochastic assumptions:

The ε_i are independent and identically distributed.   (C9a)

$E(\varepsilon_i) = 0$ and cov $\varepsilon_i = V$ for all i.   (C9b)

All X are independent of ε.   (C9c)

Here, V is a $q \times q$ positive definite matrix. As is conventional, we normalize A so $A_{jj} = 0$ for all j. Write Y_{ij} for the jth component of the row vector Y_i. Then the jth equation in the system explains Y_{ij} in terms of the other endogenous variables and the exogenous variables:

$$Y_{ij} = Y_i A_j + X_i B_j + \varepsilon_{ij}$$

where A_j is the jth column of A and B_j is the jth column of B.

In the pre-harvester tomato model, for example, i corresponds to the pairs ct of counties and years, so there are $n = 8 \times 13 = 104$ data points; $q = 2$, because there are 2 endogenous variables, price and quantity; and $p = 15$ because there are 15 exogenous variables, viz.,

8 county dummy variables

yield Y
dispersion D
fertilizer F
grower price G
wage rate W
income I
retail price R

Coming now to two-stage least squares, by conditioning on the exogenous X's, we may suppose them constant; see (C9c). Multiply (C8) by X_i^T and sum:

$$\underset{p \times q}{R} = \underset{p \times q}{R} \; \underset{q \times q}{A} + \underset{p \times p}{S} \; \underset{p \times q}{B} + \underset{p \times q}{\Delta} \tag{C10}$$

where

$$R = \sum_{i=1}^{n} X_i^T Y_i \qquad S = \sum_{i=1}^{n} X_i^T X_i \qquad \Delta = \sum_{i=1}^{n} X_i^T \varepsilon_i \tag{C11}$$

Notice that the jth column of (C10) corresponds to the jth equation in (C8).

In applications, $[A, B]$ is constrained to fall in some linear space Λ of dimension at most pq. Then A and B can be estimated from (C10) by some variant of least squares. Notice that S is constant (nonrandom), since X is. It is conventional to treat R on the right side of (C10) as constant. This may be legitimate asymptotically but is false in any finite sample. In fact, R is not only random but also correlated with Δ, and this is the source of "small-sample bias" in 2SLS estimators. In the tomato model, this small-sample bias is serious, as will be seen below.

The matrix of errors Δ on the right-hand side of (C10) does have some covariance structure, so generalized least squares is the procedure of choice. To make contact with the standard format of (C1), we stack the columns in (C10): column 1 on top of column 2, ..., on top of column q. In the stack, information corresponding to the first equation in model (C8) comes first, information about the last equation comes last.

The parameter vector β in (C1) is obtained by stacking A and B: column 1 of A, followed by column 1 of B, ..., followed by column q of A, followed by column q of B. The design matrix is obtained by writing R and S down the diagonal and padding with zeros.

At this point, the design matrix is highly singular, having dimension $pq \times (pq + q^2)$. Usually, the elements of β known a priori to vanish are suppressed, and the design matrix is adjusted accordingly by deleting the corresponding columns. An alternative approach is to use generalized inverses. In any event the left-hand side Y vector in (C1) consists of the stacked R matrix; the error vector is the stacked Δ matrix. The full system of equations (C10) is laid out in stacked form as follows, with M_j being the jth column of any matrix M:

$$
\begin{bmatrix} R_1 \\ \cdot \\ \cdot \\ R_q \end{bmatrix} = \begin{bmatrix} R & S & 0 & 0 & \cdots & 0 & 0 \\ 0 & 0 & R & S & \cdots & 0 & 0 \\ \cdot & \cdot & \cdot & \cdot & & \cdot & \cdot \\ \cdot & \cdot & \cdot & \cdot & & \cdot & \cdot \\ \cdot & \cdot & \cdot & \cdot & & \cdot & \cdot \\ 0 & 0 & 0 & 0 & \cdots & R & S \end{bmatrix} \begin{bmatrix} A_1 \\ B_1 \\ \cdot \\ \cdot \\ A_q \\ B_q \end{bmatrix} + \begin{bmatrix} \Delta_1 \\ \cdot \\ \cdot \\ \Delta_q \end{bmatrix} \tag{C12}
$$

By (C9), the covariance matrix of the error vector (the stacked matrix) is the Kronecker product

$$
\Sigma = V \otimes S = \begin{bmatrix} V_{11}S & V_{12}S & \cdots & V_{1q}S \\ V_{21}S & V_{22}S & \cdots & V_{2q}S \\ \cdot & \cdot & & \cdot \\ \cdot & \cdot & & \cdot \\ \cdot & \cdot & & \cdot \\ V_{q1}S & V_{q2}S & \cdots & V_{qq}S \end{bmatrix} \tag{C13}
$$

Pretending R is constant, we should estimate the system (C10) by generalized least squares, relative to $V \otimes S$. Of course, V is unknown and must be estimated from the data. Coming now to two-stage least squares, consider each column of (C10) in isolation. Take column j, for instance:

$$
\underset{p \times 1}{R_j} = \underset{p \times q}{R} \; \underset{q \times 1}{A_j} + \underset{p \times p}{S} \; \underset{p \times 1}{B_j} + \underset{p \times 1}{\Delta_j} \tag{C14}
$$

Now cov $\Delta_j = V_{jj} S$; the constant of proportionality V_{jj} is immaterial, so (C14) can be estimated by generalized least squares, treating R on the right as a constant. Constraints specific to the jth equation would be imposed but not the cross-equation constraints. In the tomato model, the only constraints set appropriate coefficients to zero. For example, in the pre-harvester supply equation, income I does not come in; that coefficient is set to zero.

Let $\hat{A}$ and $\hat{B}$ denote the two-stage least-squares estimate. Let

$$
\hat{\varepsilon}_i = Y_i - Y_i \hat{A} - X_i \hat{B} \tag{C15}
$$

$$
\hat{V} = \frac{1}{n} \sum_{i=1}^{n} \hat{\varepsilon}_i^T \hat{\varepsilon}_i \tag{C16}
$$

The covariance matrix of $[\hat{A}, \hat{B}]$ can be estimated from (C6), with $\hat{V}$ used to estimate V in (C13) and obtain $\hat{\Sigma}$. It is conventional, for the purpose of estimating cov$[\hat{A}, \hat{B}]$ only, to inflate $\hat{V}_{jj}$ by $n/n - r$, where r is the number of variables actually coming into the jth equation. Call this estimated covariance matrix $\widehat{\text{cov}}$.

Turn now to the bootstrap. Let $\hat{A}$ and $\hat{B}$ be the two-stage least-squares estimates for A and B in (C8). It will be assumed that $I - \hat{A}$ is invertible. The residuals are defined by (C15). These are estimates for the true disturbances in the model (C8). Let μ be the empirical distribution of the residuals, assigning mass $1/n$ to each of $\hat{\varepsilon}_1, \ldots, \hat{\varepsilon}_n$. Consider next a model such as (C8) but in which all the ingredients are known:

Set the coefficients at A and B, respectively.

Make the disturbance terms independent, with common distribution μ.

The exogenous X's are kept fixed. With this simulation model, pseudodata can be generated for $i = 1, \ldots, n$. These pseudodata will be denoted by stars:

$$Y_i^\star = (X_i \hat{B} + \varepsilon_i^\star)(I - \hat{A})^{-1} \tag{C17}$$

where the $\varepsilon^\star$'s are independent with the common distribution μ.

Now pretend the pseudodata (C17) come from a model like (C8), with unknown coefficient matrices. Since there are county-specific intercepts, assumption (C9a) is satisfied. Estimate these coefficients by two-stage least squares from the pseudodata; denote the estimates by $\hat{A}^\star$ and $\hat{B}^\star$. Also compute from the starred data the estimated covariance matrix $\widehat{\text{cov}}^\star$, using (C15) and (C16) applied to the starred data.

The distribution of the pseudoerrors

$$\hat{A}^\star - \hat{A}, \quad \hat{B}^\star - \hat{B}$$

can be computed and used to approximate the distribution of the real errors

$$\hat{A} - A, \quad \hat{B} - B$$

This approximation is the bootstrap. It is emphasized that the

calculation assumes the validity of the model (C8). The distribution
of the pseudoerrors can be computed, e.g., by Monte Carlo, simply
repeating the procedure some number of times and seeing what happens.
The distribution of the pseudoerrors is of interest only as an approximation to the distribution of the real errors.

The next object is to test plaintiffs' statistical calculations
using the bootstrap methodology. Four sets of tables are presented;
each of the four tables has four parts:

   (i)   pre-harvester supply

  (ii)   pre-harvester demand

 (iii)   post-harvester supply

  (iv)   post-harvester demand

The tables correspond to different sets of assumptions about the stochastic disturbance terms ε and δ in plaintiffs' model, as specified
in equations (B1)-(B3). In all cases, it is assumed that the 16-
vectors

$$(\varepsilon_{ct}, \; \delta_{ct} : c \text{ varies over 8 counties})$$

are independent and identically distributed in time t, with mean zero.
Furthermore, the pre- and post-harvester disturbance terms are assumed
independent.

Case 1.   The 16-vectors of disturbances have an arbitrary 16×16
           covariance matrix across equations and counties.

Case 2.   The disturbance terms are independent across equations;
           within an equation, the 8-vectors of disturbances have an
           arbitrary 8×8 covariance matrix across counties.

Case 3.   The pairs $(\varepsilon_{ct}, \; \delta_{ct})$ are independent and identically dis-
           tributed, as c and t vary; but they have an arbitrary 2×2
           covariance matrix across equations.

Case 4.   The disturbance terms are independent across equations;
           within an equation, they are independent and identically
           distributed.

A priori, Case 1 is the most plausible. Implicitly, Case 3 is assumed
in Chern and Just (1980); and the discussion of the bootstrap presented above used Case 3. Case 4 is more special than Case 3, so
plaintiffs' analysis must cover this case, too. To apply the

bootstrap in Cases 1, 2, or 4, it is necessary only to modify appropriately the distribution for the ε^*.

The main conclusions from the bootstrap analysis are as follows:

In Cases 2 and 4, there is serious "small-sample bias" in the coefficient estimates. This bias distorts the statistics in plaintiffs' favor. The bias can be seen by comparing the Parameter column with the Mean column for post-harvester demand price in Tables C2 and C4.

In Cases 1 and 2, the conventional asymptotic formulas for standard errors are off by as much as 40%, in either direction, presumably due to the covariance across counties and to the difference in variances between counties, which are ignored in plaintiffs' analysis.[13] This can be seen by comparing the SD column with the RMS nominal SE column in Tables C1 and C2, especially for pre-harvester supply price and post harvester demand price. A downward distortion of the SE's helps the plaintiffs; an upward distortion hurts.

Columns 1 and 2 in the tables show the results obtained by fitting plaintiffs' model to plaintiffs' data, with one data error corrected (see Appendix B). Pre-harvester, the results agree quite closely with those reported in Table B1; the small differences are probably the result of round-off error, since the work was done on two different computers using two different software packages. Post-harvester, the results differ appreciably from Table B1, as discussed in Appendix B.

In the tables, column 1 shows the parameter estimates; column 2, the standard errors from the conventional asymptotic formulas. These columns are the same in all four tables. Columns 3–5 give the results of a bootstrap simulation experiment. In this experiment, plaintiffs' model (B1)–(B3) is taken as true, with parameter values as given in column 1 of the tables. The disturbance terms obey the conditions laid out above; e.g., Table C1 corresponds to Case 1. On the computer, the disturbance terms are generated by resampling the residuals from the fit reported in column 1. This is done 100 times, to generate 100 simulated data sets, according to plaintiffs' theory as expressed in

---

[13]In deposition, plaintiffs' expert made an ad hoc adjustment in certain exhibits for these factors. This adjustment is highly suspect, for the reasons given in [11].

(B1)-(B3). The exogenous variables are held fixed, and the equations are solved for the endogenous price and quantity variables. For each simulated data set, we use plaintiffs' fitting procedure to estimate the parameters in the simulation model and to compute the conventional approximations to the standard errors. The mean of the 100 parameter estimates $[\hat{A}^*, \hat{B}^*]$ is shown in the Mean column; the standard deviation, in the SD column. The root-mean-square of the 100 conventional standard errors is shown in the RMS nominal SE column. This is the square root of the mean of the diagonal entries in $\widehat{\text{cov}}^*$ above.

In Table C4, for example, the post-harvester demand price coefficient is set in the simulation model as -1.68; however, in the 100 simulated data sets, the mean of the estimates done by plaintiffs' method is 0.32, so there is an upward bias of 1.68 + 0.32 = 2. Plaintiffs' method for estimating parameters is strongly biased.

Likewise, in Table C1, the standard deviation of the 100 estimates for the pre-harvester supply price is 6.58; this is a true measure of the variability in the estimates, in the simulation model. However, the root-mean-square of the 100 conventional estimates of standard error is only 4.01. Plaintiffs' method for approximating the standard error is biased downward by

$$1 - \frac{4.01}{6.58} \approx 39\%.$$

Table C1. 100 Bootstrap Replications Resampling 16–Vectors

	Pre-harvester supply				
	The fit		The bootstrap		
	Parameter	Nominal SE	Mean	SD	RMS nominal SE
P	11.6	4.11	12.5	6.58	4.01
Y	9.69	5.09	10.1	6.11	4.77
D	−6.62	11.2	−6.85	14.4	10.4
F	−2.10	2.28	−1.90	3.59	2.03
G	−91.9	45.3	−91.9	73.5	42.2
W	−328.	154.	−334.	229.	162.
San Joa	871.	335.	832.	449.	272.
Yolo	625.	336.	586.	450.	273.
Solano	316.	336.	274.	453.	274.
Sutter	295.	336.	252.	454.	273.
Sacto	378.	335.	336.	454.	272.
Stanis	300.	332.	260.	447.	270.
Santa C	273.	337.	231.	455.	274.
San Ben	227.	338.	183.	452.	275.

	Pre-harvester demand				
	The fit		The bootstrap		
	Parameter	Nominal SE	Mean	SD	RMS nominal SE
P	−14.3	5.71	−13.7	4.52	4.46
I	.700	.115	.717	.129	.118
R	132.	32.1	129.	28.4	25.2
San Joa	217.	62.2	212.	74.3	61.4
Yolo	−8.36	62.1	−11.2	75.2	61.3
Solano	−319.	62.2	−325.	72.1	61.4
Sutter	−332.	62.0	−339.	72.3	61.2
Sacto	−260	62.2	−268.	70.8	61.4
Stanis	−349.	61.9	−355.	72.4	61.2
Santa C	−381.	62.4	−387.	73.0	61.6
San Ben	−406	62.3	−412.	73.4	61.4

(continued)

Table C1. Continued

	Post-harvester supply				
	The fit		The bootstrap		
	Parameter	Nominal SE	Mean	SD	RMS nominal SE
P	4.10	1.48	3.82	2.04	1.13
Y	14.9	8.08	14.1	6.51	6.88
San Joa	263.	161.	282.	153.	145.
Yolo	500.	165.	538.	145.	149.
Solano	−145.	167.	−116.	141.	150.
Sutter	−144.	172.	−113.	146.	154.
Sacto	−291.	160.	−266.	142.	144.
Stanis	−304.	164.	−277.	146.	148.
Santa C	−321.	174.	−295.	152.	156.
San Ben	−349.	190.	−319.	164.	170.

	Post-harvester demand				
	The fit		The bootstrap		
	Parameter	Nominal SE	Mean	SD	RMS nominal SE
P	−1.68	3.98	−1.46	2.21	3.28
I	.156	.146	.155	.115	.142
R	35.2	28.5	32.7	18.7	24.4
San Joa	445.	80.2	444.	77.5	79.5
Yolo	689.	80.0	707.	83.5	79.7
Solano	48.4	80.1	58.1	56.9	79.8
Sutter	60.4	80.2	70.7	60.1	79.8
Sacto	−111.	80.1	−105.	62.4	79.7
Stanis	−111.	80.7	−104.	59.0	79.5
Santa C	−111.	80.6	−105.	60.1	79.3
San Ben	−107.	80.5	−99.5	58.2	79.6

Table C2.   100 Bootstrap Replications Resampling 8-Vectors

| | Pre-harvester supply | | | | |
| | The fit | | The bootstrap | | |
	Parameter	Nominal SE	Mean	SD	RMS nominal SE
P	11.6	4.11	11.0	5.27	3.77
Y	9.69	5.09	9.42	7.00	4.80
D	-6.62	11.2	-7.42	13.2	10.5
F	-2.10	2.28	-2.14	3.26	2.00
G	-91.9	45.3	-89.7	69.2	42.4
W	-328.	154.	-325.	184.	155.
San Joa	871.	335.	890.	483.	280.
Yolo	625.	336.	644.	480.	280.
Solano	316.	336.	334.	478.	280.
Sutter	295.	336.	312.	479.	280.
Sacto	378.	335.	395.	477.	279.
Stanis	300.	332.	318.	471.	276.
Santa C	273.	337.	289.	479.	281.
San Ben	227.	338.	244.	479.	282.

| | Pre-harvester demand | | | | |
| | The fit | | The bootstrap | | |
	Parameter	Nominal SE	Mean	SD	RMS nominal SE
P	-14.3	5.71	-11.8	3.59	3.83
I	.700	.115	.658	.108	.108
R	132.	32.1	120.	22.0	22.6
San Joa	217.	62.2	226.	63.1	59.2
Yolo	-8.36	62.1	-.747	64.0	59.3
Solano	-319.	62.2	-312.	60.5	58.7
Sutter	-332.	62.0	-325.	60.4	58.4
Sacto	-260.	62.2	-254.	59.9	58.7
Stanis	-349.	61.9	-340.	61.1	58.5
Santa C	-381.	62.4	-374.	61.7	58.8
San Ben	-406.	62.3	-399.	61.1	58.6

(continued)

Table C2.   Continued

	Post-harvester supply				
	The fit		The bootstrap		
	Parameter	Nominal SE	Mean	SD	RMS nominal SE
P	4.10	1.48	3.34	.765	.920
Y	14.9	8.08	17.5	6.71	6.06
San Joa	263.	161.	227.	148.	130.
Yolo	500.	165.	484.	170.	135.
Solano	-145.	167.	-172.	153.	135.
Sutter	-144.	172.	-172.	160.	139.
Sacto	-291.	160.	-322.	143.	129.
Stanis	-304.	164.	-334.	149.	133.
Santa C	-321.	174.	-354.	154.	140.
San Ben	-349.	190.	-386.	171.	153.

	Post-harvester demand				
	The fit		The bootstrap		
	Parameter	Nominal SE	Mean	SD	RMS nominal SE
P	-1.68	3.98	.223	1.57	2.01
I	.156	.146	.156	.0949	.115
R	35.2	28.5	23.4	14.6	18.5
San Joa	445.	80.2	438.	57.8	67.6
Yolo	689.	80.0	713.	76.1	74.5
Solano	48.4	80.1	57.9	50.0	66.6
Sutter	60.4	80.2	70.0	52.0	67.0
Sacto	-111.	80.1	-110.	53.7	66.3
Stanis	-111.	80.7	-108.	49.7	65.9
Santa C	-111.	80.6	-108.	51.0	65.9
San Ben	-107.	80.5	-102.	50.8	65.8

Table C3. 100 Bootstrap Replications Resampling 2-Vectors

	Pre-harvester supply				
	The fit		The bootstrap		
	Parameter	Nominal SE	Mean	SD	RMS nominal SE
P	11.6	4.11	12.2	3.84	3.60
Y	9.69	5.09	10.3	4.66	4.70
D	-6.62	11.2	-6.75	10.6	10.4
F	-2.10	2.28	-2.08	1.84	1.95
G	-91.9	45.3	-89.1	40.9	41.8
W	-328.	154.	-331.	167.	156.
San Joa	871.	335.	844.	250.	264.
Yolo	625.	336.	596.	250.	265.
Solano	316.	336.	288.	248.	265.
Sutter	295.	336.	269.	249.	264.
Sacto	378.	335.	354.	248.	264.
Stanis	300.	332.	272.	246.	262.
Santa C	273.	337.	246.	251.	266.
San Ben	227.	338.	198.	251.	267.

	Pre-harvester demand				
	The fit		The bootstrap		
	Parameter	Nominal SE	Mean	SD	RMS nominal SE
P	-14.3	5.71	-13.9	4.77	4.51
I	.700	.115	.686	.104	.112
R	132.	32.1	130.	27.9	25.2
San Joa	217.	62.2	221.	58.8	58.5
Yolo	-8.36	62.1	-5.42	57.4	58.4
Solano	-319.	62.2	-316.	55.0	58.5
Sutter	-332.	62.0	-328.	58.5	58.4
Sacto	-260.	62.2	-256.	58.1	58.6
Stanis	-349.	61.9	-347.	59.3	58.3
Santa C	-381.	62.4	-379.	58.7	58.7
San Ben	-406.	62.3	-402.	59.3	58.5

(continued)

Table C3. Continued

	Post-harvester supply				
	The fit		The bootstrap		
	Parameter	Nominal SE	Mean	SD	RMS nominal SE
P	4.10	1.48	3.92	1.20	1.17
Y	14.9	8.08	14.5	6.27	7.17
San Joa	263.	161.	269.	131.	150.
Yolo	500.	165.	516.	131.	154.
Solano	−145.	167.	−126.	130.	156.
Sutter	−144.	172.	−135.	129.	160.
Sacto	−291.	160.	−275.	134.	150.
Stanis	−304.	164.	−285.	134.	153.
Santa C	−321.	174.	−308.	143.	162.
San Ben	−349.	190.	−332.	151.	177.

	Post-harvester demand				
	The fit		The bootstrap		
	Parameter	Nominal SE	Mean	SD	RMS nominal SE
P	−1.68	3.98	−1.78	3.52	3.94
I	.156	.146	.150	.112	.133
R	35.2	28.5	35.2	23.4	27.7
San Joa	445.	80.2	447.	67.3	78.1
Yolo	689.	80.0	698.	66.9	78.5
Solano	48.4	80.1	60.4	69.9	78.3
Sutter	60.4	80.2	62.4	62.5	77.9
Sacto	−111.	80.1	−104.	76.4	78.8
Stanis	−111.	80.7	−98.7	73.6	77.8
Santa C	−111.	80.6	−104.	70.3	77.9
San Ben	−107.	80.5	−98.3	65.2	77.9

Table C4. 100 Bootstrap Replications Resampling 1-Vectors

Pre-harvester supply

	The fit		The bootstrap		
	Parameter	Nominal SE	Mean	SD	RMS nominal SE
P	11.6	4.11	11.2	3.06	3.69
Y	9.69	5.09	9.13	4.79	4.77
D	-6.62	11.2	-5.67	9.43	10.6
F	-2.10	2.28	-2.20	1.90	2.01
G	-91.9	45.3	-98.8	39.7	42.2
W	-328.	154.	-343.	166.	162.
San Joa	871.	335.	922.	254.	266.
Yolo	625.	336.	676.	252.	267.
Solano	316.	336.	370.	253.	267.
Sutter	295.	336.	347.	249.	266.
Sacto	378.	335.	428.	251.	266.
Stanis	300.	332.	348.	249.	264.
Santa C	273.	337.	324.	257.	268.
San Ben	227.	338.	277.	253.	269.

Pre-harvester demand

	The fit		The bootstrap		
	Parameter	Nominal SE	Mean	SD	RMS nominal SE
P	-14.3	5.71	-11.9	3.75	4.70
I	.700	.115	.676	.107	.121
R	132.	32.1	119.	22.4	26.8
San Joa	217.	62.2	221.	50.1	59.2
Yolo	-8.36	62.1	-2.35	53.6	59.3
Solano	-319.	62.2	-312.	54.0	59.6
Sutter	-332.	62.0	-325.	50.6	59.0
Sacto	-260.	62.2	-253.	51.0	59.5
Stanis	-349.	61.9	-343.	48.7	58.3
Santa C	-381.	62.4	-377.	51.5	59.8
San Ben	-406.	62.3	-398.	52.7	59.7

(continued)

Table C4.   Continued

	Post-harvester supply				
	The fit		The bootstrap		
	Parameter	Nominal SE	Mean	SD	RMS nominal SE
P	4.10	1.48	3.35	.879	1.02
Y	14.9	8.08	16.9	5.20	6.48
San Joa	263.	161.	245.	115.	136.
Yolo	500.	165.	484.	110.	139.
Solano	−145.	167.	−162.	110.	141.
Sutter	−144.	172.	−166.	117.	145.
Sacto	−291.	160.	−315.	108.	135.
Stanis	−304.	164	−313.	109.	139.
Santa C	−321.	174.	−345.	114.	146.
San Ben	−349.	190.	−372.	126.	160.

	Post-harvester demand				
	The fit		The bootstrap		
	Parameter	Nominal SE	Mean	SD	RMS nominal SE
P	−1.68	3.98	.320	1.51	2.02
I	.156	.146	.159	.0928	.120
R	35.2	28.5	22.1	12.2	15.3
San Joa	445.	80.2	454.	62.5	66.9
Yolo	689.	80.0	699.	59.1	66.2
Solano	48.4	80.1	51.9	66.8	69.1
Sutter	60.4	80.2	68.1	53.3	65.9
Sacto	−111.	80.1	−104.	59.7	66.5
Stanis	−111.	80.7	−102.	56.7	66.0
Santa C	−111.	80.6	−102.	53.1	67.9
San Ben	−107.	80.5	−98.3	60.1	67.0

REFERENCES

1. J. A. Brandt, B. C. French, and E. V. Jesse (1978). Economic performance of the processing tomato industry. Bulletin 1888, Division of Agricultural Sciences, University of California at Davis.

2. L. Breiman and D. A. Freedman (1983). How many variables should be entered in a regression equation? *Journal of the American Statistical Association*, Vol. 78, No. 381, pp. 131-136.

3. W. S. Chern and R. E. Just (1978). Econometric analysis of supply response and demand for processing tomatoes in California. Giannini Foundation Monograph No. 37, University of California at Berkeley.

4. W. S. Chern and R. E. Just (1980). Tomatoes, technology, and oligopsony. *The Bell Journal of Economics*, Vol. 11, pp. 584-602.

5. C. Christ (1975). Judging the performance of econometric models of the U.S. economy. *International Economic Review*, Vol. 16, pp. 54-74.

6. G. Debreu (1969). *Theory of Value*. Monograph 17, Cowles Foundation for Research in Economics at Yale University. (Wiley, New York)

7. M. Finkelstein (1980). The judicial reception of multiple regression studies in race and sex discrimination cases. *Columbia Law Review*, Vol. 80, pp. 737-754.

8. F. Fisher (1980). Multiple regression in legal proceedings. *Columbia Law Review*, Vol. 80, pp. 702-736.

9. D. A. Freedman (1981). Some pitfalls in large econometric models: a case study. *Journal of Business*, Vol. 54, pp. 479-500.

10. D. A. Freedman (1983). A note on screening regression equations. *American Statistician*, Vol. 37, pp. 152-155.

11. D. A. Freedman and S. Peters (1984). Bootstrapping a regression equation: some empirical results. *Journal of the American Statistical Association*, Vol. 79, pp. 97-106.

12. D. A. Freedman and S. Peters (1983). Bootstrapping an econometric model: some empirical results. *Journal of Business and Economic Statistics*, Vol. 2, pp. 150-158.

13. D. A. Freedman, T. Rothenberg, and R. Sutch (1983). On energy policy models. *Journal of Business and Economic Statistics*. Vol. 1, No. 1, pp. 24-36.

14. P. Higginbotham (1980). Opinion in Vuyanich vs. Republic National Bank of Dallas. 505 F Sup 224.

15. E. Mansfield (1982). *Microeconomics*, 4th edition. (Norton, New York)

16. P. A. Samuelson (1980). *Economics*, 11th edition. (McGraw–Hill, New York)

17. H. Theil (1971). *Principles of Econometrics*. (Wiley, New York)

18. F. M. Scherer (1980). *Industrial Market Structure and Economic Performance*, 2nd edition. (Rand–McNally, Chicago)

19. V. Zarnowitz (1979). An analysis of annual and multiperiod quarterly forecasts of aggregate income, output, and the price level. *Journal of Business*, Vol. 52, pp. 1–34.

Added in Proof

P. Meier, J. Sacks, and S. Zabell (1984). What happened in Hazelwood: statistics, employment discrimination, and the 80% rule. *American Bar Foundation Research Journal*, pp. 139–186.

D. W. Peterson, ed. (1983). Statistical inference in litigation. *Journal of Law and Contemporary Problems*, Vol. 46, No. 4.

PROBABILISTIC RISK ASSESSMENT IN THE NUCLEAR INDUSTRY: WASH-1400 AND BEYOND

T. P. SPEED   CSIRO Division of Mathematics and Statistics

ABSTRACT: In the mid-seventies the Rasmussen Report (WASH-1400) represented the state-of-the-art in applying probabilistic and statistical methods to the assessment of the risks associated with nuclear reactors. It received a great deal of criticism, much of which was well-based. Have the problems the Rasmussen team met been overcome in the intervening decade? Are the figures produced in probabilistic risk assessments of nuclear power stations of any value? After examining some of the relevant literature that has appeared since 1975 and reading the recent Sizewell B Probabilistic Safety Study, I have concluded that the answer to both these questions is **No!**

1. INTRODUCTION

Several years ago I wrote a short critique of the 1975 Reactor Safety Study (RSS), U.S. Government (1975), also known as *WASH-1400* or the *Rasmussen Report*. As a mathematical statistician, my interest in the RSS focused on the nature and quality of the data used, the probabilistic and statistical assumptions made, and the methods of analysis adopted in the study. My intention was to form an opinion on the

validity of the probability figures that were the principal conclusions of the study, and I concluded that they were totally valueless.
Not long afterwards a review of the RSS, U.S. Government (1978), concluded, among other things that

> We are unable to determine whether the absolute probabilities of
> accident sequences in WASH-1400 are high or low, but we believe
> that the error bounds on those estimates are, in general, greatly
> understated.   [p. viii]

Now, nearly a decade later, it seems that few people believe the
probability figures calculated in WASH-1400, although most would accept
that the *process* of carrying out a probabilistic risk assessment has
considerable value.   The proposal by the United Kingdom Central Electricity Generating Board (CEGB) to build a pressurized water reactor
(PWR) at Sizewell in Suffolk following the Standardized Nuclear Unit
Power Plant System designed by Bechtel for Westinghouse in the U.S.,
and the subsequent production by Westinghouse of the Sizewell B Probabilistic Safety Study (PSS) for the proposed reactor, offer us an
opportunity to see the extent to which the problems identified in the
RSS have been overcome.   Like the RSS, the Sizewell B PSS aims for,
and obtains, figures purporting to quantify the probabilities (per
year) of core melt and release from the containment, although there are
no error bounds of any kind attached to these figures.

The main aim of this paper is to examine the Sizewell B PSS in
much the same way as I did the RSS; see Speed (1979) for a Danish
version of my critique.   I sought evidence of careful data analysis,
an appreciation of the difficulties and uncertainties inherent in the
task, perhaps consideration of robustness issues and other modern
methods, but at least a critical use of traditional techniques; in
short, I sought evidence that the probabilistic and statistical
analyses of the study were carried out by competent professionals in
the area.   At the same time it was helpful to survey the literature
of probabilistic risk assessment in the nuclear and related fields
that has appeared since 1978; and, for completeness, a brief account
of the RSS and its aftermath will also be presented.   A final section draws some conclusions from this examination.

2. BRIEF ACCOUNT OF WASH-1400 AND ITS AFTERMATH

In August 1972 the U.S. Atomic Energy Commission (AEC) commissioned a
study of the safety of commercial light-water reactors, to be done by
a team of AEC employees and some outside consultants under the general
leadership of Professor Norman Rasmussen, Head of the Department of
Nuclear Engineering at M.I.T. The study group was requested to
identify reactor accident sequences having potential consequences to
the public, to estimate their probabilities of occurrence, and to
quantify their consequences in terms of the sizes of various releases
of various radioactive substances. By August 1974 a draft of this
study was ready; it consisted of some 3300 pages (Main Report and ten
appendices) resulting from about 70 man-years of work and apparently
costing more then U.S. $3 million. The draft was circulated for
comment, and in October 1975 a somewhat amended version was published,
including an extra appendix giving comments received on the draft
together with answers to these comments.

This RSS calculated that the probability of a core melt was 1 in
20,000 (per reactor year); it also gave figures for a range of possible consequences, including the much quoted 1 in 10^9, the probability of a radioactive release causing more than 1000 (early) fatalities, alleged to be comparable to an individual's risk of being killed
by a meteorite.

The period immediately following the release of the RSS saw a
large number of criticisms of its presentation, completeness, methodology, treatment of key issues, and the use to which its principal
conclusions were put in the field of public policy. Of the many critical analyses, we mention just Critchley (1976), Hubbard and Minor
(1977), and von Hippel (1977). In mid-1977, following hearings of the
House Committee of Interior and Insular Affairs and a request from the
Chairman of that Committee, the Nuclear Regulatory Commission (NRC)
convened "a special Review Group for the purpose of reviewing peer
group comments on the final report of the Reactor Safety study and
the developments in risk assessment methodology that have occurred
since the report was published," U.S. Government (1978, p. 62). This

Risk Assessment Review Group (RARG) reported in September 1978, U.S. Government (1978), and the following is a list of topics from its main criticisms of the use of probability and statistics in the RSS:

The "square-root bounding model"

The use of the log-normal distribution

The use of subjective probabilities

Variations between reactors

Completeness and relevance of data

Propagation of errors

Probability "smoothing"

Use of median instead of mean

Common cause failures

Human factor aspects, including data

These headings cover my main statistical criticisms of the RSS and, I believe, those of most other critics.

In a response to the RARG Report quoted in Monitor (1979), Professor Rasmussen is quoted as saying that he is in general agreement with most of its findings and recommendations although he feels that the uncertainty bounds are only "somewhat understated" and not "greatly understated" as the RARG believed. Rasmussen is further quoted as believing that "accumulated reactor experience to date puts an upper limit on this probability that is not much greater than the upper limit in WASH-1400."

Summarizing, the RSS is now accepted in the U.S. as having many serious defects in its methodology, and its principal numerical conclusions are no longer quoted by official U.S. Government sources such as the Nuclear Regulatory Commission. It has been left to reactor manufacturers to try to do properly what the RSS failed to do, and below we shall examine one such attempt, the Westinghouse Sizewell B Probabilistic Safety Study.

3. DEVELOPMENTS IN PROBABILISTIC RISK ASSESSMENT SINCE 1978

Perhaps the most striking development in the area of probabilistic
risk assessment since 1978 has been its enormously increased popularity; see Health and Safety Executive (1978), McCormick (1981), and
Philipson (1982) for specific examples, and Farmer (1980, 1981),
Fischoff et al. (1981), Okrent (1981), Royal Society Study Group
(1983), and Warner and Slater (1981) for reviews covering a wide
variety of fields. In many of these articles the RSS methodology and
results are quoted as though the criticisms noted above, including
those of the RARG, had never been made, much less accepted by
Rasmussen himself. For example, Farmer (1981) refers to the RSS (see
p. 106) without reservation, even reproducing (his Fig. 3) the infamous
Figure 1-1 of the Executive Summary of the WASH-1400, headed "Frequency
of Man-Caused Events Involving Fatalities." In the same vein, Lord
Rothschild (1978) quoted Lord Ashby, who was quoting the 1 in 10^9
figure from WASH-1400, and also reproduced quantitative conclusions
from the study; Jones and Akehurst (1980) actually cite these figures
from Lord Rothschild (1978) in their article purporting to explain
risk assessment!

Not all of the efforts at probabilistic risk assessment since
1978 have adopted the RSS methodology, the study Health and Safety
Executive (1978) being a notable exception. This, too, was criticized, Cremer and Warner (1980), but it stands out as a model of careful and restrained analysis. The question of whether its quantitative
conclusions are of any more value than those of other studies I mention, I leave unanswered; but it is refreshing to see a study that is
prepared to distinguish among probability evaluations

a. assessed statistically from historical data;

b. based on statistics as far as possible but with some missing
figures supplied by judgment;

c. estimated by comparison with previous cases for which fault
tree assessments have been made;

d. "dummy" figures, likely to be always uncertain, a subjective
judgment must be made;

e. fault tree synthesis, an analytically based figure that can be
independently arrived at by others.

The approach has much in common with that adopted in the excellent
paper by Fairley (1977).

Another trend, going in parallel with the increased use of probabilistic risk assessment techniques, is that of formulating *quantitative safety goals.* The Report of the RARG (p. 44) makes some comments
on the role of probabilistic methods in the regulatory process, whereas
Okrent (1981, §6) briefly describes the history of this topic in the
context of U.S. reactor safety. Marshall (1983, pp. 245-46) states
that the design of the PWR for Sizewell B was tuned to aim at a target
of "a risk factor of less than once in a million years" (for the event
of uncontrolled release of radioactivity).

Have There Been Any Major Advances in Methodology Since 1978?

Shortly after the RARG Report was released, and perhaps in response to
the criticism of the WASH-1400 use of probabilities, variously
described as "engineering judgment," "subjective," or "Bayesian," the
papers of Apostolakis on this topic appeared; see, especially, Apostolakis and Mosleh (1979). This work attempted to formalize the RSS
techniques by introducing gamma priors to incorporate expert opinion,
the use of Poisson likelihoods, Bayes' theorem, etc., all quite
familiar material to statisticians and in no way overcoming the major
objections to the use of such probabilities in a risk calculation such
as that of a reactor core melt.

If it can be described as a methodological advance, we can assert
that most of the more blatant misuses of probability and statistics
found in WASH-1400 appear to have been quietly dropped. These include
the "square-root method" for common cause failures, the Monte Carlo
method for "propagating errors" in log-normally distributed probabilities [see, however, 4(ii) below], and the technique of probability
smoothing.

The clearest evidence, however, that no major advances in methodology have been made is found by reading recent reviews of the subject
of probabilistic risk assessment in nuclear and related fields, such
as Warner and Slater (1981), Philipson (1982), and the Royal Society

Study Group (1983). In these articles we find the same basic techniques described, with or without qualifications as to their use and interpretation, and at no time are any serious doubts raised as to their validity or usefulness. The comments by a discussant (D. Andrews) to Farmer (1981) are interesting in this respect:

> It may be convenient to use quantitative methods in the manner that Professor Farmer suggests, but before doing so it is necessary to establish that the practice is valid.
>
> The man who designed the notorious pipe at Flixborough did not just guess its dimensions: he used a formula. This formula was well known and easy to understand and use. We know now that the formula used was wrong, because the pipe failed. We know also that we could have known that the formula was inappropriate because the circumstances of its application differed from the assumptions from which the formula was predicated. We have to make sure that we are not making a similar mistake ourselves with regard to quantitative assessments.
>
> In the quantitative assessment of the safety of any complete plant, however, there are two necessary types of factor about which one cannot have assurance that they satisfy these requirements, namely the factor allowed to cater for unanticipated unknowns, and the factor allowed to cater for the undependability of human performance.
>
> In respect of unanticipated unknowns, one can only guess, and guesses have no epistemological validity.
>
> In respect of human behaviour, the possibility of intelligent intervention in the selection of causes from the population of choice is specifically excluded in the archetype for which the fundamental postulates of probability theory were developed, and on which all subsequent probability theory depends. One is not allowed to mark the cards or peer in the urn. The population of causes is not constant or finite: people invent things, and have mental aberrations that extend the possibilities beyond the scope of mechanistic reliability. Moreover, there is evidence to show that human behaviour—even where the life of a person concerned may be at risk—may have no central tendency.
>
> We are entitled to use probability theory in respect of human behaviour only where the preferences of any individual member of the sample can have no significant influence on the mean response of the whole population, and where an answer in terms of the mean is adequate for the purposes intended. In respect of the assurance of safety against major dangers, we are concerned with the response of the one particular person who has the power to influence the course of events, and probability theory has nothing to say about that.
>
> Having decided that something has to be done, it is perfectly reasonable, and often desirable, to use probability theory

to determine how best to do it, and for this purpose quantitative criteria may be set. For instance, the B.C.A.R. require the probability of a hazardous effect arising from an individual failure to be less than 10^{-8} per hour of engine operation. Nevertheless, in many cases it is just not practicable to determine these probabilities, and reliance must be placed on the effectiveness of imposed controls.

Probability theory must not be allowed to become a scientific superstition. It is not an incantation for the exorcism of unthinkable events.

There is need for a forum in which the validity of proposed safety philosophies can be publicly debated. [p. 119]

It seems likely that risk analysts will ultimately recognize the probability aspects of the common risk assessment techniques such as event and fault trees for what they are, namely, frameworks for extended applications of the addition and multiplication rules of probability theory, and that their correct use involves sensible estimation of the component probabilities and a clear understanding of the associated assumptions of completeness and independence. They will give up expecting a solution to the "problem" of common-mode failures [see 4(iv) below], stop trying to account for unanticipated events, and so on, and only make probabilistic analyses in contexts in which they have good reasons to believe that the answers are of value.

4. THE SIZEWELL B PROBABILISTIC SAFETY STUDY

(i) Background and Structure of the Study

In June 1981, a task force was set up to develop firm design proposals for the pressurized water reactor that the United Kingdom Central Electricity Generating Board (CEGB) wanted to build at Sizewell in Suffolk; see Marshall (1983). In that article Marshall gives an outline of the design process followed in the U.K. leading up to the Statement of Case, Reference Design, and Pre-Construction Safety Report published by the CEGB in May 1982. As noted earlier in this paper, he states that "The target aimed for is a risk factor of less than once in a million years." At around the same time, the Sizewell B Probabilistic Safety Study, Westinghouse (1982), appeared (probably in anticipation of the public enquiry into the building of the reactor), showing that the target had been achieved. The overall conclusions of this study are

> . . . that the frequency of core melt at the Sizewell B plant from internal initiators is conservatively estimated to be 1.16×10^{-6} per year and the release frequency from the containment is 2.8×10^{-8} per year. These releases are largely dominated by either basemat failures which are not as serious as above ground failures or delayed overpressure failures which would not occur for at least 8 hours after accident initiation. [p. 5.7-1]

How did the Sizewell B PSS reach these conclusions? Although structured differently, the basic approach adopted in the study was the same as that of WASH-1400; namely,

> Component failure distributions are determined. Equipment reliability is calculated. Physical phenomena are evaluated. Finally, the probability of any accident with any number of safety system failures can be calculated along with the associated release of radioactive material. [p. 1.1-1]

The study establishes 16 *initiating events*, 26 *support states*, 45 *plant damage states*, and 12 *release categories*; the general approach is to calculate the probability $\mathrm{pr}(R_k)$ of release category k as a sum

$$\sum_i \sum_j \mathrm{pr}(E_i)\mathrm{pr}(D_j|E_i)\mathrm{pr}(R_k|D_j) \tag{1}$$

where (PSS notation) $\varphi_i = \mathrm{pr}(E_i)$ is the probability of initiating event i, $m_{ij} = \mathrm{pr}(D_j|E_i)$ is the probability of plant damage state j given initiating event i, and $c_{jk} = \mathrm{pr}(R_k|D_j)$ is the probability of release category k given plant damage state j. Each of the elements of the *plant damage matrix* $M = (m_{ij})$ is calculated as the sum

$$\mathrm{pr}(D_j|E_i) = \sum_h \mathrm{pr}(D_j|E_i \ \& \ S_h)\mathrm{pr}(S_h|E_i) \tag{2}$$

over all 26 support states S_h. The probabilities $\mathrm{pr}(S_h|E_i)$ are calculated by standard methods and then the probabilities $\mathrm{pr}(D_j|E_i \ \& \ S_h)$ are calculated as sums

$$\sum_{\mathrm{paths}} \mathrm{pr}(E_i \rightarrow D_j|E_i \ \& \ S_h) \tag{3}$$

over all paths in an *event tree* with top event E_i and terminal events the D_j. These event trees typically have 8 to 12 branch points along each path; so a typical term in (3) is a product

$$\text{pr}(I_1 | E_i \ \& \ S_h) \text{pr}(I_2 | I_1 \ \& \ E_i \ \& \ S_h) \ \dots \ \text{pr}(D_j | I_1 \ \& \ I_2 \ \& \ \dots \ \& \ I_r \ \& \ E_i \ \& \ S_h) \quad (4)$$

of $r = 8$ to 12 terms corresponding to *intermediate events* $I_1, \dots, I_r$.
Similarly, but somewhat more simply, each of the events of the
containment matrix $C = (c_{jk})$ is calculated as the sum

$$\sum_{\text{paths}} \text{pr}(D_j \rightarrow R_k | D_j) \quad (5)$$

over all paths in the *containment event tree* from D_j to R_k; and each
term in (5) should be a product of conditional probabilities

$$\text{pr}(J_1 | D_j) \text{pr}(J_2 | J_1 \ \& \ D_j) \ \dots \ \text{pr}(C_k | J_1 \ \& \ J_2 \ \& \ \dots \ \& \ J_s \ \& \ D_j) \quad (6)$$

corresponding to s intermediate events $J_1, \dots, J_s$, where here s is of
the order of 30.

This then is the structure of the probability calculations under-
taken. Just how the many probability evaluations were made will be
discussed in later subsections, and we close this introduction by
quoting from some of the study's remarks on methodology:

> The situation here is identical to that present whenever a model
> is made, whether it be mathematical, an engineering, a computer,
> or a verbal model. The point is, the model is not the entity.
> It is only a symbolic representation of the entity.

> The relevance of this point for our current work is that an event
> tree is a model of plant behavior. It can be made more accurate,
> more realistic by distinguishing more systems at the top and more
> states for each system. The price for this is greater complexity
> in the analytical work itself. The analysts' challenge is one of
> balancing the conflicting desires of realism and simplicity.
> [p. 1.1-4]

A question that is *not* addressed anywhere in this study is: What evidence is there that probabilities calculated using this model bear
any relation to the behavior of real-world PWRs and reactor accidents?

(ii) Types of Probability Evaluations Used

As with WASH-1400, the Sizewell B PSS contains an undifferentiated
mixture of probability evaluations based on hard data, apparently

quite subjective considerations, models of physical phenomena, and other methods. The conclusions of the study are that "the frequency of core melt . . . is conservatively estimated to be. . . ." It seems worthwhile to note some of these evaluations so that the reader can focus better on the question just mentioned (which is not addressed in the study).

Apart from the probabilities concerning the failure or unavailability of equipment, the main data-based probabilities in the study concern the initiating events. I comment on these in (iii) below. In Appendix 3.8.1 of the study, entitled "Hydrogen Mixing Probability Assignment," we find that the probability of a detonation event in the context of hydrogen mixing is assigned epsilon, with epsilon stated in a footnote to equal 10^{-4}. Further down, a value of 10 epsilon is used to define a related probability "in order to show that it is greater than" the former, and finally, because two "separate independent constraints must be met" the "probability of a detonation is somewhere between 10 epsilon squared, and epsilon squared, . . . In other words, . . . small enough that the possibility of a detonation need not be considered in the hydrogen burn question."

This sort of argument is reminiscent of WASH-1400 at its worst and suggests that not all the criticisms made of that report have been understood by the writers of the PSS. The lapse is not an isolated one, for the following section, headed "Probability Assignment," apparently quite arbitrarily *assigns* probabilities of 1 - epsilon, 0.95, 0.99, 0.6, 0.7, 0.95, 0.99, 0.6, 0.4, and C to a string of events associated with hydrogen mixing. When one considers the anxiety and uncertainty associated with the hydrogen bubble at Three Mile Island, U.S. Government (1979), the discussion just referred to seems far from adequate.

In §2.4 of the study, "Containment Event Tree Quantification" [cf. equation (6) above], we find the following remarks:

> The probabilities assigned to each node in the containment event trees were determined individually and were based on a number of diverse resources including auxiliary analyses by engineers; containment phenomenology computer modeling codes such as MARCH and CORCON, and the Westinghouse-developed COCOCLASS9 code;

consultants' reports and analyses; experimental investigations; containment capability analyses; and the literature.

When the events specified by the paths of the containment event tree were analyzed, certain branches for a path were found to be highly unlikely or even physically impossible. Rather than assign a zero probability to these unlikely branches, a very small probability, often called epsilon ε (usually 10^{-4}) was assigned. [p. 2.4-1]

It is difficult indeed for a statistical reader of the study to assess these probability evaluations or "assignments." Very few details are given, although the programs referred to are from WASH-1400, and that source could be consulted. The end result, the containment matrix C, is given (p. 2.4-8 or p. 5.3-9) as a 23×12, not a 45×12, matrix corresponding (with one exception) to *only* those plant damage states with *successful* containment isolation (cf. p. 1.6.4.1-7). The rows of this matrix add to unity, as they should, and one wonders what happened to the other 22 rows corresponding to plant damage states for which containment isolation *failed*.

Turning to the actual figures c_{jk} presented, one finds that six rows are dominated by a figure of 0.9882 (two) or 0.9886 (four) corresponding to $k = 12$, whereas a further four each give 0.9732 to the same column. Of the other twelve rows, four are identical with non-trivial probability in three columns, whereas the remaining six have 0.9832 (two), 0.9835 (two), or 0.9732 (two) in column $k = 11$, and the remaining two having 1.0000 in columns 1 and 2, respectively. This is claiming a high order of accuracy when, speaking elsewhere of one of the computer models used in their calculation, we find

In a study of this type, MARCH has a number of limitations. It will not predict the exact course a particular type of core meltdown will take. This is, in part, because the models used in MARCH are simplified to increase computational speed, because MARCH does not model certain physical phenomena, and because the exact characteristics of an LWR core melt are not known. [p. 3.2-4]

No such reservations are voiced about the COCOCLASS9 code which is described as "a sophisticated mathematical model of a generalized reactor containment." The record of this model in calculating

probabilities that can be related to the real world is nowhere
mentioned.

(iii) Data and Data Analyses Performed

The data used in the Sizewell B PSS concerns initiating events, hardware (components), and human factors. I shall comment briefly on each
of the last two below, my main remarks being directed at the study's
use of the data embodied in Tables 1.2-13 and 1.2-14. It is on the
basis of data in these tables that the study team evaluated the probabilities φ_i of initiating events, and their methods of doing so seem
to be at least as arbitrary as some of those used in WASH-1400.

On what data are the PSS probabilities based? This is not
entirely clear, for although they reproduce a 30 × 14 table of population event data, our Table 1, 30 PWRs with a combined operating experience of 131 years and 14 types of events, their later calculations use
figures described in Table 1.2-15 as *Sizewell B Type Plant Specific*.
We quote:

> In recognition that the Sizewell B plant has no operating history
> to extract plant specific data a "like" plant was substituted.
> The selection of a "like" plant was necessary, since generic
> plant data would cover too broad a spectrum and vintage of plants
> (i.e., two, three, and four loop plants, etc.). The lack of a
> sufficient four loop plant specific data base prevented a generic
> four loop approach. The Zion plants of the Commonwealth Edison
> Company were chosen as the like plant from which to extract plant
> specific data. [p. 1.2-5]

It would seem, then, that probability evaluations relating to the Sizewell B PWR should be based upon the experience of the Zion plants.

What can we make of the population data? Denoting by λ_{ir} and
n_{ir} the *rate* and *observed number* of initiating events of type i and
by t_r the number of operational years associated with PWR r, $i = 1$,
..., 14, $r = 1$, ..., 30, we can consider as plausible the model that
has the n_{ir} independently distributed as Poisson variates with expectation $\mathbb{E}(n_{ir}) = \lambda_{ir} t_r$. One reasonable interpretation of the term
"population" would be that the λ_{ir} do not depend on r, i.e., are
homogeneous:

Table 1. PSS Population Event Data

PWR	SLOCA	S/GTR	LOFF	COOM	LOPF	CPI	TT	SSI	RT	TTLOOP	OY
1. YR	0	0	4	0	3	0	8	0	46	9	15
2. IP1	1	0	40	0	7	0	21	0	143	3	12
3. SO	0	0	0	0	0	0	8	0	11	0	8
4. CY	0	0	10	2	2	0	16	0	32	4	8
5. REG	0	0	14	4	0	0	8	0	11	1	6
6. HBR	0	0	31	1	2	0	32	1	37	1	5
7. PB1	0	1	2	1	0	0	7	0	14	3	5
8. PB2	0	0	4	0	0	0	7	0	11	0	4
9. P	0	2	13	0	1	0	4	0	15	0	4
10. MY	0	0	4	0	1	1	3	0	6	0	3
11. S1	1	2	27	3	2	1	17	2	11	2	4
12. S2	0	1	14	7	0	0	10	0	6	1	4
13. O1	0	0	10	0	0	0	18	0	10	0	4
14. O2	0	0	7	0	0	0	7	0	4	0	2
15. O3	0	0	4	0	0	1	13	0	3	0	3
16. FC	0	0	3	0	2	0	2	0	7	4	3
17. K	0	0	11	5	0	0	16	0	11	0	2
18. A1	0	1	1	0	0	0	4	0	10	2	2
19. TMI	0	0	0	0	0	0	3	0	2	0	2
20. CC	0	0	3	0	1	1	5	0	4	0	1
21. T	0	0	5	0	1	0	7	1	4	0	1
22. M2	0	0	6	0	2	0	17	0	11	0	1
23. Z1 + Z2	1	0	58	3	5	0	41	8	42	0	11
24. IP2	0	0	34	5	2	0	32	1	42	1	5
25. IP3	0	0	12	0	0	0	4	0	10	3	3
26. DCC1	0	0	1	0	0	0	3	0	3	0	1
27. PI1	0	0	10	5	1	0	11	0	12	0	3
28. PI2	0	0	5	2	1	0	11	0	7	0	2
29. TP3	0	0	16	1	5	0	18	1	22	0	4
30. TP4	0	0	14	0	0	0	18	0	4	0	3
Total no. of events	3	7	363	39	38	4	371	14	551	34	131

Source: Adapted from Tables 1.2–13 and 1.2–14 of PSS.

KEY:
YR: YANKEE ROW	S1: SURRY 1	T: TROJAN
IP1: INDIAN POINT 1	S2: SURRY 2	M2: MILLSTONE 2
SO: SAN ONOFRE	O1: OCONEE 1	Z1 + Z2: ZION 1 PLUS ZION 2
CY: CONN. YANKEE	O2: OCONEE 2	IP2: INDIAN POINT 2
REG: R.E. GINNA	O3: OCONEE 3	IP3: INDIAN POINT 3
HBR: H.B. ROBINSON	FC: FORT COLHOUN	DCC1: D.C. COOK 1
PB1: POINT BEACH 1	K: KEWAUNEE	PI1: PRAIRIE ISLAND 1
PB2: POINT BEACH 2	A1: ARKANSAS 1	PI2: PRAIRIE ISLAND 2
P: PALISADES	TMI: THREE MILE ISLAND	TP3: TURKEY POINT 3
MY: MAINE YANKEE	CC: CALVERT CLIFFS	TP4: TURKEY POINT 4

SLOCA: SMALL LOCA	TT: TURBINE TRIP
S/GTR: S/G TUBE RUPTURE	SSI: SPURIOUS SAFETY INJECTION
LOFF: LOSS OF FEEDWATER FLOW	RT: REACTOR TRIP
COOM: CLOSURE OF ONE MSIV	TTLOOP: TURBINE TRIP LOSS OF
LOPF: LOSS OF PRIMARY FLOW	OFFSITE POWER
CPI: CORE POWER INCREASES	OY: OPERATIONAL YEARS

$$\mathbb{E}(n_{ir}) = \lambda_i t_r \tag{H}$$

and one can readily fit and test this model. Indeed, one can also fit
and test the *multiplicative* model

$$\mathbb{E}(n_{ir}) = \alpha_i \beta_r t_r \tag{M}$$

with some convention identifying the parameters α_i and β_r. These
models are readily fitted (omitting the four events with no occurrences to date), and the likelihood-ratio (approximate) chi-square
test statistics for them are 527 on 261 d.f. for (M) and 996 on 290
d.f. for (H). Clearly neither is a plausible model for the data.
How then should we summarize it? One could, preferably with further
information, group the reactors into more homogeneous subsets and pool
across these, perhaps assigning the probabilities of one of these
groups to the proposed Sizewell B reactor. Alternatively, one could
simply leave the population as it is and bear in mind that any pooled
estimates of rates (probabilities) would have greater than the usual
(Poisson) uncertainty associated with them.

The study does neither of these things. It presents Table
1.2-15, our Table 2, having given in the text the following almost
incomprehensible explanation:

> The plant-specific and the population data is summarized in Table
> 1.2-13 by plant and by initiating event. Table 1.2-14 shows the
> number of operation years in the data base by plant.
>
> It should be noted that these frequencies are derived using
> Bayesian techniques (1-3) and represent a conservative approach
> to initiating event qualification. This is particularly true
> for rare events. Rare events are defined as initiators which
> have not occurred within the ~200 years of PWR operating experi-
> ence.
>
> Table 1.2-15 shows the probability of occurrence of each initiat-
> ing event expressed in terms of various lognormal parameters for
> the plant-specific and PWR population generic data. A review of
> Sizewell B design against typical PWR design and review of con-
> sequential failures was performed to identify inconsistencies
> within the data base. This review indicated that the small LOCA
> frequency as an initiator was very conservative, most small LOCAs
> have occurred as consequential failures (pump seal due to loss of
> cooling or PORVs sticking when challenged). Since these conse-
> quential failures are explicitly treated via looping functions as
> described in Section 1.6.2, the small LOCA frequency as an

Table 2.   Initiating Event Occurrence Probability per Plant Year

Initiating event category	Sizewell B type plant specific						PWR population generic	
	5%	Median	95%	Range factor	Mean	Variance	Mean	Variance
1	3.33-5	3.44-4	3.55-3	1.03+1	9.40-4	5.74-6	1.01-3	6.37-6
2	3.33-5	3.44-4	3.55-3	1.03+1	9.40-4	5.74-6	1.01-3	6.37-6
3*	3.35-5	3.44-4	3.55-3	1.03+1	9.90-4	5.74-6	1.01-3	6.37-6
4	2.84-3	1.48-2	7.68-2	5.20	2.44-2	1.03-3	8.75-2	1.40-1
5*	3.33-5	3.44-4	3.55-3	1.03+1	9.40-4	5.74-6	1.01-3	6.37-6
6*	3.33-5	3.44-4	3.55-3	1.03+1	9.40-4	5.74-6	1.01-3	6.37-6
7	4.14	5.13	6.35	1.24	5.17	4.55-1	3.41	2.02+1
8	9.36-2	2.20-1	5.18-1	2.35	2.52-1	1.97-2	6.00-1	1.07+1
9	1.90-1	3.37-1	5.98-1	1.78	3.58-1	1.66-2	3.21-1	9.57-2
10	4.65-3	1.68-2	6.07-2	3.61	2.28-2	4.37-4	4.77-2	8.13-2
11	2.84	3.65	4.69	1.29	3.69	3.21-1	4.00	1.29+1
12	3.29-1	5.96-1	1.08	1.81	6.36-1	5.62-2	1.59-1	4.02-1
13	2.94	3.73	4.74	1.27	3.77	3.03-1	4.11	1.00+1
14†								
V‡								

Source: From Table 1.2-15 of PSS.

Note: Values are presented in an abbreviated scientific notation, e.g., 1.11-5 = 1.11×10^{-5}.

*Conditional probabilities (looping functions) not included.
†ATWS ($\emptyset_{14}$) is not an initiating event, rather it is an initiator plus failure to trip.
‡V sequence quantification treated in Section 1.6.4.1-7.

initiator, alone, is set consistent with the medium and large
LOCAs. [pp. 1.2-6, 1.2-7]

Let us look closely at what these (presumably) Bayesian techniques, the use of the log-normal distribution and review of the
Sizewell B design have done with the data. We shall compare them
with results from a classical statistical analysis, assuming a Poisson
process for the events and homogeneity for the population data; see
Table 3 for selected events. Incorporating the evident inhomogeneity
would have little effect on our estimated rates but would increase
their standard errors.

There are several points to note in connection with our Table 3.
First, the obvious classical estimates for the rate of small loss
of coolant accident (LOCA) are either 1 in 10 (specific) or 1 in 50
(population); the PSS uses 1 in 1000 and asserts this with a degree of
confidence of which the $97\frac{1}{2}\%$ point is still 1 in 200! The same figures
are used for large LOCAs, whereas the upper limit of a classical 95%
confidence interval for this rate, based on no events observed, is
either 1 in 3 (specific) or 1 in 40 population. Second, we observe
that there is some consistency between the two approaches for turbine
trips (specific), although a big difference for the population rates.
And here the numbers are quite large. It is apparent that some very
unusual statistical methods are being used. Third, we note that in
10 out of 13 events whose probabilities are evaluated in the PSS Table
1.2-15, our Table 2, the *precision* claimed for the plant specific estimate is *greater* than that claimed for the population generic, a quite
impossible conclusion to draw from data alone. Clearly the review of
the Sizewell B design has produced these figures, not any statistical
analysis, Bayesian or otherwise.

The reliability data on *components* is stated as being from
Masarik (1981), which I have been unable to consult, from other
(unspecified) sources, or from engineering judgment. It is still
quite disconcerting for a statistician to come across the following
(in Table 1.3-1 headed *Component Failure Data*, p. 1.3-6):

Manual valve, normally closed: ε.

Table 3. Estimation of Probabilities of Certain Initiating Events

| | | | | Classical statistics | | Sizewell B PSS | |
| | | | | | 95% confidence | | |
E	S	n	T	MLE	interval	Mean	95% interval
(2)	P	3	131	22.9×10^{-3}	$[4.6, 67.2] \times 10^{-3}$	1×10^{-3}	$[.02, 5.9] \times 10^{-3}$
	S	1	11	90.9×10^{-3}	$[9.1, 509] \times 10^{-3}$	1×10^{-3}	$[.02, 5.5] \times 10^{-3}$
(1)	P	0	131	0.000	$[0.0, 28.2] \times 10^{-3}$	1×10^{-3}	$[.02, 5.9] \times 10^{-3}$
	S	0	11	0.000	$[0.0, 336.4] \times 10^{-3}$	1×10^{-3}	$[.02, 5.5] \times 10^{-3}$
(11)	P	371	131	2.83	$[2.54, 3.12]$	4.00	$[0.66, 13.43]$
	S	41	11	3.73	$[2.67, 5.05]$	3.69	$[2.70, 4.92]$

Key: E: events (2): small LOCA
         (1): large LOCA
        (11): turbine trip

n: number of events
MLE: maximum likelihood estimate

S: source of data
P: population generic
S: Sizewell specific = Zion 1 + Zion 2
T: number of operating years
95% confidence interval:

From Johnson and Leone (1977, p. 525) the 95% confidence interval for a Poisson rate
with n events observed in T units of time is

n	Lower	Upper
0	0	$3.7/T$
1	$0.1/T$	$5.6/T$
3	$0.6/T$	$8.8/T$
41	$29.4/T$	$55.6/T$

When $n > 50$, a normal approximation was used.
For log-normal distribution with mean $\alpha = \exp(\mu + \tfrac{1}{2}\sigma^2)$ and variance $\beta = \alpha^2[\exp(\sigma^2) - 1]$, the median is $\exp(\mu)$ and scale factors giving the 2½% and 97½% probability points are

$$\exp \mp \left[1.96 \left\{ \log\left(\frac{\beta}{\alpha^2} + 1\right) \right\}^{\frac{1}{2}} \right].$$

Elsewhere (p. 1.5-8), we find the same, namely,

> Assumption 3: All manual valves in an in-service train will
> have negligible (ε) failure probability.

The value of ε is not specified in this context; but elsewhere, as we have already noted, ε is taken as 10^{-4}. However, there are other probabilities of this magnitude or smaller in Table 1.3-1.

The data on human errors have also been inaccessible to me, being taking from Commonwealth Edison Company (1981) and U.S. Government (1980). Much has been said about human aspects of reactor accidents, see especially pp. 29-30 of the RARG Report and the report on the accident at Three Mile Island, U.S. Government (1979). I have nothing to say on this topic other than to observe that it is apparent that no wholly satisfactory way has yet been developed of managing human beings and, accordingly, predicting and quantifying their behavior in reactor accidents; see Denton (1983):

> . . . still worried about the safety of the nation's 78 operating
> reactors. . . . The real touchstone of reactor safety is the
> human element and not the hardware. . . . We dream up these tech-
> nologicically sophisticated machines and forget the humans.
> [p. 13]

(iv) Multiplication of Probabilities, Including Treatment of Common-Mode Failures

The general form of the multiplication rule for probabilities is

$$\text{pr}(A \ \& \ B \ \& \ C \ \& \ \dots \,|\, H) = \text{pr}(A\,|\,H)\text{pr}(B\,|\,A \ \& \ H)\text{pr}(C\,|\,A \ \& \ B \ \& \ H) \ \dots \ (7)$$

and in the present context, a large number of such probabilities of conjunctions of many terms are being calculated. To what extent is (7) fully appreciated and used in the Sizewell B PSS? In my opinion, it is probably not properly understood. There are so many tacit probabilistic independence assumptions and inconsistencies involving (7) that an explicit concern for common-mode failures (see below) seems quite beside the point.

Consider the basic structure (1) of the calculation of the proba-bility $\text{pr}(R_k)$ or release category k. It expresses the answer as a sum of terms, each of which *should* be of the form

$$\text{pr}(E_i)\text{pr}(D_j|E_i)\text{pr}(R_k|D_j \text{ \& } E_i),$$

i.e., it assumes that for all i, j, and k we have

$$\text{pr}(R_j|D_k \text{ \& } E_i) = \text{pr}(R_k|D_j). \tag{8}$$

Expressed another way, (8) states that release events are probabilistically independent of initiating events given any of the plant damage states for which (8) is nonzero. This might sound plausible when said in isolation, but it seems to contradict quite explicitly the discussion in 2.2, "Selection of Key Accident Sequences." There we find the initiating events playing a role (see especially p. 2.2-4 and Table 2.2-3) in the evaluation of probabilities associated with the various release categories, although these have not yet been defined at this point in the study. Thus the basic structure of the study calculation appears to be flawed because of a lack of appreciation of (7).

There is some awareness of the relevance of conditional probabilities in long products such as (4) or (6):

> Generally, the probabilities at each nodal branch are conditional on predecessor events. For example, the probability of hydrogen burn late in a sequence may be influenced by the occurrence of a prior hydrogen burn because hydrogen concentration depends upon previous history. [p. 2.1-1]

However, this knowledge has not influenced practice for many probabilities calculated using data from the various data sources noted in (iii) above. Not only are the probabilities there simply tabulated without qualification—unconditional probabilities, one might say— but an examination of the numbers used in event tree calculations, such as that displayed in Tables 1.6.4.1.4-2, 1.6.4.2.4-2, etc., shows that the same figures are used, apart from a few exceptions, across all support states and across all initiating events. The exceptions are all *logical* consequences of the conditions defining the support states; different probabilities are not evaluated.

How much credibility can we place in probabilities calculated as products of 8, 10, 12, or more terms, assuming independence all along the path? In the PSS, figures of the order 10^{-15}, 10^{-17}, and 10^{-19} are obtained in this way. I would suggest that there is little or no reliability experience with such complex systems. I know of none in which this sort of calculation had led to results that related reasonably to subsequent experience. Is it not reasonable to expect the PSS to convince us that these methods will be effective in *their* case? Or should we be content with the knowledge that they tried as hard as they could to bridge the gaps in their understanding?

The issue of common-mode failure probabilities [How do we calculate $\mathrm{pr}(A \,\&\, B | H)$ when we only know $\mathrm{pr}(A | H)$ and $\mathrm{pr}(B | H)$?] is raised in this study as it was in WASH-1400. Of course, all that can be said without extra information is

$$\max(1 - \mathrm{pr}(A | H) - \mathrm{pr}(B | H), \ 0) \leqslant \mathrm{pr}(A \,\&\, B | H) \leqslant \min(\mathrm{pr}(A | H), \ \mathrm{pr}(B | H)).$$

The Sizewell B rejects the thoroughly discredited WASH-1400 "solution" to the "problem" of common-mode failures and embraces a different false technique, the *additive cut-off approach*:

> The additive cut-off approach limits the system unavailability to a pre-determined minimum (i.e., "cut-off") value. However, since there is no statistical basis for the choice of the cut-off value, it must be selected entirely on engineering judgment.

> The choice of the cut-off value is no easier than the choice of the approach. The CEGB Design Safety Guidelines require that the CMF probability shall not be assumed to be less than 1.(-5) and the system unreliability shall not exceed 1.(-3) overall.

> Thus a value of 1.(-4) per demand was chosen for the CMF cut-off contribution for all systems, with the exceptions noted below. [p. 1.3-3]

The arbitrariness of the procedure is quite evident, and a simple illustration of the silliness of this rule can be found in the analysis of the *Power State Probability Model*, §1.5.1.3. Four diesel generators are to be operated when off-site power fails, and the notion of support state involves in part the number of these generators that *succeed*, i.e., that start up and continue for three hours when called upon to do so. When calculating the probabilities of all four, three only out of the four, two only out of the four, and one generator only succeeding, the component probabilities are essentially multiplied as though mutual independence obtained. (Here $p = 0.9642$). However, the probability of all four failing under the same assumption would be $(1 - p)^4 = 1.64 \times 10^{-6}$ which, being smaller than the figure of 1.0×10^{-4}, must be replaced by the latter. This would result in the probabilities summing to more than unity. Accordingly, the other probabilities appear to have been juggled slightly to restore their sum to unity. Table 4 compares my calculations with those found in Table 1.5-3 of the PSS.

Table 4. Power State Probabilities During LOOP

Power state	Probability from PSS	Probability under independence
0	0.0000	0.0000
1	0.8230	0.8290
2	0.0825	0.0796
3	0.0825	0.0796
4	0.0075	0.0076
5	0.0019	0.0019
6	0.0019	0.0019
7	0.0017	0.0002
8	0.0017	0.0002
9	0.0001	0.0000

If the explanation offered above is not correct, then I am quite unable to explain how the results in Table 1.5-3 of the PSS follow from the discussion in §1.5.1.

Finally, we observe that no such additive cut-off was used in the probability multiplications carried out on the various event trees. Surely the case for the use of such a rule there would be stronger; 10^{-4} instead of 10^{-19} in such examples would certainly be more worthy of the description *conservative*. It is ironic that in such calculations a probability of zero is distinguished from one of 10^{-17}, 10^{-18}, or 10^{-19}.

(v) Completeness of the Study

Is it possible that the members of the study group identified *all* accident sequences that could contribute significantly to the risk? The discussion in the RARG Section III is relevant here, although essentially leaving the question unanswered, but in a footnote we find the opinion:

> One of us (F. v. H.) questions whether, for a system as complex as a nuclear power plant, the methodology can be implemented to give such a high level of confidence that the summed probability of many known and unknown accident sequences leading to an end point such as a core melt is well below the limit set by experience. [p. 15]

Experience with actual reactor accidents (Browns Ferry, Three Mile Island) would seem to support this view and it appears to be shared by Critchley (1976):

> No high-risk, major-hazard, safety-assured plant like a nuclear reactor should be built unless it is so well designed, constructed and operated that disastrous failure cannot be foreseen in the anticipated circumstances of its existence; that is, such an event must be "incredible." Thus, the permitted net chance of occurrence of a catastrophic radiation accident arising from any envisaged cause must tend to be vanishingly small. A risk so forecast cannot be true. The true hazard is given by the summation of the occurrence probabilities of all accident-producing causes which includes an almost infinite spectrum of unexpected, unusual or highly improbable though possible happenings or coincidences. At the present time, at least, the task of catching such a large number of rare, random and diverse things is Sisyphean. There is, thus, a severe limitation on the input data which vitiates any quantitative predictions, and such serious accidents as might occur will be most likely to be "rogue" events which would not be identified in the quantifier's philosophy. [p. 18]

Although this issue is not, strictly speaking, a statistical one, it becomes one when we recall our earlier question.
Is there a body of experience that compels us to believe, or even suggests to us, that probabilities based on such potentially incomplete analyses can be accepted with confidence as having a useful frequency interpretation? I know of no such experience; all that I have read concerning actual nuclear reactor accidents suggests just the opposite.

(vi) Sensitivity Analyses Carried Out

It is common in the discussion of a complex mathematical model for which any check on its realism and predictive power seems out of the question to vary some of its parameters or assumptions and note the

effect on certain overall features of the model. Exactly what we
learn about the validity or usefulness of a model by doing this is far
from clear, but such a practice can certainly help to isolate unsatisfactory features of the model which might otherwise escape notice.

A very perfunctory sensitivity analysis of the kind just described
was carried out in the Sizewell B PSS. Bleed and feed techniques feature in a number of the plant event trees (see, e.g., §§1.6.4.4,
1.6.4.5, and 1.6.4.6), and two re-analyses are done assuming restricted
or no bleed and feed capability for any event. These changed assumptions alter the total core melt frequency from 1.16 to 1.9 and
2.16×10^{-6} per year, respectively. A second analysis excludes the
possibility of power recovery after a turbine trip due to loss of off-
site power and finds that this has no effect whatsoever on the core
melt frequency. Finally, the assumption of quarterly testing of sump
recirculation valves is changed to annual testing; in this case the
base figure 1.16×10^{-6} changes to 2.7×10^{-6}, at the same time,
apparently, *decreasing* the frequency of two types of plant damage
states. Just how this theoretically impossible conclusion resulted is
difficult to discover; the discussion in §1.5.9 is too brief to be
helpful so that Table 1.5-17 cannot be checked (Table 1.5-17A seems
not to be mentioned anywhere).

Even without considering this last point, I was left wondering
precisely how much more confidence I should have in the overall analysis *after* reading the sensitivity analysis than I had before I did so.
In reality its very perfunctoriness, compounded with the final puzzle,
left me feeling generally more critical of the study than I did initially. The topics discussed seemed well away from the real problems
associated with such an ambitious undertaking as this Probabilistic
Safety Analysis.

5. CONCLUSIONS

Having read a good deal of the Sizewell B PSS fairly carefully, having
found clear evidence that the study group failed to appreciate a number of important points of probability and statistics and that they

failed to use clear and accepted methods of statistical analysis with
their data, and having seen no evidence of worthwhile advances on
problems in the area well known to be difficult, if not insurmountable,
I find myself concluding that there is no reason at all for me to
accept their final probability figures as having any value. Many of
the reasons that have led me to this conclusion have been detailed
above; many others have been omitted.

Do I think that with careful, competent statistical analysis this
approach *could* yield probability figures that I might believe? The
answer is again *no*, I do not, for much the same reasons as those put
forward by D. Andrews or O. H. Critchley in the extracts quoted above.
I think that at many important points in the analysis of accidents,
probability methods are quite irrelevant, and their use might give a
dangerous sense of security to analysts. The WASH-1400 discussion of
the Browns Ferry fire illustrates this point very well; see the comments on it (p. 25) in the RARG Report.

The fact that the groups who write reports such as the Sizewell B
PSS or WASH-1400 invariably show such a lack of appreciation of the
subtleties of probability and statistics I see as intimately connected
with their obvious belief and hope that the approach will yield worthwhile probability figures; if they understood the subject far better,
they would expect far less from it. If actual figures are really
required, workers in the field of reactor safety should turn to the
calculation of·the well-based and realistic risk assessment which,
although they might not be as low, would be believable and should also
be closer to the truth. There are many statisticians who would be willing and able to collaborate in such an exercise.

ACKNOWLEDGMENTS

I am very grateful to William Cannell for supplying me with a copy of
the Sizewell B PSS and for his general encouragement, to Marj Johnson
for so willingly chasing up references for me, to Camilla Fazekas for
analyzing the population event data, and to Pam Carriage for typing
the manuscript so quickly and efficiently. Special thanks are also

due to the organizers of the Neyman-Kiefer Memorial Conference from the Department of Statistics, University of California at Berkeley, for giving me the opportunity to speak there on this topic.

REFERENCES

Articles listed as "cited in . . ." have not been consulted.

Apostolakis, G., and Mosleh, A. (1979). Expert opinion and statistical evidence: an application to reactor core melt frequency. *Nucl. Sci. Eng. 70*, 135-149.

Commonwealth Edison Company (1981). *Zion Probabilistic Safety Study.* [Cited in Westinghouse Electric Corporation (1982).]

Cremer and Warner (1980). *An Analysis of the Canvey Report.* London: Oyez Publishing Limited.

Critchley, O. H. (1976). Risk prediction, safety analysis and quantitative probability methods—a caveat. *J. Br. Nucl. Energy Soc. 15*, 18-20.

Denton, H. (1983). Quoted in an interview in "Update," *Newsweek*, June 27, p. 13.

Editorial (1978). *Nature 276*, 429.

Electric Power Research Institute (1978). ATWS: A Reappraisal, Part III: Frequency and Anticipated Transients: EPRI NP-801. [Cited in Westinghouse Electric Corporation (1982).]

Fairley, W. B. (1977). Evaluating the 'small' probability of a catastrophic accident from the marine transportation of liquefied natural gas. In *Statistics and Public Policy.* W. B. Fairley and F. Mosteller, eds. Reading, Mass.: Addison-Wesley, pp. 331-353.

Farmer, F. R. (1980). The risk assessment of major hazards. *Bull. Inst. Math. Appl. 16*, 114-115. [From Symposium on "Mathematical Modelling of Large Scale Accidents and the Environment," held at the University of Cambridge, March 20, 1979.]

Farmer, F. R. (1981). Experience in the quantification of engineering risks. *Proc. Roy. Soc.* (A) *376*, 103-120.

Fischoff, B., Lichtenstein, S., Slovic, P., Derby, S. L., and Keeney, R. L. (1981). *Acceptable Risk.* London: Cambridge University Press.

Health and Safety Executive (1978). *Canvey: An Investigation of Potential Hazards from Operations in the Canvey Island/Thurrock Area.* London: Her Majesty's Stationary Office.

Hubbard, R. B., and Minor, G. C. (1977), eds. *The Risks of Nuclear Power Reactors. A Review of the NRC Reactor Safety Study*

WASH-1400. NUREG-75/014. Cambridge, Mass.: Union of Concerned Scientists.

Johnson, N. L., and Leone, F. C. (1977). *Statistics and Experimental Design in Engineering and the Physical Sciences*. 2nd ed. New York: Wiley.

Jones, D. R., and Akehurst, R. L. (1980). Risk assessment: an outline. *Bull. Inst. Math. Appl. 16*, 252-258.

Marshall, Sir Walter (1983). Design and safety of the Sizewell pressurized water reactor. *Proc. Roy. Soc. Lond.* (A) *385*, 241-251.

Masarik, R. J. (1981). Selected failure-rate data for Westinghouse NSS components. Westinghouse Electric Corporation. [Cited in Westinghouse Electric Corporation (1982).]

McCormick, Norman J. (1981). *Reliability and Risk Analysis: Methods and Nuclear Power Applications*. New York: Academic Press.

Monitor (1979). Does reactor safety need reassessment? *Nuclear Engineering International*, May 1979, pp. 47-49.

Okrent, D. (1981). Industrial risks. *Proc. Roy. Soc. Lond.* (A) *376*, 133-149.

Philipson, L. L. (1982). General risk analysis methodological implications to explosives risk management systems. In *Twentieth DOD Explosives Safety Seminar*, Norfolk, Virginia, 24-26 August 1982, pp. 703-724.

PSS. See Westinghouse Electric Corporation (1982).

RARG. See U.S. Government (1978).

Rothschild, Lord (1978). Risk. *The Listener*, November 30, pp. 715-718.

Rothschild, Lord (1978a). Correspondance. *Nature 276*, 555.

Royal Society Study Group (1983). *Risk Assessment*. London: The Royal Society.

RSS. See U.S. Government (1975).

Speed, T. P. (1979). Forsvindende sandsynligheder og atomkraftsikkerhed: et nyt misbrug af sandsynlighedsregningen? *RAMA 1*, University of Copenhagen. [Danish translation of "Negligible probabilities and nuclear reactor safety: another misuse of probability?" Department of Mathematics, University of Western Australia (1977).]

U.S. Government (1975). *Reactor Safety Study WASH-1400*. NUREG-75/014. Washington, D.C.: NRC. [Also known as the *Rasmussen Report*.]

U.S. Government (1978). *Risk Assessment Review Group Report to the U.S. Nuclear Regulatory Commission*. NUREG/CR-0400. Washington, D.C.: NRC. [Also known as the *Lewis Report*.]

U.S. Government (1979). *The Need for Change: The Legacy of TMI*. Report of The President's Commission on the Accident at Three Mile Island. New York: Pergamon Press.

U.S. Government (1980). *Generic Evaluation of Feedwater Transients and Small Break Loss-of-Coolant Accidents in Westinghouse Designed Operating Plants.* NUREG-0611. [Cited in Westinghouse Electric Corporation (1982).]

Von Hippel, F. (1977). Looking back on the Rasmussen Report. *Bull. of the Atomic Scientists 33*, 42–47.

Warner, Sir Frederick, and Slater, D. H. (1981). The assessment and perception of risk: a discussion. *Proc. Roy. Soc.* (A) *376*, 1–206.

Westinghouse Electric Corporation (1982). *Sizewell B Probabilistic Safety Study.* Westinghouse Electric Corporation WCAP 9991 Rev. 1. [Referred to in the text as Sizewell B PSS or just PSS.]

NAIL FINDERS, EDIFICES, AND OZ

LEO BREIMAN University of California, Berkeley

The thesis of this paper is that many, if not most, statisticians in government and industry are poorly trained for their profession and are consequently poor problem solvers in terms of public policy decisions. Since this thesis is illustrated by anecdotal materials in which I often emerge as the hero, I begin with a situation in which my failure was undeniable.

In the late sixties I was hired as a consultant to the defense in the famous Ellsberg trial. The defense was interested in bringing a challenge to the court concerning the composition of the empaneled jurors. A Supreme Court ruling had held that empaneled jurors should reflect the characteristics of the population in which the particular court had jurisdiction. In the Federal Court that was hearing the Ellsberg case, the jurisdiction consisted of Los Angeles County and four surrounding counties. The Ellsberg defense was specifically concerned with under-representation of blacks.

The basic data consisted of the questionnaires that were mailed to the people selected for jury duty, filled in by them, and returned to the Jury Commissioner. We were given access to the about 6000 questionnaires that had been received over the past year. A large majority of these were from people who subsequently had been excused from jury duty because of some stated hardship.

These 6000 questionnaires were in boxes, which I took to a commercial card punching company together with a coding for various questions and answers. After the data were put on the cards, I started the tabulations that were to be presented to the court.

There were two fatal errors on my part. The first came out when the Jury Clerk, in her testimony, disputed the numbers in my tabulation. She had gone through the questionnaires and counted the number in a certain category. My count was much higher than hers.

With a sinking feeling I started checking through the output in detail and finally realized what had happened. The card punchers had punched in one box of questionnaires *twice*! My first failure was that I did not ensure

GOOD QUALITY CONTROL ON THE DATA.

The second error was even more fundamental. The critical question on the questionnaire about the respondant's race included the comment that answering it was optional. About 15% of the questionnaires submitted by people who were later empaneled were missing these answers. As I recall, about 5% of those answering the question were black, and blacks made up about 13% of the total population in the court's jurisdiction. There was no way in which those data could be used to determine whether blacks were under-represented. I had failed to ask the fundamental question

CAN THE RELEVANT QUESTIONS BE ANSWERED BY THE DATA?

In my defense, there was a mitigating circumstance. I had just come out of the university and started my consulting career. I learned and did not repeat those early mistakes too often.

My experience in meeting, working with, and reviewing the work of many statisticians in the field practice is that they generally suffer from one or the other of the three following complexes.

THE FIND-THE-NAIL COMPLEX

Jerome Friedman has a lovely saying (source unknown):

> If all you have is a hammer, then every problem will look
> like a nail.

Applied to statisticians, this refers to absorption with the technique rather than the problem; to the failure to see the problem whole; to the failure to ask, "Does it all make sense?"

THE EDIFICE COMPLEX

This refers to the building of a large, elaborate, and many layered statistical analysis that covers up the simple and obvious.

THE WIZARD OF OZ COMPLEX

The exploitation of the mysteries of statistics to dazzle and mystify the less knowledgeable.

Here are some recent illustrations.

THE ASA AD HOC ADVISORY COMMITTEE ON NUCLEAR REGULATORY RESEARCH

This committee was formed by the ASA in 1980. For background, I include the relevant sections of a June 30, 1980 memo from Fred C. Leone, Executive Director of the ASA.

> The board of Directors of the American Statistical Asso-
> ciation is establishing an Ad Hoc Advisory Committee on Nuclear
> Regulatory Research. This is a result of negotiation between
> members of the Nuclear Regulatory Commission and the American
> Statistical Association, followed by an invitation from the
> Director of the Office of Nuclear Regulatory Research. This is
> a major step which the ASA Board has taken and, hence, it is
> especially important that the advisory committee be very strong
> and have the necessary balance to be most effective. The terms
> of reference (charge) of the Advisory Committee are stated in
> the accompanying sheet.

Terms of Reference (Charge) of ASA Ad Hoc Advisory Committee
on Nuclear Regulatory Research

This Committee will provide advice and peer review with
respect to programs of the Office of Nuclear Regulatory Research
of the U.S. Nuclear Regulatory Commission which involve or
require statistical and probabilistic techniques and approaches.
In particular, it will

1. take a broad responsibility for the review of the statistical and probabilistic technique proposed for the Numerical
Risk Criteria Project and assessment of statistical contributions to risk assessment procedures and applications.

2. Review and comment as requested on statistical and probabilistic approaches or techniques proposed together with other
nuclear regulatory research and development programs.

3. Define (a) monographs and guideline documents on existing
statistical and probabilistic topics and techniques and
(b) areas of statistical and probabilistic research and
development, that are needed to further the effective use of
statistics in connection with nuclear regulatory procedures
and programs.

The committee first met on August 1, 1980. It consisted of 18
members of whom 17 were statisticians, the majority academics. It
interfaced to the Office of Nuclear Regulatory Research (NRR).

The NRR was committed to the probabilistic risk assessment (PRA)
methodology. This methodology was initiated in the Rasmussen Report
(WASH-1400, 1975). An NRC requirement as of 1978-79 was that each new
nuclear power plant conduct a PRA prior to operation.

A PRA starts with each basic component in the plant, e.g., pipes,
valves, and diesel generators, and estimates a failure rate for each
such component. Then it attempts to construct all possible sequences
of events leading to severe core damage or meltdown. In some way, a
probability is assigned to each sequence. Then these probabilities
are combined to give an overall probability per year of severe core
damage.

A PRA is a serious and extensive undertaking. The original volumes of the WASH-1400 report form a stack almost a foot high. The
cost of a PRA for a power plant is several million dollars. The
superstructure of a PRA is extremely large and elaborate, and is constructed with a variety of tenuous assumptions. After a long climb it

finally emerges at the top with "the bottom-line number," i.e., the
probability per year of severe core damage. This is the number
reported to the press and bandied about in the NRC licensing decisions.

After the original Rasmussen report came under fire, the Lewis
Committee was appointed to review it. The Lewis Committee report
approved the basic methodology but had a number of criticisms, including lack of adequate peer review of statistical methodology. Against
this background the ASA committee was formed with travel and other
expenses funded by the NRR.

Here are two examples of what happened at committee meetings.

COMMITTEE REPORT, MAY 15, 1981

2. Modeling component failure and reactor error sequences. In
    general, members of the Committee felt that significantly
    more attention should be paid to the use of methodology from
    stochastic processes in this area. In particular, it was
    felt that the methodology of renewal processes and non-
    homogeneous Poisson processes could be useful in producing
    mathematical models which would describe certain observed
    phenomena more closely than models currently in use. To be
    more specific, a more flexible but more complicated model
    for the binomial failure rate common cause model which could
    account for the differing lifetimes of different components
    could be constructed using non-homogeneities in both the
    Poisson process and the binomial probabilities. For some
    types of equipment the use of renewal processes may prove
    fruitful in modeling such events as non-catastrophic break-
    downs.

MINUTES, APRIL 25-27, 1982 MEETING

Suggestions and issues raised in general discussion included:

a. A discussion of the positive and negative aspects of Bayesian
    methodology when applied in risk estimation.

b. Questions and responses on the meaning and implications of the
    phrase "uncertainty propogation" and on the communications
    problems engendered in its use.

c. The suggestion that a more thorough survey of the literature
    and greater methodological adaptation be attempted in address-
    ing NRC RES's statistical problems. Such a review and com-
    parison of methods and problems should yield areas for further
    statistical research by individuals in academic or other
    research institutions.

These excerpts are, admittedly, not a random selection. But they illustrate what was often happening. In the course of their meetings, many committee members were operating in the *find-the-nail* mode, trying to find technical pieces of the problem that could be dealt with by known statistical methods.

The fundamental issue that we should have been addressing from the start was:

DOES THE WHOLE IDEA OF PRA's MAKE SENSE?

My conclusion is *absolutely not*, at least in terms of producing believable estimates of risk. In mid 1982, I suggested recommendations to the ASA committee for submission to the NRR. These are contained in the appendix to this paper. To quote:

> The opinion of the ASA Ad Hoc Committee is that "bottom line" estimates of severe core damage are misleading and inaccurate. The continued focus on them is harmful to the goal of nuclear reactor safety.
> . . .that overall risk assessments of severe core damage be based on the analysis of past nuclear reactor operating experience.

The NRR, faced with budget problems, cut out the funding for the ASA committee in late 1982, and meetings stopped before there was any careful consideration of these recommendations. However, the NRR had already started a precursor study, which obtained an estimate of risk by looking at all records of serious power plant incidents and applying the PRA methodology to get estimates that the incident might have led to severe core damage. The advantage is that one starts far along in the sequence of events. This study resulted in a risk estimate two orders of magnitude higher than that given in the Rasmussen report.

If it had looked at the problem whole instead of finding nails, the ASA committee might have realized that PRA's are an outrageous misuse of statistics, probability, and common sense.

THE EPA CRITERION DOCUMENT ON TSP

TSP stands for *total suspended particulates*, the particles in air that
are small enough to remain suspended. The larger of these are kept
from entering the lungs by the body's defense system. The smaller
microscopic particles can get through, lodge in the lung tissue, and
cause some damage. Because of this, EPA requires TSP monitoring,
which is done for 24 hours every 6 days at about 4000 sites in the U.S.

The measurement is done by a Hi-Vol sampler. This instrument
sucks a measured volume of air through a filter for 24 hours. The
filter is weighed before and after. The weight difference (in micro-
grams) is divided by the volume of air (in meters^3) to give the TSP
reading.

The current standards are

annual geometric mean less than 75
2nd highest 24 hour reading in the year less than 260

By the provisions of the Clean Air Act, the EPA preiodically
reviews all research relevant to the standards and, based on its
review, decides whether to change the standard.

The procedure is the the EPA puts together a criterion document
that summarizes all available relevant information and serves as the
basis for its decision. This document is then submitted for review
and written public comments before it becomes final.

Among others, I reviewed the draft criterion document, the written public comments, and a review of both by an EPA contractor. The
fundamental issue is

WHAT EVIDENCE OF HEALTH EFFECT EXISTS THAT IS RELEVANT TO
CURRENT STANDARDS?

The criterion document discussed a number of epidemiological
studies. The most important of these were done in England, because
of the concern there over high smoke levels. Most of the urban particulate matter in England consists of fine carbon particles generated by
the use of coal in house heating and factories. Because of this, they

use the BS method of measurement, which is based on optical reflectiv-
ity of the filter instead of the total mass of the deposited particles.

The British have had extremely high smoke episodes in some of
their cities. More than 20 years ago, London experiences an episode
that was estimated to have caused several thousand deaths. The data
from such episodes and from persistently high smoke areas give the
most clear cut evidence of particulate health effects.

The written public comments consisted of over 200 pages. Well
over half were filled with discussions of statistical techniques.
The correct way to model with multiple time series, transformation of
variables, lags, standard errors, confidence intervals, etc., were
subjects written about by a number of very eminent statisticians. It
constitutes a very nice example of nail finding and edifice building.
Here are some excerpts from my review:

> The statistician's first question when faced with data must be
> "is this data capable of answering the questions I am interested
> in?" No amount of fancy statistical footwork can make up for
> unsuitable data. All of the statisticians who are bemoaning the
> fact that some data sets have not been analyzed using high-powered
> time series analysis are missing the fundamental point—to wit YOU
> CAN'T MAKE A SILK PURSE OUT OF A SOW'S EAR. After reading over
> some of the fundamental epidemiological papers and many descrip-
> tions and criticisms of other studies, the clear fact emerges
> that there is very little or no data suitable for setting TP or
> TSP standards in reference to health hazards. . . . The big
> problem is not lack of appropriate statistical technique, but lack
> of good data. Give me a well thought out and carefully executed
> experiment resulting in good data and I will tell you the names
> of at least a dozen statisticians around the country who will do
> a very credible job in the analysis of the data.

In brief, the problem was that the British studies were virtually
useless for two reasons. First, the health effects observed took
place at particulate levels that, by any standard of comparison, were
much higher than our current EPA standards. Second, because there is
no site-independent method to reliably convert BS measurements to TSP.
The few U.S. studies had other flaws.

THE ETHYL CORPORATION MMT APPLICATION

Ethyl Corporation manufactures a lead-based additive to increase the octane rating of gasoline. With all of the new cars using unleaded gas, Ethyl devised a new additive MMT based on manganese (Mn). Use of an additive is a serious affair. Quoting from Ethyl Corporation's Reapplication for MMT Waiver (May 22,1981), "use of a 1/64 g. Mn/gal. in unleaded . . . indicate a savings of 35,000 B. oil per day and $350 million in processing facilities."

To get EPA permission to use a new additive in gasoline, a company must present proof that use of the additive will not increase automotive exhaust pipe levels of nitrous oxides (NOX), carbon monoxide (CO), and hydrocarbons (HC).

Ethyl Corporation funded a study by the Corrdinating Research Council. It used 63 cars of 7 different types, i.e., Ford, GM, Chrysler, and two foreign. The 9 cars of each type were divided into 3 groups of 3 each. The first group used unleaded gas with no MMT, the second group used gas with 1/32 g/gal of MMT, and the third used 1/16 g/gal of the additive. Each car was driven 50,000 miles, and the NOX, CO, and HC levels were checked at 0, 3, 5, 10, 15, 22.5, 30, 37.5, 45, and 50 thousand miles.

The 1979 report of the Coordinating Research Council states:

> The results of this study indicate that the use of MMT at either test concentration increases both engine and tailpipe hydrocarbon emissions compared to clear fuel.

The EPA disallowed the waiver. But on May 22, 1981, Ethyl Corporation reapplied for a MMT waiver for 1/64 g/gal. Their application states:

> Interpolating the available emission data for clear fuel and 1/32 and 1/16 g./gal. Mn to 1/64 g./gal. Mn shows no significant effect of MMT on emissions at this low concentration.

Over half of the reapplication document was devoted to a summary (50 pages) of a statistical analysis that "proved" the above statement. The analysis is a marvelous exercise in edifice building and statistical wizardry.

To begin with, the analysis dropped three car types from the analysis (the reasons were interesting), leaving a total of 360 recorded HC values. A total of 105 regression equations were fit to these 360 data points. There were various discussions of significance tests for rejecting outliers, for linearity, of degrees of freedom, of reduced variance, etc.

At the end of this long and complex analysis, they produced prediction equations for HC at any mileage for any given amount of MMT additive. An outcome was that their equations predicted lower HC emissions at 1/64 g/gal MMT than for clear gas for all four automobile types tested.

I was asked to review their analysis and requested the original data. Perhaps the most telling point emerged when I used their equations to predict actual data values. Averaged over the four types, at 30,000 miles, here are the results:

	HC emissions		
	0 MMT	1/32 MMT	Increase
Ethyl prediction	.410	.422	3%
Actual data	.400	.455	14%

The statistical jargon and complexity of the analysis make it hard to penetrate, but this magic edifice had the effect of using the actual data with a 14% increase at the bottom and producing a predicted 3% increase at the top.

There are many other illustrations that could be given, but these three are, I think, enough to get me to my major point:

BECAUSE OF THEIR FAILURE TO TREAT A PROBLEM WHOLE, MANY STATISTICIANS ARE POOR SERVANTS OF PUBLIC POLICY.

In practice, the primary issues are

1. Problem formulation: What are the right questions?
2. Data
    a. How to gather data capable of answering the relevant questions,

   b. Assessing whether the data at hand are capable of answering
       the questions,

   c. Understanding the measurement methods that produce the data,

   d. Data quality.

3. Analysis—interpretation

   a. An analysis <u>appropriate</u> to the data,

   b. Sensible interpretation of results.

Succeeding in 1 and 2 is three-quarters of the battle, yet these
issues are rarely adressed in formal statistical training.

> I KNOW OF NO FIELD IN WHICH THERE IS SUCH A LARGE DIVERGENCE
> BETWEEN WHAT IS NEEDED IN PRACTICE AND THE TEACHING AND RESEARCH
> OF THE UNIVERSITIES.

We do not encourage

   CAREFUL THINKING

   INTELLIGENT FORMULATION

   COMMON SENSE

Instead, statisticians are equipped with a narrow and often inappli-
cable methodology that produces

   LIMITED VISION

   WIZARD-OF-OZISM

   EDIFICE BUILDING

That the impact of statisticians on public policy has not been larger
and that statisticians are distrusted is due, to a good extent, not to
our stars, dear statisticians, but to ourselves.

APPENDIX: Suggested Committee Recommendation on PRAs (1982)

Probabilistic Risk Assessment (PRA) has two major functions:

First. It forces an extensive engineering analysis of the system,
starting at the component level and working its way up. By isolating
higher probability paths, it focuses attention on the critical parts of
the system and can lead to corrective action.

Second. It produces "bottom line" estimates of the probability of severe core damage. These estimates appear prominently in the summary. They are widely circulated to the public and used by the Commission in their licensing decisions.

The opinion of the ASA Ad Hoc committee is that the "bottom line" estimates of severe core damage are misleading and inaccurate. The continued focus on them is harmful to the goal of nuclear reactor safety. The reasons for this opinion will be expanded below.

Use of PRAs in their first function, as an engineering systems analysis tool, does provide valuable information concerning the failure modes of the system. Therefore, we recommend that

RECOMMENDATION: *That PRAs make no attempts to estimate overall probabilities of severe core damage. Instead of numerical assignments, paths to failure should be ranked as High, Medium and Low Probability. That licensing decisions not be based on numerical estimates of core damage, but instead on whether the best current safety standards have been met by the plant.*

We note, to begin, that neither the committee nor any of the technical staff of the NRC with whom the committee has been in contact have any belief in the scientific merit or accuracy of the "bottom line" estimates.

The major reason for this disbelief is inherent in the structure of the fault/event tree analysis. At each stage in the tree construction, questionable estimates or questionable methods of combining previous estimates are introduced. Errors are compounded and propagate upward. The final estimates have many sources of error, some of which are difficult, if not impossible, to quantify.

Two particularly weak places in the analysis are:

I. The impossibility of quantifying human error probabilities to within several orders of magnitude. This problem has been seen repeatedly in the various precursor events involving surprising and unanticipated modes of human failure.

II. The similarly difficult problem of assessing probabilities of common mode failures. The estimated probability for the simultaneous occurrence of two events can differ by several orders of magnitude depending on whether the events are assumed independent or have a common cause origin.

In addition, there are numbers of other quite questionable assumptions used in PRAs to arrive at the final estimate.

It is sound and accepted statistical practice to always compute error bounds for any estimate. In view of the methodological obstacles mentioned above, realistic error bounds on estimated probability of core damage would be so wide as to make the estimates useless for decision making. For instance, we do not consider it unlikely that error bounds on a 10^{-6} estimate might be a lower bound of 10^{-2} and an upper bound of 10^{-10}.

Continued use and emphasis on these "bottom line" estimates has some harful effects.

First. Since many of the NRC's own technical staff and much of the outside scientific community do not place any credibility in these numbers, an atmosphere of cynicism and frustration is created. For the sake of public relations, suspension of sound judgment is required. Not only is this harmful to internal morale, but it also exposes the NRC to justifiable external criticism.

Second. Because of the focus on the overall estimates, otherwise important engineering information may be distorted. We find it hard to believe that a PRA analysis carried out by a consulting firm hired by the utility will produce unacceptably high overall risk estimates. The emphasis is not only distorting in this way, but also it diverts technical time and funding away from the more important engineering systems analysis aspects, both in terms of NRC technical staff and of the direction of research carried out by subcontractors. If the emphasis were removed from the overall risk assessment and PRAs viewed instead as an engineering analysis tool, this might open the way to significant technical improvements; a much more realistic set of goals would be set; and attention and research directed at those goals.

If overall risk assessments are needed, then a much sounder approach is the statistical analysis of the precursor events generated over the history of many hundreds of reactor-years of ordering experience. The committee commends the NRC for moving in this direction and recommends

RECOMMENDATION: *That overall risk assessments of severe core damage be based on the analysis of past nuclear operating experience. Furthermore, that the accuracy of past PRAs in locating high probability paths be retrospectively assessed in terms of the history of precursor events.*

Adoption of the recommendations will help in re-establishing the credibility of the NRC risk assessment program and place it on a more honest and realistic statistical footing.

ON INFERENCE FOR DEMOGRAPHIC PROJECTION
OF SMALL POPULATIONS

C. C. HEYDE   University of Melbourne

1. INTRODUCTION

In this paper we shall be concerned with age-structured populations
whose vital rates vary stochastically in time. Let Y_t be the (column)
vector consisting of the numbers of individuals in each of K age
classes at time t. We shall assume an evolution of the kind

$$Y_{t+1} = X_{t+1} Y_t + \varepsilon_{t+1} \tag{1}$$

where $\{X_i\}$ is a stationary ergodic sequence of random matrices of
vital rates, and $\{\varepsilon_i\}$ is a sequence of independent and identically
distributed stochastic disturbance vectors with $E\varepsilon_1 = 0$ such that
$\{\varepsilon_i\}$ is independent of the σ-fields $\mathscr{G}$ generated by Y_0, X_1, X_2, ...

The model (1) in the degenerate case $\varepsilon_t \equiv 0$ for all t, and under
various conditions on the X_t, has been much discussed in the literature, particularly with a view to establishing asymptotic results on
the long-term population growth rate [e.g., Tuljapurkar and Orzack
(1980) and references therein]. Further, an asymptotic inferential

From *Proceedings of the Berkeley Conference in Honor of Jerzy Neyman
and Jack Kiefer*, Volume I, Lucien M. Le Cam and Richard A. Olshen,
eds., copyright © 1985 by Wadsworth, Inc. All rights reserved.

theory has recently been developed by Heyde and Cohen (1984). This
allows, under mild regularity conditions, for the testing of hypotheses about, and the construction of confidence intervals for, the growth
rate and population size.

In the case $\varepsilon_t \equiv 0$, the emphasis in the model (1) is on the
effects of variations in the vital rates themselves, and the variability of births and deaths conditional on given vital rates is suppressed. However, for small populations this variability can be
crucial in determining survival or extinction. The full model (1) is
necessary to cope with this situation. Fortunately, it turns out that,
conditional on nonextinction $\cap_{k \geqslant 0} \{Y_k > 0\}$, the inferential theory of
Heyde and Cohen (1984) developed for the case $\varepsilon_t \equiv 0$ continues to hold
for the full model (1). It is the purpose of this paper to establish
this robustness result. An application to the assessment of the long-
term viability of the endagnered North American whooping crane is given
to illustrate the use of the theory. This application involves small
population sizes for which the full model (1) is essential.

2. THEORETICAL RESULTS

Let $\{X_i\}$ be a stationary ergodic random sequence of $K \times K$ matrices
with nonnegative elements. Following Heyde and Cohen (1984) (hence-
forth denoted by HC) we shall suppose that the $\{X_i\}$ satisfy the two
assumptions:

A1. There exists an integer n_0 such that any product $X_{j+n_0} \cdots X_{j+1} X_j$
     of n_0 of the matrices has all its elements positive with
     probability one.

A2. For some constant C, $1 < C < \infty$, and each matrix X_i,
$$1 \leqslant M(X_i)/m(X_i) \leqslant C$$
with probability one, where $M(X)$ and $m(X)$ are, respectively, the
maximum and minimum *positive* elements of X.

The significance of these assumptions is discussed in HC; see also
Seneta (1980, Sections 3.1, 3.2).

Next, suppose that the process $\{X_i\}$ is defined on a probability space $(\Omega, \mathscr{F}, P)$. Write $\mathscr{M}_a^b$ for the σ-field generated by X_a, ..., X_b and let

$$\phi(n) = \sup_{k \geqslant 0}\{\,|P(B|A) - P(B)|\,;\ A \in \mathscr{M}_0^k,\ B \in \mathscr{M}_{k+n}^\infty,\ P(A) > 0\}.$$

The uniform mixing assumption is also used in HC and is one of the standard conditions of asymptotic independence. It is convenient to frame the theorem below in terms of conditions on the rate at which $\phi(n) \to 0$ although many other variants could equally be proposed.

The generalization of the inferential result of HC is provided by the following theorem.

Theorem: *Let* $Z_t = \alpha' Y_t$ *where* $\{Y_t\}$ *satisfies* (1) *and* α *is a nonzero vector of nonnegative elements. Conditional on nonextinction* $\cap_{k \geqslant 0}\{Y_k > 0\}$ *the following results hold:*

(a) *If A1 and A2 are satisfied and* $E|\log M(X_1)| < \infty$, *then*

$$\lim_{t \to \infty} t^{-1} \log Z_t = \log \lambda \quad \text{a.s.} \tag{2}$$

(b) *If the additional conditions* $E[\log M(X_1)]^2 < \infty$ *and* $\sum_{n=1}^\infty \phi^{1/2}(n) < \infty$ *hold, then*

$$\lim_{t \to \infty} t^{-1/2}\, E\big|\log(X_t \cdots X_1)_{11} - t \log \lambda\big| = \sigma(2/\pi)^{1/2} \tag{3}$$

exists for $0 \leqslant \sigma < \infty$ *and if* $\sigma > 0$

$$(t\sigma^2)^{-1/2}\{\log Z_t - t \log \lambda\} \xrightarrow{d} N(0, 1) \quad \text{(mixing)} \tag{4}$$

as $t \to \infty$, *the convergence being mixing in the sense of Rényi. Furthermore, under the same conditions as* (3) *and* (4),

$$(\log t)^{-1} \sum_{i=1}^{t} \big|\log Z_i - i \log \lambda\big| i^{-3/2} \xrightarrow{p} \sigma(2/\pi)^{1/2} \tag{5}$$

as $t \to \infty$ *and*

$$(\log t)^{-1} \sum_{i=1}^{t} \big|\log Z_i - i \log \hat{\lambda}\big| i^{-3/2} \xrightarrow{p} \sigma(1/\pi)^{1/2} \tag{6}$$

as $t \to \infty$ *where* $\log \lambda = t^{-1} \log Z_t$.

These results offer the possibility of constructing approximate confidence intervals for population growth and growth rates and also for the testing of hypotheses about these quantities.   The methods are identical to those proposed in HC for the case $\varepsilon_t \equiv 0$.

3. PROOF OF THE THEOREM

Iteration of (1) gives

$$Y_t = X_t \ \ldots \ X_1 Y_0 + \delta_t \tag{7}$$

where $\delta_t = \varepsilon_t + X_t \varepsilon_{t-1} + X_t X_{t-1} \varepsilon_{t-2} + \cdots + X_t \ \cdots \ X_2 \varepsilon_1$, so that on $\{Y_t > 0\}$,

$$\log Z_t = \log \alpha' Y_t = \log \alpha' X_t \ \ldots \ X_1 Y_0 + \beta_t \tag{8}$$

where

$$\beta_t = \log \left(1 + \frac{\alpha' \, \delta_t}{\alpha' X_t \ \cdots \ X_1 Y_0} \right).$$

We shall show that $\alpha' \delta_t / \alpha' X_t \ \ldots \ X_1 Y_0 \xrightarrow{\text{a.s.}} Z$ as $t \to \infty$ for some random variable Z.   To do this it is necessary to apply the series version of the dominated convergence theorem.

Recalling that $M(X)$ and $m(X)$ denote, respectively, the maximum and minimum positive elements of a matrix X, we have

$$|\alpha' \delta_t| \leq (1' \alpha) \left[\sum_{\ell=2}^{t} (1' |\varepsilon_{\ell-1}|) M(X_t \ \ldots \ X_\ell) + 1' |\varepsilon_t| \right]$$

and using Lemma 2' of HC,

$$M(X_t \ \ldots \ X_\ell) \leq (KC)^{2n_0} m(X_t \ \ldots \ X_\ell),$$

so that

$$|\alpha' \delta_t| \leq (1' \alpha)(KC)^{2n_0} \left[\sum_{\ell=2}^{t} (1' |\varepsilon_{\ell-1}|) m(X_t \ \ldots \ X_\ell) + 1' |\varepsilon_t| \right].$$

Furthermore,

$$\alpha' X_t \ \ldots \ X_1 Y_0 \geq (1' \alpha)(1' Y_0) m(X_t \ \ldots \ X_1)$$
$$\geq (1' \alpha)(1' Y_0) m(X_t \ \ldots \ X_\ell) m(X_{\ell-1} \ \ldots \ X_1)$$

for $2 \leq \ell < t$, and hence

$$\frac{|\alpha' \delta_t|}{\alpha' X_t \ \ldots \ X_1 Y_0} \leq (KC)^{2n_0} (1' Y_0)^{-1} \left[\sum_{\ell=1}^{t} \frac{1' |\varepsilon_\ell|}{m(X_\ell \ \ldots \ X_1)} \right], \tag{9}$$

while

$$\sum_{\ell=1}^{\infty} 1' |\varepsilon_\ell| / m(X_\ell \ \cdots \ X_1) < \infty \ \text{a.s.} \tag{10}$$

since the ε_ℓ are i.i.d. with $E|\varepsilon_\ell| < \infty$ and $\log m(X_\ell \ \cdots \ X_1) \sim \ell \log \lambda$ a.s. as $\ell \to \infty$ using Theorem 1 and Lemma $2'$ of HC.   Then, defining

$$a_{t\ell} = \frac{\alpha' X_t \ \cdots \ X_\ell \varepsilon_{\ell-1}}{\alpha' X_t \ \cdots \ X_1 Y_0}, \ 2 \leqslant \ell \leqslant t,$$

$$a_{t1} = \frac{\alpha' \varepsilon_t}{\alpha' X_t \ \cdots \ X_1 Y_0},$$

and $a_{t\ell} = 0$ for $\ell > t$, we have

$$\frac{\alpha' \delta_t}{\alpha' X_t \ \cdots \ X_1 Y_0} = \sum_{\ell=1}^{\infty} a_{t\ell}$$

and, in view of (9) and (10), the dominated convergence theorem gives

$$\lim_{t \to \infty} \frac{\alpha' \delta_t}{\alpha' X_t \ \cdots \ X_1 Y_0} = \sum_{\ell=1}^{\infty} \lim a_{t\ell}. \tag{11}$$

Now, using Theorem 3.3, p. 87 of Seneta (1980), we obtain under the assumptions A1 and A2 that, for any $1 \leqslant i, j, k \leqslant K$ and fixed ℓ,

$$\frac{(X_t \ \cdots \ X_\ell)_{ij}}{(X_t \ \cdots \ X_\ell)_{ik}} \xrightarrow{\text{a.s.}} V_{jk}^{(\ell)} > 0$$

as $t \to \infty$, the $V_{jk}^{(\ell)}$ being certain positive quantities that do not depend on i.   Consequently, if $\alpha = (\alpha_j)$,

$$(\alpha' X_t \ \cdots \ X_\ell)_i = \sum_{j=1}^{K} \alpha_j (X_t \ \cdots \ X_\ell)_{ji}$$

$$= \sum_{j=1}^{K} \alpha_j \frac{(X_t \ \cdots \ X_\ell)_{ji}}{(X_t \ \cdots \ X_\ell)_{j1}} (X_t \ \cdots \ X_\ell)_{j1}$$

$$\sim \sum_{j=1}^{K} \alpha_j V_{i1}^{(\ell)} (X_t \ \cdots \ X_\ell)_{j1}$$

$$= V_{i1}^{(\ell)} (\alpha' X_t \ \cdots \ X_\ell)_1$$

as $t \to \infty$.   Then, for $\ell \geqslant 2$,

$$a_{t\ell} = \frac{\alpha' X_t \cdots X_\ell \varepsilon_{\ell-1}}{\alpha' X_t \cdots X_1 Y_0} = \frac{\sum_{i=1}^{K} (\alpha' X_t \cdots X_\ell)_i (\varepsilon_{\ell-1})_i}{\sum_{i=1}^{K} (\alpha' X_t \cdots X_\ell)_i (X_{\ell-1} \cdots X_1 Y_0)_i}$$

$$\sim \frac{\sum_{i=1}^{K} V_{i1}^{(\ell)} (\varepsilon_{\ell-1})_i}{\sum_{i=1}^{K} V_{i1}^{(\ell)} (X_{\ell-1} \cdots X_1 Y_0)_i}$$

$$= \frac{(V^{(\ell)})' \varepsilon_{\ell-1}}{(V^{(\ell)})' X_{\ell-1} \cdots X_1 Y_0}$$

as $t \to \infty$ where $v^{(\ell)} = (V_{i1}^{(\ell)})$, while using Theorem 1 of HC, $\lim_{t \to \infty} a_{t1} = 0$ a.s. and $\lim_{\ell \to \infty} \lim_{t \to \infty} a_{t\ell} = 0$ a.s. It then follows from (11) that

$$\lim_{t \to \infty} \frac{\alpha' \delta_t}{\alpha' X_t \cdots X_1 Y_0} = \sum_{\ell=2}^{\infty} \frac{(V^{(\ell)})' \varepsilon_{\ell-1}}{(V^{(\ell)})' X_{\ell-1} \cdots X_1 Y_0} \quad \text{a.s.}$$

$$= Z, \text{ say.}$$

Next we need to show that Z has a continuous distribution. Take $\mathcal{G} = \sigma(Y_0, X_1, X_2, \ldots)$ and recall that the ε_i are i.i.d. and independent of $\mathcal{G}$. An application of Theorem 13.1.1, p. 537 of Kawata (1972) gives immediately that the distribution of Z conditional on $\mathcal{G}$ must be continuous and this forces the distribution of Z itself to be continuous.

Then, on the nonextinction set $\mathscr{E} = \cap_{k \geq 0} \{Y_k > 0\}$ we have positivity of

$$\alpha' Y_t = \alpha' X_t \cdots X_1 Y_0 \left(1 + \frac{\alpha' \delta_t}{\alpha' X_t \cdots X_1 Y_0} \right)$$

$$\sim \alpha' X_t \cdots X_1 Y_0 (1 + Z^*) \quad \text{a.s.}$$

where Z^* is Z conditional on $Z \geq -1$, and since Z has a continuous distribution, $P(1 + Z^* = 0) = 0$. Note that the probability of nonextinction is $\lim_{t \to \infty} P(Y_t > 0) = P(Z > -1)$. Thus, on $\mathscr{E}$,

$$\log Z_t = \log \alpha' X_t \cdots X_1 Y_0 + \beta_t \tag{12}$$

where $\beta_t \xrightarrow{\text{a.s.}} \log(1 + Z^*)$ as $t \to \infty$.

Further, using Lemma 2' of HC we have for $t \geq n_0$,

$$(KC)^{-2n_0} \leq (X_t \cdots X_1)_{ij} / (X_t \cdots X_1)_{11} \leq (KC)^{2n_0}$$

and hence

$$
(KC)^{-2n_0} (X_t \ \dots \ X_1)_{11} (1'\alpha)(1'Y_0) \leqslant \alpha'X_t \ \dots \ X_1 Y_0
$$
$$
\leqslant (KC)^{2n_0} (X_t \ \dots \ X_1)_{11} (1'\alpha)(1'Y_0) \tag{13}
$$

so that on $\mathscr{E}$ we have from (12) that for $t \geqslant n_0$

$$
\left| \log Z_t - \log(X_t \ \dots \ X_1)_{11} \right| \leqslant \log(1'\alpha)(1'Y_0)
$$
$$
+ 2n_0 \log KC + |\beta_t|. \tag{14}
$$

The result (3) is established in Theorem 1 of HC, whereas (2), (4), (5), (6) follow immediately from (13), (14), and Theorems 1 and 3 of HC. This completes the proof.

4. AN APPLICATION

As an illustration of the results, we shall consider their application to the assessment of the long-term viability of the endangered North American whooping crane. This is a rare migratory bird species that breeds in the Wood Buffalo Park, Northwest Territories, Canada, and winters in the Aransas National Wildlife Refuge, Texas, USA.

Annual counts of whooping cranes arriving at the Aransas National Park in the fall for the years 1938 to 1972 have been published by Miller, Botkin, and Mendelssohn (1974), and their data are given in the following table. Recent observations are not available. Birds born in a particular year have juvenile plumage and are referred to as *young*; the rest are called *adults*. These data have been tentatively discussed by Miller et al. (1974) using a linear birth-and-death process model and by Keiding (1976) using a birth-immigration-death model. These models, however, assume a constant environment and are consequently less biologically plausible than an approach based on the model (1).

It should be remarked that some of the earlier counts are known to be incomplete. In particular, those of 1938 and 1945 are clearly deficient because the total number of birds was less than the number of adults in the following year.

Adult and Young Whooping Cranes Counted at the Aransas National
Wildlife Refuge from 1938 to 1972

Year	Adult	Young	Total	Year	Adult	Young	Total
1938	10	4	14	1956	22	2	24
1939	16	6	22	1957	22	4	26
1940	21	5	26	1958	23	9	32
1941	13	2	15	1959	31	2	33
1942	15	4	19	1960	30	6	36
1943	16	5	21	1961	33	5	38
1944	15	3	18	1962	32	0	32
1945	14	3	17	1963	26	7	33
1946	22	3	25	1964	32	10	42
1947	25	6	31	1965	36	8	44
1948	27	3	30	1966	38	5	43
1949	30	4	34	1967	39	9	48
1950	26	5	31	1968	44	6	50
1951	20	5	25	1969	48	8	56
1952	19	2	21	1970	51	6	57
1953	21	3	24	1971	51	5	56
1954	21	0	21	1972	46	5	51
1955	20	8	28				

The early population counts are erratic, but a trend is apparent
from 1952. Consequently, three analyses have been done. The first is
on the data set from 1952 to 1972 inclusive. The second is for the
full set (however, we have adjusted the totals for 1938 and 1945 to
16 and 22, respectively, these being the minimum possible values).
The third is for the set from 1939 to 1972 (again with the adjusted
total for 1945). Population totals in each year are used, and the
sexual composition of the population is ignored (which amounts to
assuming that a fixed proportion of the population are females). No
specific assumptions about the matrices X_i of vital rates are required
in addition to those specified in the theory developed earlier in the
paper.

Period 1952-1972 inclusive. Using the estimators (11) and (12) of HC,
which in our context come from (2) and (6), we find that

$$\log \hat{\lambda} = 0.044, \quad \hat{\sigma} = 0.089.$$

Consequently, the approximate 95% confidence interval for $\log \lambda$ is

0.004 ± 0.039, and log λ is just significantly different from zero at the 5% significance level. This positive result should, however, be interpreted with caution. The time interval over which data have been collected is small for the use of asymptotic theory, and the estimators are not rapidly convergent.

Periods 1938-1972 inclusive and 1939-1972 inclusive. Again using the estimators (11) and (12) of HC (and with adjusted data for 1938 and 1945), we find that for 1938-1972

$$\log \hat{\lambda} = 0.034, \quad \hat{\sigma} = 0.198$$

giving an approximate $100(1 - \alpha)$% confidence interval for log λ of $0.034 \pm 0.034 z_{\alpha/2}$, whereas for the period 1939-1972

$$\log \hat{\lambda} = 0.025, \quad \hat{\sigma} = 0.166$$

leading to an approximate $100(1 - \alpha)$% confidence interval of $0.025 \pm 0.029 z_{\alpha/2}$. The instability of the population numbers over the period 1938-1951 produces inconclusively wide confidence intervals for log λ, and the point estimate of log λ is quite sensitive to changes in the initial population size.

REFERENCES

Heyde, C. C., and Cohen, J. E. (1984). Confidence intervals for demographic projections based on products of random matrices. *Theor. Pop. Biol.* (to appear).

Kawata, T. (1972). *Fourier Analysis in Probability Theory*. Academic Press, New York.

Keiding, N. (1976). Population growth and branching processes in random environments. In *Proc. 9th Int. Biom. Conf. Inv. Papers II*, pp. 149-165.

Miller, R. S., Botkin, D. B., and Mendelssohn, R. (1974). The whooping crane (*Grus Americana*) population of North America. *Bio. Conserv. 6*, 106-111.

Seneta, E. (1980). *Non-negative Matrices and Markov Chains*, 2nd ed. Springer-Verlag, New York.

Tuljapurkar, S. D., and Orzack, S. H. (1980). Population dynamics in variable environments I. Long-run growth rates and extinction. *Theor. Pop. Biol. 18*, 314-342.

MAXIMAL SEGMENTAL MATCH LENGTH AMONG RANDOM SEQUENCES FROM A FINITE ALPHABET

SAMUEL KARLIN Stanford University
FRIEDEMANN OST Technical University of Munich

ABSTRACT: The distributional properties of the maximal match length (identity blocks) among multiple random sequences composed of letters from a finite alphabet are described. The letter sequences are generated allowing for Markov dependence of any finite order. Applications are given to a variety of comparisons of actual DNA sequences.

1. INTRODUCTION

The results of this paper are motivated in part by the recent dramatic technical advances in rapid DNA sequencing methodology. These have been used to generate large data bases that currently total five million nucleotides corresponding to many DNA pieces of chromosomes from many different species; see Smith and Burke (1983) and Karlin et al. (1983) for references. The problem of classifying nucleotide patterns and relationships within and between multiple sequences is of paramount importance. The ascertainment of long homologies (matching blocks) across different sequences or repeats of a segment on a single

From *Proceedings of the Berkeley Conference in Honor of Jerzy Neyman and Jack Kiefer*, Volume I, Lucien M. Le Cam and Richard A. Olshen, eds., copyright © 1985 by Wadsworth, Inc. All rights reserved.

sequence are of particular interest. Such homologies may reflect regions that are conserved because of their biological function, or they may be evolutionary remnants of ancestral DNA segments. Distinguishing nonrandom sequence patterns from chance configurations is important in DNA and protein sequence comparisons.

Consider an alphabet $\mathcal{A}$ of m letters ℓ_1, ℓ_2, $\ldots$, ℓ_m (e.g., $m = 4$ for DNA sequences, $m = 20$ in the amino acid alphabet). A *word* of length k is a set of k consecutive letters of the sequence. The following *independence random model* can often serve as a standard by which to assess statistical significance of various DNA patterns and structures. In this purview, we generate s sequences of lengths N_1, $\ldots$, N_s, respectively, where the successive letters of the νth sequence are sampled independently, realizing letter ℓ_i with probability $p_i^{(\nu)}$. In the case of DNA, the components of $p^{(\nu)}$ are usually taken as the actual nucleotide (A, T, C, G) frequencies of the original sequences.

A more elaborate random model that allows for first-order neighbor dependencies is constructed as follows. In the νth sequence, we specify $p_{ij}^{(\nu)}$ as the conditional probability of sampling letter ℓ_j following letter ℓ_i. The generation of successive letters of the νth sequence is now governed by the transition probability matrix $\| p_{ij}^{(\nu)} \|$. (In the case of DNA, $p_{ij}^{(\nu)}$ usually correspond to the dinucleotide frequencies of the original sequences.) We refer to this sequence model as the *first-order Markov dependent random model*. Other versions can be constructed to allow for higher-order Markov dependence or stationary sequences obeying uniform mixing conditions.

This paper describes several theoretical results on maximal homology lengths for these random sequence models. For a randomly generated sequence $\mathcal{S}$ of N letters based on the alphabet $\mathcal{A}$, an *r*-fold *repeat of length k* is a word of k letters appearing in $\mathcal{S}$ at least r times. Let $L_2^{(N)}$ be the length of the longest repeat ($r = 2$) contained in $\mathcal{S}$. The growth of $L_2^{(N)}$ is of asymotitic order $\log\binom{N}{2}/(-\log \lambda^{[2]})$, see Theorem 3.1 later, where $\lambda^{[2]} = \sum_{i=1}^{m} p_i^2$ for the independence random model having probability p_i of sampling letter ℓ_i. We also describe the order growth of the maximal length of a word in $\mathscr{S}$ that occurs r or more times. More generally, we shall examine s random sequences each of length essentially N, and determine the distributional properties of $K_{r,s}^{(N)}$, defined as the length of the longest word that is common to r or more of the s sequences.

The distributional properties of other word relationships and patterns in $\mathscr{S}$ can be ascertained. For example, let $I^{(N)}$ be the length of the longest word in $\mathscr{S}$ whose direction-reversed word (i.e., the word with the letters arranged in reverse order) also occurs in $\mathscr{S}$. For the case in which the letters are independently generated, the growth of $I^{(N)}$ is also of order $\log N$, but on the average $I^{(N)} < L_2^{(N)}$.

Section 2 formalizes these concepts and problems in mathematical terms, and a number of results are described in Sections 3 and 4. Several comparisons with real data examples are highlighted in Section 5. Although we emphasized applications to DNA and protein sequences, these same models bear on a variety of problems in informational sciences concerned with comparisons between texts of any kind.

There is considerable literature on problems of estimating the expected length of the longest subsequence (not necessarily composed from contiguous elements) that is common to two sequences, e.g., Chvatal and Sankoff (1975), Deken (1979), Steele (1982). The asymptotic order in this case is a constant multiple of N, whereas the variable $K_{r,s}^{(N)}$ entertains growth order $\log N$. Zubkov and Mikhailov (1974) established a limit theorem pertaining to the length of the longest repeat on a single sequence of independently generated letters; see also Mikhailov (1974) and Karnin (1983) for repeats of symmetric binary sequences.

2. FORMULATIONS AND PROBLEMS

Consider an alphabet $\mathscr{A}$ of m letters and s independent sequences (strings of letters from $\mathscr{A}$), $\mathscr{S}_1$, $\mathscr{S}_2$, ..., $\mathscr{S}_s$ of lengths N_1, N_2, ..., N_s, respectively. We consider the following mechanisms for generating the elements of $\mathscr{S}_\nu$:

(i) *Independence model.* The consecutive letters of $\mathscr{S}_\nu$ consist of independent, identically distributed (i.i.d.) samples of the alphabet, in which the letter ℓ_j is realized with probability p_j, $j = 1, 2, \ldots, m$.

(ii) *Markov dependence model.* The successive letters of each $\mathscr{S}_\nu$ are generated as a realization of a Markov chain following the transition probability matrix $\mathbf{P} = \| p_{ij} \|$.

We seek to characterize long word patterns within and between the sequences. In this context, we focus on the following class of random variables. Let

$K_{r,s}^{(N)}$ be the length of the largest common word appearing in at least r out of the s sequences $\mathscr{S}_1, \mathscr{S}_2, \ldots, \mathscr{S}_s$ where the components of $N = (N_1, \ldots, N_s)$ refer, respectively, to their lengths. (2.1)

For a single sequence $\mathscr{S}$ of length N we concentrate on the random variable

$L_r^{(N)}$ equal to the length of the longest word occurring at least r rimes in $\mathscr{S}$. (2.2)

Functionals of Word Relationships. We begin with the simpler case of comparisons between two words. For definiteness, consider two random strings of letters (from the alphabet $\mathscr{A}$), the realization of $\mathscr{S}_1$ governed by the Markov transition matrix P with the realization of $\mathscr{S}_2$ governed by the Markov transition matrix Q. $\mathscr{S}_1$ and $\mathscr{S}_2$ are assumed to be generated independently with lengths N_1 and N_2, respectively. Let

$W_{P,Q} = W_{P,Q}^{(N_1, N_2)}$ be the length of the longest word common to both $\mathscr{S}_1$ and $\mathscr{S}_2$. (2.3)

Of particular biological interest is the case in which $Q = P^\star$ such that the transition matrix Q is that of the time reversed Markov chain to P. Another relevant class of examples stipulates $Q = \Pi'P\Pi$ with Π a prescribed permutation matrix, where Π' stands for the transposed matrix to Π. For example, when $\Pi = \| \pi_{ij} \|$ such that $\pi_{ij} = 1$ for $j = m + 1 - i$ and 0 otherwise, then $W_{P, \Pi'P\Pi}$ assesses the length of the longest word in $\mathscr{S}_1$ whose letter inverted word appears in $\mathscr{S}_2$.

It is useful to distinguish two categories of permutations. The first refers to the letter alphabet delineating a one-to-one mapping f

of $\mathscr{A}$ to $\mathscr{A}$, which is equivalently characterized by a permutation of the integers 1, 2, ..., m (m being the size of the alphabet) to itself. The second consists of a coherent family of permutation mappings $\{\sigma_k\}$ applied to the positions of a word, where for each integer k the permutation σ_k maps the indices (positions) 1, 2, ..., k of a k-word to itself. Let $\mathscr{S}$ be a random string of letters (based on $\mathscr{A}$) generated as a Markov dependent sequence. We define

$Z_{f,\sigma}^{(N)}$ as the length of the longest word

$$w = (a_{t+1}, a_{t+2}, \ldots, a_{t+\ell})$$

of $\mathscr{S}$ for which there also exists another word $\qquad$ (2.4)

$$w^* = (a_{t^*+1}, a_{t^*+2}, \ldots, a_{t^*+\ell})$$

in $\mathscr{S}$ with the property that

$$a_{t^*+\mu} = f(a_{t+\sigma_\ell(\mu)}), \quad \mu = 1, 2, \ldots, \ell.$$

The following example arises naturally in DNA sequence analysis. In this case $\mathscr{A} = (A, T, C, G)$. Set f to be the complementary map that sends $A \to T$, $T \to A$, $C \to G$, $G \to C$; i.e., every nucleotide maps into its complementary bonding nucleotide. Let $\boldsymbol{\sigma} = \{\sigma_k\}$ comprise the collection of inversion permutations $\sigma_k(1, 2, \ldots, k) = (k, k-1, \ldots, 1)$. The combined operations f and $\boldsymbol{\sigma}$ produce

$$\text{GACTTCCAA} \xrightarrow{f \circ \boldsymbol{\sigma}} \text{TTGGAAGTC.}$$

The two words above are said to be in *dyad symmetry relation*. An invariant word under this transformation is called *self-dyad*; e.g.,

$$\text{GACGTC} \xrightarrow{f \circ \boldsymbol{\sigma}} \text{GACGTC.}$$

Dyad symmetry combinations are biologically important because they may reveal the possible secondary structures (potential stem-loop formations) of DNA sequences; cf. Karlin et al. (1983).

3. RESULTS (GROWTH ORDERS)

We describe the asymptotic growth properties of the random variables

$K_{r,s}^{(N)}$ [defined in (2.1)], $L_r^{(N)}$ [defined in (2.2)], and $\qquad$ (3.1)
$W_{P,Q}^{(N_1, N_2)}$ [defined in (2.3)].

For definiteness, we focus on the sequence models in which the letters
are generated and governed by the common transition probability matrix
P. Let $\lambda^{[r]}$ be the principal eigenvalue (the spectral radius) of
$P^{[r]} = P \circ P \circ \cdots \circ P$, the rth Schur product of P. Recall that the
Schur product of $A = \| a_{ij} \|$ and $B = \| b_{ij} \|$ is $A \circ B = \| a_{ij} b_{ij} \|$. In
particular, the entries of $P^{[r]}$ are the rth powers of the elements of
P. The special case of i.i.d. sequences arise from the Markov matrix
$\Gamma = \| p_j e_i \|$, $(e_i \equiv 1)$, such that p_j is the probability of obtaining the
letter ℓ_j. In this case we have $\lambda^{[r]} = \lambda(\Gamma^{[r]}) = \sum_{j=1}^{m} p_j^r$.

When dealing with s sequences generated by Markov transition probability matrices P_1, P_2, ..., P_s, respectively, we define

$$\lambda^{[r,\,s]} = \max_{1 \leqslant \nu_1 < \nu_2 < \cdots < \nu_r \leqslant s} \lambda(P_{\nu_1} \circ P_{\nu_2} \circ \cdots \circ P_{\nu_r}) \tag{3.2}$$

where $\lambda(A)$ is the spectral radius of the matrix A.

Theorems 3.1-3.3 indicate the order growth of some of the random
variables listed in (3.1). We assume hereafter that the lengths N_i of
the sequence $\mathcal{S}_i$ increase to infinity at the same rate, meaning that
the ratios N_i/N_j for all i and j are bounded away from 0 and ∞.

Theorem 3.1: *Let* $K_{r,\,s}^{(N)}$ *be the length of the longest common word occur-*
ring in at least r *among the* s *sequences* $\{\mathcal{S}_\nu\}_1^s$. *The letters of each*
$\mathcal{S}_\nu$ *are assumed to be generated by the Markov chain matrix*
$P_\nu = \| p_{ij}^{(\nu)} \|$ *where the* P_ν *are assumed not too deviant from each other. Let*

$$n(N) = \sum_{(i_1, i_2, \ldots, i_r)} \left(\prod_{\nu=1}^{r} N_{i_\nu} \right) \tag{3.3}$$

where the sum extends over all r*-tuples of indices* $(i_1, i_2, \ldots, i_r)$,
$1 \leqslant i_1 < i_2 < \cdots < i_r \leqslant s.$ *Then with* $\lambda^{[r,\,s]}$ *defined in (3.2) we have*

$$K_{r,\,s}^{(N)} \approx \frac{\log n(N)}{-\log(\lambda^{[r,\,s]})} = u(N,\, r,\, s) \tag{3.4}$$

meaning that provided $n(N) \to \infty$,

$$\frac{K_{r,\,s}^{(N)}}{\log n(N)} \to -\log(\lambda^{[r,\,s]}) \quad \text{in probability.} \tag{3.5}$$

For multiple repeats within a single sequence, we have

Theorem 3.2: *The length* $L_r^{(N)}$ *of the longest word repeated at least* r *times within a single sequence has asymptotic growth*

$$L_r^{(N)} \approx \frac{\log\binom{N}{r}}{-\log(\lambda^{[r]})} \quad \text{with } \lambda^{[r]} = \lambda(P^{[r]}).$$ (3.6)

Theorem 3.3: *Consider letter strings* $\mathscr{S}_1$ *and* $\mathscr{S}_2$ *of lengths* N_1 *and* N_2, *respectively, generated as respective realizations governed by the Markov chain matrices* P *and* Q. *Let* $W_{P,Q}^{(N_1, N_2)}$ *be the length of the longest shared word between* $\mathscr{S}_1$ *and* $\mathscr{S}_2$. *Then*

$$W_{P,Q}^{(N_1, N_2)} \approx \frac{\log(N_1 N_2)}{-\log(\lambda_{P,Q})}$$ (3.7)

where $\lambda_{P,Q} = \lambda(P \circ Q)$ *provided* $\lambda(P \circ Q^{[m]})$ *and* $\lambda(P^{[m]} \circ Q) < (\lambda_{P,Q})^{(m+1)/2}$ *holds for all* m = 1, 2, ...

It can be proved [see Karlin and Ost (1985)] that

$$\lambda(P \circ Q) \leqslant \sqrt{\lambda(P \circ P)} \sqrt{\lambda(Q \circ Q)}.$$ (3.8)

Consider $Q = P^{\star}$ as the time reversed Markov chain matrix to P [see Karlin and Taylor (1982), Chap. 11] such that

$$P^{\star} = \frac{\pi_j P_{ji}}{\pi_i} = D_{\pi}^{-1}(P')D_{\pi},$$

where D_{π} is the diagonal matrix diag(π_1, π_2, ..., π_m) and $\boldsymbol{\pi}$ is the stationary frequency vector of P, then $\lambda(P^{\star} \circ P^{\star}) = \lambda(P \circ P)$ and (3.8) in this context reduces to

$$\lambda(P \circ P^{\star}) \leqslant \lambda(P \circ P).$$ (3.9)

On the basis of (3.9), we obtain

Corollary 3.1: *The following asymptotic order comparison applies*

$$E\left[W_{P,P^{\star}}^{(N_1, N_2)}\right] \leqslant E\left[W_{P,P}^{(N_1, N_2)}\right] \quad \text{as } N_1, N_2 \to \infty,$$

where E *denotes the expectation of the indicated variable.*

4. THE PARENTAL DISTRIBUTION OF MATCHES

Concerning common occurrences of a word among r out of s independently generated stationary sequences, there is a basic distribution for the size of a local match that underlies the distribution of the length of the longest global match.

For r independent stationary sequences generated by the processes, $\mathscr{P}_1$, $\mathscr{P}_2$, ..., $\mathscr{P}_r$, we let $p_\rho^t(w)$ be the probability of observing the specific word w (of length k) in the process $\mathscr{P}_\rho$ at position t. The probability of the joint event is $\prod_{\rho=1}^r p_\rho^{t_\rho}(w)$. The probability of some common word of length k localized to the positions t_1, ..., t_r is

$$\sum_{w \in \mathscr{W}_k} \prod_{\rho=1}^r p_\rho^{t_\rho}(w), \tag{4.1}$$

where $\mathscr{W}_k$ stands for the collection of all possible words of length k. When each sequence is generated by the same stationary probability law, (4.1) reduces to $\sum_{w \in \mathscr{W}_k}[p(w)]^r$ where $p(w)$ is the probability of the word w. In many cases the following limit exists:

$$\lim_{k \to \infty}\left(\sum_{w \in \mathscr{W}_k}[p(w)]^r\right)^{1/k} = \lambda^{[r]}. \tag{4.2}$$

It is instructive to indicate some examples. Where the probabilities $p(w)$ are induced by a common stationary finite Markov chain process with underlying transition Markov matrix $P = \|p_{ij}\|$ having stationary frequency vector $\boldsymbol{\pi}$, the expression (4.1) is explicitly

$$\sum_{w \in \mathscr{W}_k}[p(w)]^r = \sum_{j_1, j_2, \ldots, j_k}[\pi_{j_1} p_{j_1 j_2} p_{j_2 j_3} \cdots p_{j_{k-1} j_k}]^r$$

which can be compactly written in the inner product form (using the notation $\langle \mathbf{x}, \mathbf{y} \rangle = \Sigma x_i y_i$)

$$\langle \boldsymbol{\pi}^{[r]}, (P^{[r]})^{k-1}\mathbf{u}^{[r]} \rangle$$

with $P^{[r]}$ as the rth Schur power of the matrix P, $\boldsymbol{\pi}^{[r]} = (\pi_1^r, \ldots, \pi_m^r)$, $\mathbf{u} = (1, 1, \ldots, 1)$, and $(P^{[r]})^{k-1}$ the $(k-1)^{\text{st}}$ matrix product of $P^{[r]}$. If $\lambda^{[r]}$ is the principal eigenvalue of $P^{[r]}$, we find that for every positive $\tilde{\boldsymbol{\pi}}$,

$$\lim_{k \to \infty}\left(\langle \tilde{\boldsymbol{\pi}}, (P^{[r]})^{k-1}\mathbf{u} \rangle\right)^{1/k} = \lambda^{[r]}. \tag{4.3}$$

Generally, for r independent stationary Markov chain sequences following the transition probability matrices $P(\rho)$, $\rho = 1, 2, \ldots, r$, respectively, and letting now $p^{(r)}(w)$ evaluate the joint probability of observing the same word w of length k (k-word) at positions $t_1, t_2, \ldots, t_r$ in sequences $\mathscr{S}_1, \ldots, \mathscr{S}_r$, respectively, we ascertain the asymptotic geometric probability of a match to be

$$\lim_{k \to \infty} \left(\sum_{w \in \mathscr{W}_k} p^{(r)}(w) \right)^{1/k} = \lambda^{[r]}$$

where now $\lambda^{[r]} = \lambda(P(1) \circ P(2) \circ \cdots \circ P(r))$ is the spectral radius of the Schur product matrix $P(1) \circ P(2) \circ \cdots \circ P(r)$.

5. LIMIT THEOREMS FOR LONG HOMOLOGIES

The scope of the results is amply illustrated by concentrating on the random variable $K_{r,s}^{(N)}$ [see (2.1)] with all N_ν equal to N. From Theorem 3.1 we know the order growth. The asymptotic variance is bounded as implied in Theorem 5.1 below. Consider now the translated random variable

$$\hat{K}_{r,s}^{(N)} = K_{r,s}^{(N)} - \frac{\log\left(\binom{s}{r} N^r\right)}{-\log \lambda^{[r]}} \tag{5.1}$$

having subtracted the mean growth term. The results of this section concern the limit probabilities ($N \to \infty$)

$$Pr\{\hat{K}_{r,s}^{(N)} \leqslant z\} \quad \text{for all real } z. \tag{5.2}$$

The computations are based on the parental match distribution. In this context, for every choice of r out of s sequences and collection of positions $\mathbf{j} = (j_1, j_2, \ldots, j_r)$, j_ν specifying an index position in the νth sequence, $1 \leqslant j_\nu \leqslant N$, we let $\alpha_{\mathbf{j}}(k)$ be the probability of a common match of size $\geqslant k$ at the positions embodied in $\mathbf{j}$.

For all sequences generated as i.i.d. letters or as a stationary Markov chain, the collection of probabilities $\{\alpha_{\mathbf{j}}(k)\}$ are independent of $\mathbf{j}$ and possesses the form

$$\alpha_{\mathbf{j}}(k) = (\lambda^{[r]})^{k-1}\gamma + \delta(k), \tag{5.3}$$

where γ is a positive constant (depending on the components of the principal eigenvectors of $P^{[r]}$) and δ, although a complicated quantity depending on r, s and the parameters of P, satisfies $\dfrac{\delta(k)}{(\lambda^{[r]})^k} \to 0$ at a geometric rate as $k \to \infty$.

More explicitly, for the case of r sequences of letters generated independently governed by the Markov chain matrix P (assumed to be irreducible aperiodic), we have

$$\gamma = \left(\sum_{i=1}^{m} (\pi_i)^r \eta_i \right) \left(\sum_{i=1}^{m} \xi_i \right), \tag{5.4}$$

where $\boldsymbol{\xi} = (\xi_1, \ldots, \xi_m)$ and $\boldsymbol{\eta} = (\eta_1, \ldots, \eta_m)$ are the principal left and right eigenvectors of $\mathbf{P}^{[r]}$, associated with $\lambda^{[r]}$ normalized such that $\sum_{i=1}^{m} \xi_i \eta_i = 1$ and $\boldsymbol{\pi} = (\pi_1, \ldots, \pi_m)$ is the stationary frequency vector of P and $(\pi_i)^r$ is the rth power of the ith component.

The parental match probabilities $\alpha_1(k)$ in the stationary case behave asymptotically $(k \to \infty)$ as $(\lambda^{[r]})^{k-1} \gamma$.

In seeking the limit distribution of $\hat{K}_{r,s}^{(N)}$, we find it convenient to modify the model from a discrete to a continuous match length by assigning to each r-tuple of positions the probability of a match of length $\geqslant z$ for at least r out of s sequences as equal to $(\lambda^{[r]})^{z-1} \gamma$. This converts the probabilities for integer k to a continuous variable z. In this extended framework, we designate the corresponding random variable of the longest r out of s match by $\tilde{K}_{r,s}^{(N)}$ where, as in (5.1), the order term $\log\left(N^r \binom{s}{r} \right) / (-\log \lambda^{[r]})$ is subtracted away.

For a parental match distribution such as

$$\alpha_1(z) = (\lambda^{[r]})^{z-1} \gamma \tag{5.5}$$

we can prove

Theorem 5.1: *For the modified matching process subject to (5.5) as* $N \to \infty$,

$$\mathrm{Prob}\{\tilde{K}_{r,s}^{(N)} \leqslant z + 1\} \to \exp[-(1-\lambda)\lambda^z \gamma] = G_{\lambda,\gamma}(z) \tag{5.6}$$

where $\lambda = \lambda^{[r]}$ *and* γ *is an appropriate constant depending on the parameters of the Markov transition matrix;* $\gamma = \lambda$ *in the independence case.*

Remark: The cumulative distribution $H_\lambda(z) = \exp\{-\lambda^z\}$, $-\infty < z < \infty$, $0 < \lambda < 1$ is recognized as one of the classical extremal distributions.

Observe that the cumulative distribution $G_{\lambda,\gamma}(z)$ differs from $H_\lambda(z)$ by a translation constant since by setting $\gamma(1 - \lambda) = \lambda^\beta$ we have $G_{\lambda,\gamma}(z) = H_\lambda(z + \beta)$.

A stochastic ordering analysis establishes the following comparison between $\tilde{K}^{(N)}_{r,s}$ and $\hat{K}^{(N)}_{r,s}$.

$$\varlimsup_{N \to \infty} \Pr\{\hat{K}^{(N)}_{r,s} \leqslant z\} \leqslant G_{\lambda,\gamma}(z)$$

$$\varliminf_{N \to \infty} \Pr\{\hat{K}^{(N)}_{r,s} \leqslant z\} \geqslant G_{\lambda,\gamma}(z - 1). \tag{5.7}$$

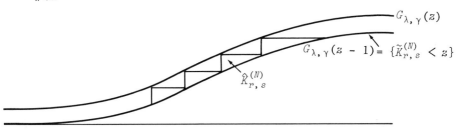

It is essential to emphasize that the random variable $\hat{K}^{(N)}_{r,s}$, defined in (5.1), cannot have a limit distribution because it is discrete. However, by virtue of the relations (5.7), close bounds on all the moments of $\hat{K}^{(N)}_{r,s}$ are available. Note first that the mean of the distribution $G_{\lambda,\gamma}(z)$ is

$$a(\lambda, \gamma) = -\left[\frac{\log(1 - \lambda) + \log \gamma}{\log \lambda} + \frac{0.5772}{\log \lambda}\right]$$

where 0.5772 is Euler's constant, and the variance is $1.645(1/\log \lambda)^2$. The limiting mean of $\hat{K}^{(N)}_{r,s}$ in the independence case ($\gamma = \lambda$) is

$$E(K^{(N)}_{r,s}) = \frac{r \log N}{-\log \lambda^{[r]}} + \frac{\log\left(\dfrac{s}{r}\right)}{-\log \lambda^{[r]}} \tag{5.8}$$

$$+ a(\lambda^{[r]}, \lambda) + .5 + \varepsilon_N(\lambda^{[r]}),$$

where $\varepsilon_N(\lambda)$ satisfies $|\varepsilon_N(\lambda)| < 1$. More exactly, when $\lambda = .25$, we ascertain $\varepsilon_N(\lambda)$ between the bounds $\pm.0014$. We can also establish

$$\mathrm{Var}[K^{(N)}_{r,s}] = 1.645\left(\frac{1}{\log \lambda^{[r]}}\right)^2 + \delta_N(\lambda^{[r]}). \tag{5.9}$$

For $\lambda = .25$ we estimate numerically $\delta_N(\lambda)$ in the interval $(.1091,$ $.1104)$ such that as $N \to \infty$, $\text{Var}(K_{r,s}^{(N)}) \in (.933, .945)$.

The limit distributional behavior of the random variables $L_r^{(N)}$, $W_{P,Q}^{(N_1, N_2)}$, defined in Section 2 have similar representations as in Theorem 5.1. More specifically,

Theorem 5.2:

$$G_{\lambda, \gamma}(z - 1) \leqslant \varliminf_{N \to \infty}\{\ \} \leqslant \varlimsup_{N \to \infty} \text{Pr}\left\{ L_r^{(N)} - \frac{\log\binom{N}{r}}{-\log(\lambda^{[r]})} \leqslant z \right\} \leqslant G_{\lambda, \gamma}(z).$$

The asymptotic mean and variance are

$$E(L_r^{(N)}) = \frac{\log\binom{N}{r}}{-\log \lambda^{[r]}} + a(\lambda^{[r]}, \gamma) + .5 + \varepsilon_N(\lambda^{[r]})$$

and

$$\text{Var}(L_r^{(N)}) = 1.645\left(\frac{1}{\log(\lambda^{[r]})}\right)^2 + \delta_N(\lambda^{[r]}),$$

respectively, where ε_N and δ_N are confined by the same bounds as in Theorem 5.1.

Theorem 5.3: *Consdier two letter sequences $\mathscr{S}_1$ and $\mathscr{S}_2$ of lengths N_1 and N_2 governed by the Markov transition matrices P and Q, respectively. Then*

$$G_{\lambda, \gamma}(z - 1) \leqslant \lim_{N_1, N_2 \to \infty}\{\cdot\} \leqslant \varlimsup_{N_1, N_2 \to \infty} \text{Pr}\left\{ W_{P,Q}^{(N_1, N_2)} - \frac{\log N_1 N_2}{-\log \lambda} \leqslant z \right\}$$
$$\leqslant G_{\lambda, \gamma}(z)$$

with $\lambda = \lambda(P \circ Q)$ and γ is a constant depending on P and Q. The asymptotic mean is

$$E[W_{P,Q}^{(N_1, N_2)}] = \frac{\log N_1 N_2}{-\log \lambda} + a(\lambda, \gamma) + .5 + \varepsilon(\lambda)$$

and $\text{Var}[W_{P,Q}^{(N_1 N_2)}]$ is as in (5.9) with $\lambda^{[r]}$ replaced by $\lambda = \lambda(P \circ Q)$.

All the limit distributions are basically of the same form $G_{\lambda, \gamma}(z)$, where the λ and γ parameters reflect differences in the sequence structures and the homology model specified. The critical parameter is the characteristic value λ. As remarked earlier, the parameter γ can be incorporated into the distribution function $H_\lambda(z) = \exp[-\lambda^z]$ as a translation constant.

6. EXAMPLES OF HOMOLOGY IN DNA SEQUENCE DATA

In this section we shall compare the theoretical results of Sections 3–5 with the observed length of the longest multiple repeat in a single extended DNA sequence and the length of the longest block match among several sequences. The data sets to be considered are

(i)   *Papovaviruses*. These are a class of carcinogenic viruses consisting of a single circular double stranded piece of DNA approximately 5200 base pairs (bp) long, which live in certain higher mammals. Three prominent representatives are SV-40 (Simian Virus-40) whose host is the green monkey, BKV-Dun which live in humans, and Polyoma which live in mice.

(ii)   *Papilloma viruses* (wart viruses). About 9000 bp of circular structure. We shall compare HPV in humans and BPV in bovines.

(iii) *Lambda-phage*. This is a well structured virus that replicates after integrating itself into the bacterium E-Coli.

(iv)   *Mitochondrial genomes*. The genomes to be compared are of human, mouse, bovine, and Xenopus (frog).

(v)   *Some immunoglobulin gene pieces*. An important gene involved in the immune system is the κ-(kappa-)light chain immunoglobulin gene. We shall consider its representatives in human, mouse, and rabbit.

(vi)   *Two Drosophila genes*. The gene E-74 (stimulated by a growth hormone ecdysome) and the AdH (alcohol dehydrogenase) enzyme gene, partly involved in metabolizing alcohol.

(vii) Finally, we consider three cases of hepatitis B virus living in humans, duck, and ground squirrel.

The above examples represent a meager spectrum of the DNA sequences currently available in DNA data banks. Detailed analysis of these and other sequences will be dealt with elsewhere; cf. Karlin et al. (1985a, b).

The third column of Table 1 presents the theoretical expectation for the size of the longest word that occurs at least twice for the random sequence model with i.i.d. letters standardized to the same total length N and with the letter frequencies of A, T, C, G coincident to those of the nascent DNA sequence. The same column indicates the theoretical variance for this model. The last column is the longest repeat size L^* [$(L_2^{(N)})$ in the notation of (2.2)] observed for the data.

The theoretical formulas for the mean of $L_2^{(N)}$ are calculated from (5.8) and that of $\mathrm{Var}(L^{(N)})$ from (5.9). In each case the associated parameter $\lambda^{[2]} = (\mathrm{freq\ A})^2 + (\mathrm{freq\ C})^2 + (\mathrm{freq\ G})^2 + (\mathrm{freq\ T})^2$ is based on the live nucleotide sequence.

Table 1.

DNA Sequence (N = Size)	Nucleotide Frequencies				Theoretical $E(L_2^n)$, $[\mathrm{Var}(L_2^n)]$	Observed L_2^n
	A	C	G	T		
Papovaviruses						
SV-40 N = 5243	.30	.21	.19	.30	11.85 [.91]	72[†]
BKV-Dun N = 5153	.30	.20	.20	.30	11.88 [.91]	68[†]
Polyoma N = 5293	.26	.24	.24	.26	11.61 [.86]	16[†]
Papilloma Viruses						
HPV N = 7811	.31	.19	.21	.29	12.50 [.93]	16[†]
BPV N = 7945	.29	.21	.24	.26	12.32 [.93]	15[*]
λ-Phage						
N = 48502	.25	.23	.27	.25	14.80 [.86]	15
Ig-κ-Gene						
Human N = 5019	.30	.20	.21	.29	11.78 [.91]	18[†]
Mouse N = 5473	.29	.20	.21	.30	11.94 [.91]	20[†]
Rabbit N = 5525	.30	.20	.20	.30	11.84 [.92]	18[†]
Mitochondrial Genomes						
Human N = 16569	.31	.31	.13	.25	14.10 [.99]	15
Mouse N = 16295	.35	.24	.12	.29	14.27 [.99]	16

(continued)

Table 1. Continued

DNA Sequence (N = Size)	Nucleotide Frequencies				Theoretical $E(L_2^n)$, [$\text{Var}(L_2^n)$]	Observed L_2^n
	A	C	G	T		
Mitochondrial Genomes (continued)						
Bovine N = 16338	.33	.26	.14	.27	14.16 [.99]	16
Xenopus N = 17550	.33	.24	.13	.30	13.06 [.97]	45†
E-74 Gene						
N = 7169	.29	.20	.21	.30	13.12 [.87]	23†
AdH						
N = 2724	.31	.23	.22	.24	10.77 [.94]	12
Hepatitus Viruses						
Human N = 3182	.23	.27	.22	.28	10.47 [.87]	11
Duck N = 3021	.30	.22	.21	.27	10.39 [.88]	12
Ground Squirrel N = 3311	.27	.23	.20	.30	10.53 [.89]	12

An asterisk (*) signifies that the observed longest repeat length would occur by chance according to the independence model with probability $\leqslant$ .01. A dagger (†) means the observed length would occur by chance with probability $\leqslant 10^{-4}$.

Table 1 reveals that the observed length of the longest repeat is extraordinarily significant for papovaviruses, immunoglobulins, and the E-74 gene in Drosophila. Significantly long repeats also occur in the Papilloma viruses HPV and BPV. On the other hand, in the mitochondrial sets, human, mouse, bovine (with exception, Xenopus), and λ-phage, the lengths of the longest observed repeat is not significantly different from chance occurrence.

The observed multiple repeats continue to be significantly long when compared with the model having dependence as described by a corresponding Markov chain. More specifically, we have considered the dinucleotide frequencies, i.e., the frequencies of the 16 pairs AA, AC, AG. AT, CA, ..., TT, and construct thereby the Markov chain based on these dinucleotide frequencies. Random letter sequences are then generated following these Markov transition matrices. The repeat lengths in this model persistently fall several standard deviations below those occurring with the actual sequences. Corresponding to 2nd- to 4th-order Markov dependence comparisons also fail to explain the substantially long word repeats observed.

We next highlight several examples of the longest word matches between related DNA sequences. Table 2 indicates the longest shared word lengths in four examples. The middle column indicates the expected length of the longest word match occurring in at least r out of s sequences when each sequence is generated i.i.d.

7. GENERALIZATIONS AND REMARKS

I. There are many extensions and refinements on the ideas and constructions of Sections 2-5. We describe two generalizations of the variables $L_r^{(N)}$ and $K_{r,s}^{(N)}$, respectively.

Consider s sequences $\mathscr{S}_1$, $\mathscr{S}_2$, ..., $\mathscr{S}_s$ composed of letters from the alphabet $\mathscr{A}$, in which $\mathscr{S}_\nu$ of length N_ν is generated, governed by the transition Markov matrix P. We specify for $\mathbf{N} = (N_1, N_2, ..., N_s)$ and $\mathbf{r} = (r_1, r_2, ..., r_s)$, the random variable $L_{\mathbf{r},s}^{\mathbf{N}}$ as the length of the longest common word over the aggregate of all $\{\mathscr{S}_\nu\}$ that appears at least r_ν times in $\mathscr{S}_\nu$. The characteristic value of the appropriate λ is

$$\lim_{k \to \infty} \left(\sum_{w \in \mathscr{U}_k} [p(w)]^r \right)^{1/k} = \lambda^{[r]} = \lambda(p^{[r]}),$$

where $r = r_1 + r_2 + \cdots + r_s$ and $n(N_1, ..., N_s) = \binom{N_1}{r_1}\binom{N_2}{r_2} \cdots \binom{N_s}{r_s}$.

II. We nest prescribe $K_{\mathbf{R},s}^{\mathbf{N}}$ to be the length of the longest common word that appears at least R_1 times in some unspecified sequence among $\{\mathscr{S}_\nu\}$, at least R_2 times in a distinct sequence, etc., and at least R_s in a different unspecified sequence in which

Table 2.

DNA Sequences (s = Number of Sequences)	r	Theoretical Expectation [Variance]	Observed Length of Longest Homology in r out of s	
Papovaviruses			SV-40, BKV-Dun	32[†]
(s = 3)			SV-40, Polyoma	14
SV-40	2	13.53 [.87]	BKV-Dun, Polyoma	15
BKV-Dun				
Polyoma	3	9.43 [.21]	12 (2 Distinct Groups)[†]	
Ig-κ-Gene			Human, Mouse	19[†]
(s = 3)			Human, Rabbit	28[†]
Human	2	12.17 [.89]	Mouse, Rabbit	19[†]
Mouse				
Rabbit	3	8.72 [.23]		14[†]
Hepatitus Viruses			Human, Duck	18*
(s = 3)			Human, Ground Squirrel	37[†]
Human	2	12.29 [.88]	Duck, Ground Squirrel	13
Duck				
Ground Squirrel		8.51 [.22]		12[†]
Mitochondrial Genomes			Human, Mouse, Bovine	52[†]
(s = 4)			Human, Mouse, Xenopus	21[†]
Human	3	11.71 [.25]	Human, Bovine, Xenopus	38*
Mouse			Mouse, Bovine, Xenopus	25*
Bovine				
Xenopus	4	9.90 [.11]		30[†]

An asterisk (*) signifies that the observed longest common word would occur by chance according to the independence model with probability $\leqslant$ .01. A dagger (†) means the observed length would occur by chance with probability $\leqslant 10^{-4}$.

$$r^{\star} = \sum_{\nu=1}^{s} R_{\nu}$$

is a fixed number. For the special choice $R_{\nu} = 1$, $\nu = 1, 2, \ldots, r^{\star}$, and $R_{\nu} = 0$ for $\nu = r^{\star} + 1, \ldots,$, we recover the random variable $K^{N}_{r^{\star}, s}$ defined in (2.1). The corresponding characteristic value is $\lambda = \lambda^{[r^{\star}]}$. However, we now have

$$n(\mathbf{N}) = \sum_{(r_1, \ldots, r_s)} \prod_{\nu=1}^{s} \binom{N_{\nu}}{r_{\nu}}$$

where $(r_1, \ldots, r_s)$ runs through all permutations of $(R_1, \ldots, R_s)$.

III. The results of Theorems 3.1 and 5.1–5.3 can be extended to count-ably infinite alphabets in which the sequences are generated by a positive recurrent Markov transition matrix.

IV. The ascertainment of the longest block identity can be adapted to allow *for a prescribed number of mismatches*, say, *e*. The analog of Theorem 3.1 now has the order growth

$$K^{(N)}_{2, 2}(e) \approx \frac{\log n(N) + e \log \log n(N)}{-\log \lambda}$$

where $n(N) = N^2$ in the case of unrestricted matches for two sequences. As contrasted with (3.4), the second-order term, unlike the case of no mismatches, is unbounded as it increases to infinity with N at the very slow rate $\log \log n(N)$.

Added in Proof. We have learned, from the referee, of independent work focusing on strong laws for the maximal match length between inde-pendent letter sequences (Arratia and Waterman, "An Erdos Renyi Law with Shifts," to appear 1985 in *Adv. in Applied Math.*; see also Erdos and Renyi, 1970, "On a New Law of Large," *J. Analyse Math.* 22, 103–111).

REFERENCES

Chvatal, V., and Sankoff, D. (1975). Longest common subsequences of two random sequences. *J. Appl. Prob. 12*, 306–315.

Deken, G. A. (1979). Some limit results for longest common sub-sequences. *Discrete Mathematics 26*, 17–31.

Karlin, S., Ghandour, G., Ost, F., Tavaré, S., and Korn, L. J. (1983). New approaches for computer analysis of nucleic acid sequences. *Proc. Natl. Adac. Sci. USA 80*, 5660-5664.

Karlin, S., and Ghandour, G. (1985a). Alignment maps and homology analysis of the J-C intron in human, mouse and rabbit immunoglobulin kappa gene. *Molecular Biology and Evolution* 2:31-41.

Karlin, S., Ghandour, G., and Foulser, D. E. (1985b). DNA sequence comparisons of the human, mouse and rabbit immunoglobulin kappa gene. *Molecular Biology and Evolution* 2:18-30.

Karlin, S., and Ost, F. (1984). Some monotonicity properties of Schur powers of matrices and related inequalities. *Linear Algebra and Its Appls.* (in press).

Karlin S., and Taylor, H. M. (1981). *A second Course in Stochastic Processes.* Academic Pres, New York.

Karnin, E. D. (1983). The first repetition of a pattern in a symmetric Bernoulli sequence. *J. Appl. Prob. 20*, 413-418.

Mikhailov, V. G. (1974). Limit distributions of random variables associated with multiple long duplications in a sequence of independent trials. *Theor. Prob. and Appls. 19*, 180-184.

Smith, T. F., and Burks, C. (1983). Searching for sequence similarities. *Nature 301*, 194.

Steele, J. M. (1982). Long common subsequences and the probability of two random strings. *SIAM J. Appl. Math. 42*, 731-737.

Zubkov, A. M., and Mikhailov, V. G. (1974). Limit distributions of random variables associated with long duplications in a sequence of independent trials. *Theor. Prob. Appls. 19*, 172-179.

TWELVE-MONTH PROGNOSIS FOLLOWING MYOCARDIAL INFARCTION: CLASSIFICATION TREES, LOGISTIC REGRESSION, AND STEPWISE LINEAR DISCRIMINATION

RICHARD A. OLSHEN University of California, San Diego

ELIZABETH A. GILPIN University of California, San Diego

HARTMUT HENNING Vancouver General Hospital and The University of British Columbia

MARTIN L. LeWINTER Veterans Administration Hospital and University of California, San Diego

DANIEL COLLINS United States Naval Regional Medical Center, San Diego

JOHN ROSS, JR. University of California, San Diego

ABSTRACT: Recursive partitioning, logistic regression, and stepwise linear discrimination have been applied to one-year prognoses for 1494 patients who suffered myocardial infarctions and survived to leave the hospital. Data are derived from several San Diego hospitals and one in Vancouver, British Columbia. Schemes developed on data from one location are applied to test samples from the other. Cross-validation is employed, and issues regarding stratification are addressed. In general, for our prognostic problem, it appears to be difficult to reduce the misclassification cost below 60% of that of the no data Bayes rule.

1. INTRODUCTION

For the most part, this paper is a report on the application of recursive partitioning, logistic regression, and stepwise linear discrimination to the problem of one-year prognoses for patients who have suffered a myocardal infarction (MI) and survived to leave the

Supported by NIH Research Grants HL 17682, Ischemic SCOR awarded by the NHLBI PHS/DHHS, CA 26666, and NSF MCS 83-02188.

hospital. For this problem is is difficult to reduce the misclassification cost below 60% of what it is in the absence of data. We
also address the question of stratification in cross-validation.
Though our empirical studies are inconclusive, one very special
theoretical contribution is reported; its conclusion is that stratification is always preferable.

Our data were gathered from medical histories and noninvasively
(without use of a catheter) from 1494 patients during 24-hour
periods following their admission into hospitals in San Diego and
Vancouver, British Columbia.* Prognostic schemes developed at one
location were applied to patients at the other. Error rates by technology have been broken down by San Diego Hospital and are different
in interesting ways. Tenfold cross-validation (see Breiman et al.,
1984) was employed to estimate how various schemes for classification
would perform in a prospective mode.

The study of one-year survivorship following MI is viewed here
as a problem in classification, as in Breiman et al. (1984) and Gilpin
et al. (1983). The survivors are termed *class 1*, with prior proba-
bility $\pi(1)$ and misclassification cost $C(2|1)$; $\pi(2)$ and $C(1|2)$ are
defined by analogy for the *class 2* patients, the deaths. Throughout
the paper we take $\pi(1) = .87 = 1 - \pi(2)$. The Cs are chosen so that
$\pi(1)C(2|1) = \pi(2)C(1|2) = 1/2$. Thus, for a measurement vector $\mathbf{x}$ from
a patient of unknown class membership, and estimates $\hat{f}_j$ ($j = 1, 2$) of
the conditional densities (with respect to any dominating measure) of
$\mathbf{x}$ given that the patient is of class j, the estimated Bayes classification rule we employ classifies the patient as class 1 if
$\hat{f}_1(\mathbf{x}) \geqslant \hat{f}_2(\mathbf{x})$; otherwise the patient is classified as class 2 (see
Breiman et al., 1984, p. 14).

If one bears in mind that the identification of patients at high
risk of death is a contribution toward improved and cost-effective

---

*These efforts are a project of the Specialized Center for Research
on Ischemic Heart Disease (SCOR) at the University of California,
San Diego.

medical care for all patients, then our choice that $C(1|2)$ is about seven times $C(2|1)$ should be acceptable. Moreover, simple chi-square or other analyses of Table 1 in Section 2 provide compelling evidence that our choice of priors is acceptable for prognosis by location, regardless of whether data are complete. In San Diego hospitals, the empirical prior probabilities $\hat{\pi}(i)$ vary considerably. This variation and other considerations pose a challenge regarding combining technologies; this challenge is discussed here but not met.

To stratify m-fold cross-validation in a classification problem is to require that each of the classes be, as nearly as possible, equally represented in each of the m randomly selected subsets of the learning sample, whose definition is the first step of the cross-validatory process. We shall give considerable attention to the question of whether or not it is better to stratify. Intuitition suggests that in retrospective studies it is preferable to stratify. In studies such as the one being reported, in which the mode is prospective, the choice is perhaps less clear. As was mentioned, the theoretical result presented here for a split-sample scheme in recursive partitioning concludes, like Breiman et al. (1984, pp. 80, 179, 245-247), that to stratify seems always to be better. Empirical comparisons are made of stratified and unstratified cross-validation in predicting the performance of schemes developed at San Diego on a Vancouver test sample and vice versa. This part of our work does not shed light on the issue of whether it is preferable to stratify.

A number of studies of prognosis following MI have preceded ours. The first, which used a combination of prognostic factors, was by Peel et al. (1962); their paper introduced the Peel Index. That index was criticized by Norris et al. (1969), who introduced the Norris Index. The work by Helmers (1973) was directed toward two-year prognoses; its scheme involved multiple linear regression. The work of Luria et al. (1979) involved 24-hour dynamic monitoring. Note our previous collaborations, which were part of the cited SCOR project: Henning et al. (1979); Battler et al. (1980); Ryan et al. (1981); Madsen, Gilpen, and Henning (1983); the same authors (1984); Gilpin et al. (1983); and Madsen et al. (1984). Our brief survey of the literature

concludes with mention of the recent work on risk stratification by
the Multicenter Postinfarction Group (1983), and the study of ventricular arrhythmias, left ventricular dysfunction, and two-year
mortality following MI by Bigger et al. (1984). The papers that have
been cited are, by and large, replete with long lists of references.
However, too often the authors of a study view their contribution as
the latest in a long line of work which is almost disjoint from the
references of a closely related paper that appeared at almost the
same time. We intend to publish a bibliographic study in which diverse
technologies and results on prognosis following MI are compared and
related. We believe that efforts here are noteworthy because they
focus on one-year survival; separate technologies are compared within
locations by cross-validation and across them; and institutions in
the San Diego location are compared and contrasted.

2. PATIENT POPULATIONS

The study population consisted of 1494 patients who suffered MIs and
survived to leave the hospital. Complete data on 34 variables
selected for use in our analyses were available for 1145 of the
patients. The diagnosis of MI was established by two of three
criteria: (a) characteristic chest pain; (2) electrocardiographic
changes with evolution of Q-waves (transmural infarction); (3) eleva-
tion of creatine kinase, an enzyme released by damaged heart muscle.
Nontransmural infarction was diagnosed by typical ST-segment and
T-wave changes accompanied by (3). All patients were admitted to the
hospital within 24 hours after onset of symptoms, but there were no
further eligibility restrictions. Patients who died of noncardiac
causes were excluded.

Since 1969, a computerized data base on patients who have suffered MIs has been maintained as part of the Myocardial Infarction
Research Unit and the cited SCOR project at the University of California, San Diego. In 1976, an exhaustive univariate screening of
over 150 variables was carried out for the purpose of identifying

those features prognostic of 30-day mortality. The results of the
screening, along with definitions and procedures for gathering data,
were reported by Henning et al. (1979). A new data base, which is
focused on a subset of approximately 100 variables proven to have
prognostic value for early or late mortality (in the previous data
base) and which includes variables reported in other studies, has
been used since late 1978. Some data that describe each patient's
history, therapy during hospitalization, and complications were
retained to provide a fairly complete clinical description, even
though these variables have not been related to prognosis. Between
1977 and 1980, data on patients from Vancouver General Hospital were
entered into the previous data base. In 1978, several San Diego
hospitals besides what was then called University Hospital—the San
Diego Veterans Administration Mecical Center, the United States
Naval Regional Medical Center, and Mercy Hospital—joined the study.
Mercy Hospital was dropped in 1981 because of its low recruitment
of patients.

Not all patients with confirmed diagnoses of MI agreed to have
their data entered into the data base. At times, personnel were not
available to collect requisite data. However, over half of all
eligible patients from all centers (except Mercy Hospital) did agree,
and their data have been used. Examination of coronary care unit
(CCU) registries shows that in-hospital mortality for patients
entered and not entered are nearly equal. Thus, we believe that the
results being reported are not affected by a selection bias regarding entry to the study.

Table 1 summarizes the patients in our analyses. The previous
and present data bases yield a combined data base of 652 patients
(91 deaths) from San Diego and 842 patients (100 deaths) from Vancouver.
Of the 652 patients from San Diego, we obtained complete data for the
restricted list of 34 variables for 499 patients (70 deaths); correspondingly, we had complete data for 646 Vancouver patients (69 deaths).
By location for the combined data base, $\hat{\pi}(2)$ (the empirical prior probability of death within one year) varies from a low of .11 for the complete Vancouver data to .14 for both San Diego data sets. It was

Table 1. Sources of Data

	Time Period	All Patients			Complete Data		
		Survivors	Deaths	Total	Survivors	Deaths	Total
		Previous Data Base					
UCSDMC	1969-1978	197	32	229	109	22	131
VGH	1977-1980	487	72	559	346	46	392
		Present Data Base					
UCSDMC	1978-1983	82	19	101	68	17	85
MH	1978-1980	30	3	33	26	3	29
SDVAH	1978-1983	95	21	116	77	16	93
USNRMC	1978-1983	158	15	173	149	12	161
		365	58	423	320	48	368
VGH	1980-1983	255	28	283	231	23	254
		Combined Data Base					
San Diego		562	90	652	429	70	499
Vancouver		742	100	842	577	69	646
		1304	190	1494	1006	139	1145

Key: UCSDMC = UC San Diego Medical Center; VGH = Vancouver General
Hospital; MH = Mercy Hospital; SDVAH = San Diego Veterans Administration Hospital; USNRMC = US Naval Regional Medical Center.

noted that by various criteria these differences are insignificant.
Within the San Diego institutions and the present data base, the story
is quite different. For all patients, the values of $\hat{\pi}(2)$ vary from
.09 for the United States Naval Regional Medical Center to .19 for the
UC San Diego Medical Center. The three degrees of freedom chi-square
statistic for the comparison of proportions is 8.41 ($.025 < p < .050$).
For those patients with complete data, the low and high values for
$\hat{\pi}(2)$ are .08 and .20, and the corresponding chi-square statistic is
9.67 ($.010 < p < .025$). The variation in mortality among San Diego
institutions is discussed later in regard to results, and also in
regard to our plans for combining procedures and competing with the
prognoses of physicians. Finally, note that both San Diego and Vancouver show insignificant differences in mortality in old and new data
bases, for all patients and for those with complete data. No one
degree of freedom chi-square statistic for a comparison exceeds 1.2.

3. VARIABLE SELECTION AND TECHNOLOGIES

The variables of the new data base were screened extensively with respect to their prediction of mortality at six and twelve months. Recursive partitioning, by its very nature, and logistic regression, as we generally employ it, use synergistic effects of variables, and our process of variable selection might have utilized procedures that employ pairs or triples of variables. However, our rules for inclusion (attained significance level less than .10 for some suitable comparison of class 1 and class 2 patients) were so generous, and our experience in prognosis sufficiently substantial by now, that we are confident that all useful variables from among those available were included. The final list of variables has 34 entries, including features extracted from medical histories, physical examinations, laboratory findings, chest x-rays, and electrocardiograms. For definitions and details see Madsen et al. (1984).

The software for recursive partitioning was an adaptation of an early version of the CART* program described by Breiman et al. (1984). The Gini splitting criterion was used with a minimum node content of five observations. All 34 retained variables were used in all analyses. For each data set and location, from among the nested sequence of optimally pruned trees, we report (in the next section) results on the tree that gave the smallest overall misclassification cost for the corresponding test sample from the other location. The software for stepwise linear discrimination (SLD) was that written by Jennrich and Sampson (1981) and described in *BMDP Statistical Software 1981*.[†] Again, all 34 variables were used in all analyses. A variable was not entered, or was removed, if its F-ratio was less than 4.0. No further trimming of the list of 34 variables was used in analyses with these procedures, because both CART (implicitly) and SLD (explicitly) have their own built-in variable selection schemes. Our software for logistic regression was DuMouchel's DREG program (see DuMouchel, 1981). This program, as we used it, has no automated variable selection. However, it does report t-like statistics for each

---

*CART is a trademark of California Statistical Software, Inc.
†BMDP is a trademark of BMDP Statistical Software, Inc.

variable based on the asymptotic distribution of the maximum–likelihood estimates of the coefficients. So a preliminary analysis was
done on each complete, combined data set by location, and only variables for which the t-like statistic exceeded 1.6 in absolute value
were retained for subsequent analyses by logistic regression. (This
admittedly ad hoc procedure has worked well in the past and seemed
satisfactory here.) The union of the two sets of variables retained
includes only 11 variables.

Both in Gilpin et al. (1983, p. 75) and in Breiman et al. (1984,
p. 181) interactions suggested by successive CART splits were
employed in logistic regression. That approach was attempted in the
present study, too. However, when a classification rule, including
interactions, was developed on a learning sample from one location
and applied to a test sample from the other, no improvement in test
sample classification resulted, so the idea was discarded. In particular, logistic regression with interactions was not cross–validated
within locations.

Though the results of our analyses will be reported in the next
section, we note here that only CART, through its use of surrogate
variables (see Breiman et al., 1984, pp. 140–146), permitted us to
analyze the full data sets, including those patients missing some
data on the retained 34 variables. No imputation schemes were
employed with SLD or with DREG.

Three sets of preliminary analyses using SLD were carried out
on patients in the present data base. The first used historical
variables and data obtained throughout the hospital stay and at discharge. The second used data from the history and throughout the
hospital stay up to but not including discharge studies. The third
analysis used historical data and clinical variables gathered only
during the first 24 hours after admission. The prediction of outcome
up to one year was about as accurate with the third (restricted) set
as with the first two (see Madsen et al., 1984). Therefore, we
enlarged the population for the study beyond the present data base
by including patients from the previous data base, for whom only
early data are available. In fact, these earlier data were collected
only during the CCU period (first 3 to 5 days in the hospital).

Table 2 lists variables by their "importance" for each location. It is clear from the footnotes of the table that the notions of importance differ by technology. However, age and maximum blood urea nitrogen (a measure of kidney function and indirectly of cardiac function) are important prognostic factors by any criterion, and likewise for a history of previous MI. Note that previous angina was important for all technologies for Vancouver, but not for San Diego. This might be related to the fact that generally the Vancouver patients were older but less severely ill. They are significantly older for both class 1 and class 2 subjects, have had significantly less congestive heart failure for both groups (both chi-square statistics exceed 10), have had fewer previous MIs in the class 1 group (chi-square 8.89, $p < .01$), and have significantly lower maximum heart rates for class 1 patients. Table 3 summarizes average values and standard deviations for continuous variables by location and class, and reports frequencies for dichotomous variables.

Maximum creatine kinase was selected as a splitting variable by CART in one analysis of the San Diego data, but it produced a counterintuitive split, as did another variable, *recent history of angina*. (By *counterintuitive*, we mean that well-established relationships between values of these variables and class membership were reversed in the resubstitution application of the tree to its San Diego learning set.) Closer examination of these variables revealed some problems concerning nonstandardization of measurement and definition between the previous and present data bases. CART's use in detecting faulty data is one of its most salutary aspects; see, for example, Section 6.4 of Breiman et al. (1984).

One fact that is apparent from Tables 2, 3, and 4 and Figure 1 is that there is no unique set of variables that contains all the information regarding classification. There seem to be substantial redundancies.

Table 2. Variables Important in Each Prognostic Scheme: Complete Data

Variables	Stepwise Linear Discrimination (SLD)*		Logistic Regression (DREG)†		Recursive Partitioning (CART)‡	
	SD	VANC	SD	VANC	SD	VANC
Age	6.6	17.7	2.75	2.91	419 (2)	298 (2,4)
Previous myocardial infarction	10.4	5.8	1.84	2.08		149
Previous hypertension						119 (5)
Previous angina		26.7		3.18		578 (1)
Maximum heart rate					189	256 (5,6)
Minimum systolic blood pressure					146	170
Maximum respiration rate						167 (3)
Maximum blood urea nitrogen	43.3	7.8	3.67	2.00	153 (1)	182 (3,5,6)
Abnormal apex		5.1		1.81		(3)
Rales above the scapulae	4.6					
Systolic murmur		4.7			146	160
Maximum PR interval						107 (2)
Maximum QRS duration		4.9				116
Location of myocardial infarction		17.8	2.61			
Sinus tachycardia	7.8					(4)
Sinus brachycardia	5.3					
Old bundle branch block					(3)	
More than 6 ventricular premature beats/minute	4.5					

*For SLD, the F-ratio to enter or remove the variable is shown. A variable was not entered, or was removed, if its F-ratio was less than 4.0.

†For DREG, the ratio of the coefficient for the variable to its standard error is shown. Variables with a ratio less than 1.8 are not included.

‡For CART, the relative coordinate importance shown is the original times 10^4; all values that exceed 100 are given. Also, the level of any split on this variable is shown in parentheses.

Table 3. Important Prognostic Variables by Location and Class: Complete Data*

	Vancouver			San Diego		
	Survivors Class 1 (577)	Deaths Class 2 (69)	P	Survivors Class 1 (429)	Deaths Class 2 (70)	P
Age	63 ± 12	69 ± 11	< .001	58 ± 12	65 ± 11	< .001
Previous MI	.20	.42	< .05	.28	.54	< .001
Previous hypertension	.30	.38		.42	.46	
Previous angina	.33	.64	< .001	.39	.62	< .001
Maximum heart rate	84 ± 18	90 ± 17	< .005	89 ± 17	94 ± 21	< .002
Minimum systolic blood pressure	121 ± 20	121 ± 20		120 ± 21	118 ± 24	
Maximum respiration rate	21 ± 4	23 ± 5		20 ± 4	21 ± 4	< .10
Maximum blood urea nitrogen	19 ± 7	23 ± 8	< .001	18 ± 7	25 ± 15	< .001
Rales above the scapulae	.05	.07		.05	.08	
Maximum PR interval	.173 ± .032	.184 ± .039	< .01	.175 ± .030	.182 ± .039	< .10
Maximum QRS duration	.080 ± .015	.089 ± .024	< .001	.087 ± .020	.094 ± .029	< .02

*Continuous variables are given as averages with standard deviations. Frequencies of occurrence are noted for dichotomous variables. Attained significance levels are for comparisons of classes within locations; one degree of freedom chi-squared statistics or normal approximations to the distributions of pooled t or Behrens-Fisher statistics were used in the calculations.

4. ERROR RATES, MISCLASSIFICATION COSTS, AND PLANS FOR THE FUTURE

The topics of this section are the error rates and overall costs of
misclassification of the three techniques. They are evaluated by
resubstitution, 10-fold cross-validation, and test sample estimation.
Table 4 is one summary. For each location, procedure, and class, the
table gives the error rates for the associated approximate Bayes rule
applied to the specified data, both for resubstitution and for the
test sample from the other location. The results of 10-fold (stratified) cross-validation were obtained as described in the first section and in Breiman et al. (1984). The overall misclassification
cost R is an average of the two error rates, with the ith class rate
weighted by $\pi(i)C(j|i)$, $j \neq i$; of course, with our choice in Section 1
the weights are .5 each, and the no data optimal rule has misclassification cost .5. (In general, a no data optimal rule assigns any
observation to a class j, which minimizes $\sum_i \pi(i)C(j|i)$; see Breiman
et al., 1984, p. 178.) Sample standard errors for the cross-
validation are part of Table 6. The three San Diego complete data
decision rules are given in Figure 1. Note their complexities and
also that all splits and signs of coefficients are in keeping with
intuition except the coefficient of SB in logistic regression.

How to interpret differences in cross-validated, let alone test-
sample, estimates of misclassification cost is unclear. Simple
t-like comparisons based on normal theory approximations seem to be
prone to excessive Type I errors (see Breiman et al., Section 11.5).
If that is in fact the case, then all cross-validated estimates of
differences in R by technology within location are at the noise level.
However, there seem to be real differences in procedures within
classes.

Among comparisons of technologies applied to the Vancouver data,
DREG does better than other procedures for resubstitution, cross-
validation, and test-sample estimation for the survivors. For the
deaths, CART all cata seems best by cross-validation, but only at
the noise level over SLD. (Naturally, a procedure that does as well
for all data as another that applies only to complete data is to be

Table 4. Summary of Error Rates and Misclassification Costs R, 10-Fold Stratified Cross-Validation

	Vancouver			San Diego		
	R	Survivors Class 1	Deaths Class 2	R	Survivors Class 1	Deaths Class 2
Stepwise linear discrimination						
Resubstitution	.26	.25	.28	.29	.21	.36
Cross-validation	.29	.24	.35	.40	.26	.54
Applied to other population	.37	.26	.47	.32	.21	.44
Logistic regression						
Resubstitution	.31	.15	.48	.34	.22	.47
Cross-validation	.32	.12	.52	.38	.24	.51
Applied to other population	.34	.15	.53	.34	.18	.49
CART, complete data	16 terminal nodes			4 terminal nodes		
Resubstitution	.15	.18	.13	.33	.38	.29
Cross-validation	.36	.33	.41	.37	.32	.43
Applied to other population	.39	.40	.39	.39	.27	.51
CART, all data	5 terminal nodes			3 terminal nodes		
Resubstitution	.28	.39	.17	.33	.25	.42
Cross-validation	.32	.33	.31	.36	.28	.43
Applied to other population	.39	.35	.42	.34	.19	.49

preferred unless other considerations mitigate against it.) Overall, SLD does better for cross-validation and DREG for test-sample estimation, both at the noise level. For the San Diego data, DREG again outperforms the others for the survivors, whereas comparisons of the deaths are unclear. Overall, SLD is the test sample winner, but only at the noise level over DREG and all data CART. Why SLD works relatively well on data for which its underlying assumptions are obviously false is not completely understood. We are far from the first to observe this. See, for example, Henning et al. (1979) and Lachenbruch (1982).

CART Tree

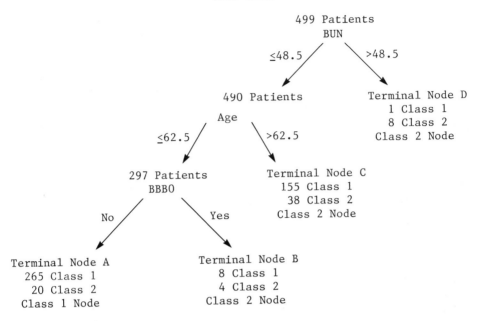

Stepwise Linear Discriminant

$$-7533 + 32(AGE) + 750(HXMI) + 82(BUN) + 935(SM) + 1246(VPBGT6)$$
$$+ 2700(SB) + 1686(BBBO);$$

if ≥ 0, classify as Class 2; otherwise classify as Class 1.

Logistic Regression

$$-7319 + 338(AGE) + 5645(HXMI) + 1198(HXANG) + 50(BUN) + 503(APEX)$$
$$+ 2855(QRS) + 76(LOCCAT) - 1075(SB);$$

if ≥ 0, classify as Class 2; otherwise classify as Class 1.

Figure 1. Three decision rules based on San Diego patients with complete data. HXMI means *previous MI*; HXANG means *previous angina*; APEX means *abnormal apex*; SM means *systolic murmur*; VPBGT6 means *more than 6 ventricular premature beats per minute*; SB means *sinus brachy-cardia*; BBBO means *old bundle branch block*. In each instance, *yes* is indicated by a higher value than is *no*. LOCCAT refers to the location of the infarction. High values go with nontransmural or indeterminate, both of which are associated with poorer prognosis than is transmural. BUN means *maximum blood urea nitrogen* during the 24-hour period follow-ing admission. QRS means *maximum QRS duration*.

Though Table 4 renders comparisons of technologies unclear
(except that DREG does better for survivors and worse for deaths
than the others), what is clear is our inability to reduce misclassification costs on test sample data below about 60% of those of the
no data Bayes rule, if data within locations are pooled. When
San Diego test sample data are broken down by location, an interesting, if confusing, picture emerges (see Table 5). Obviously the
Vancouver rules do spectacularly well on the 29 patients from Mercy
Hospital. However, the three key rows of Table 5 are those corresponding to the UCSD present data base, the VA Hospital, and the US
Naval Regional Medical Center. Recall that $\hat{\pi}(2)$ for these institutions
varies from a high of .20 for UCSD to a low of .08 for the generally
healthier and younger Naval Medical Center group. So, for some
institutions, either the πs or the Cs are "wrong." Note that for
the UCSD present data base, DREG's error rate for the survivors is
only 16%, and that all three technologies have 24% error rates
for the deaths (but by no means the same 24% of the patients!). CART
does well on the VA survivors, whereas SLD does much better on VA
deaths. Naval Medical Center deaths are difficult to find for all
technologies, but somewhat easier for SLD.

One fact that seems to emerge from our analyses is that the
different procedures do differentially well in different regions of
the feature space. Perhaps combining them in some fashion is advis-
able. An obvious approach is simply to average their estimation of
the conditional *densities* $\hat{f}_j$ or their posterior probabilities of
class membership. Especially with CART, there are questions of bias
correction that must be addressed if the averaging is implemented.
Alternatively, the approach should be that of class probability
estimation. In a pilot study (see Gilpin et al., 1983, p. 78), we
found that the crude combining by voting of the estimated Bayes rules
generally did well.

Against our computer-derived prognoses, perhaps the real bench
mark should not be the no data misclassification cost, but rather the
prognoses of expert physicians. CART fared well in a related
comparison of the diagnosis (of MI) by emergency-room physicians for

Table 5. Misclassification Costs R for Vancouver Schemes by San Diego Institution: Complete Data

	Stepwise Linear Discrimination			Logistic Regression			CART		
	R	Survivors Class 1	Deaths Class 2	R	Survivors Class 1	Deaths Class 2	R	Survivors Class 1	Deaths Class 2
Overall	.37	.26	.47	.34	.15	.53	.39	.40	.39
UCSD (old data base)	.47	.35	.59	.37	.20	.55	.39	.46	.32
UCSD (present data base)	.27	.29	.24	.20	.16	.24	.24	.25	.24
Veterans Administration Hospital	.41	.38	.44	.44	.25	.63	.50	.12	.88
Mercy Hospital	.04	.08	0	.06	.12	0	.10	.19	0
US Naval Regional Medical Center	.45	.15	.75	.49	.07	.92	.49	.05	.92

patients whose chief complaint was acute chest pain (see Goldman et al., 1982). As a future task of the cited SCOR project, the individual technologies plus one other nonparametric approach, and some combinations of technologies, will be compared with the prognoses of three experts. For a contrasting of data-driven diagnosis and the approach of so-called artificial intelligence, see Section 6.5 of Breiman et al. (1984).

5. STRATIFICATION IN CROSS-VALIDATION

The question of stratifying cross-validation is discussed in Breiman et al. (1984) and in the first section of this paper. In general, these investigations—largely by simulation—suggest that it is never worse to stratify, and typically it is better. Tenfold cross-validation, stratified and not, was carried out for all technologies. The criterion for comparison was the accuracy in predicting test sample misclassification costs. Results are summarized in Table 6. They are not informative, because the unstratified approach produced more accurate estimates for all four comparisons for Vancouver data, whereas just the opposite is true for San Diego. Moreover, comparisons seem to be at the noise level. We believe that our choice of equal values of $\hat{\pi}(i)C(j|i)$ made our task particularly difficult.

The remainder of this section is a small theoretical contribution to the study of cross-validation of tree-structured methods. Its results suggest that for recursive partitioning the stratified approach to cross-validation is preferable.

To fix notation, assume that the learning sample $\mathscr{L}$ consists of (X_1, Y_1), ..., (X_N, Y_N), which are independent and identically distributed, each distributed as (X, Y). Assume further that $X \in \mathbb{R}^d$ and has coordinatewise continuous marginal distributions; so without loss of generality assume that all X_i are distinct. Suppose that Y denotes class membership, say 1 and 2, in a two-class problem, and that X and Y are in fact independent.

Table 6. Stratified versus Unstratified 10-Fold Cross-Validation*

	Stepwise Linear Discrimination			Logistic Regression			CART Complete Data			CART All Data		
	R	Class 1 Survivors	Class 2 Deaths	R	Class 1 Survivors	Class 2 Deaths	R	Class 1 Survivors	Class 2 Deaths	R	Class 1 Survivors	Class 2 Deaths
Vancouver												
Stratified	.29 (.03)	.24 (.01)	.35 (.05)	.32 (.04)	.12 (.02)	.52 (.04)	.36 (.03)	.33 (.04)	.41 (.06)	.32 (.02)	.33 (.02)	.31 (.04)
Unstratified	.33 (.03)	.26 (.02)	.40 (.06)	.33 (.02)	.15 (.01)	.50 (.04)	.34 (.03)	.27 (.03)	.41 (.06)	.36 (.03)	.33 (.04)	.41 (.08)
San Diego												
Stratified	.40 (.02)	.26 (.03)	.54 (.04)	.38 (.02)	.24 (.03)	.51 (.04)	.37 (.02)	.32 (.02)	.43 (.04)	.36 (.02)	.28 (.03)	.43 (.07)
Unstratified	.41 (.03)	.26 (.02)	.56 (.06)	.39 (.03)	.18 (.03)	.59 (.06)	.33 (.03)	.22 (.03)	.44 (.08)	.41 (.02)	.29 (.02)	.52 (.04)

*Unstratified cross-validation produced more accurate estimates of test sample misclassification cost (data from the other location) for all four comparisons for Vancouver; just the opposite is true for San Diego. Parenthetical numbers are sample standard errors.

A tree T is grown according to any splitting rule that, like Anderson's rules based on statistically equivalent blocks (1966), partitions a node t based only on $\{X_i : X_i \in t\}$, with the proviso that each terminal node has exactly one member of $\mathscr{L}$. If X (for which Y is unknown) belongs to terminal node t, it is classified by rule $d = d(X)$ as Y_i, where $(X_i, Y_i) \in \mathscr{L}$ and $X_i \in t$. Suppose that misclassification costs are 0 for a correct classification and 1 for a mistake and that the known prior probabilities are: $P(Y = 1) = \alpha$, $P(Y = 2) = 1 - \alpha$ for some $0 < \alpha < 1$. It is straightforward to see that the overall misclassification cost R^* of the described classification rule, that is, its expected cost of misclassification, is $2\alpha(1 - \alpha)$. This is the Bayes risk of a single nearest neighbor rule.

Our comparison of stratified and unstratified *cross-validation* (really *test sample estimation*) is limited to the simplest possible case, the estimation of R^* by a single split-sample scheme.

For expository convenience we take N to be even. Denote by N_j, $j = 1, 2$, the number of Y_i, $i \leqslant N$, for which $Y_i = j$; N_1 has a binomial distribution with parameters N and α $[N_1 \sim B(N, \alpha)]$, while $N_2 = N - N_1 \sim B(N, 1 - \alpha)$. It is assumed that $\mathscr{L}$ is divided randomly into two subsamples, $\mathscr{L}_1$ and $\mathscr{L}_2$, each of size $N/2$. An algorithm meeting the specifications of the previous paragraph is applied to $\mathscr{L}_1$, and the data from $\mathscr{L}_2$ are classified by the resulting tree. For $j = 1, 2$ and $k = 1, 2$, we denote by $N_j^{(k)}$ the number of $Y_i \in \mathscr{L}_k$ for which $Y_i = j$, and by R the misclassification cost of $d(X)$ estimated from the test sample $\mathscr{L}_2$. We shall compute the mean-squared error $E\{[R - 2\alpha(1 - \alpha)]^2\}$ under two different schemes. The first corresponds to the unstratified case in which $N_1^{(2)}$ given N_1 and N_2 has a hypergeometric distribution with (in an obvious notation) parameters $N/2$, N_1, and N_2. The second corresponds to the stratified case in which $N_1^{(2)}$ is the integer part of $N_1/2$. The limiting inverse ratio of these two mean-squared errors, as N grows without bound, is what we define to be the *efficiency* of unstratified to stratified test sample estimation.

Let $N_{ij}^{(2)}$, for $i = 1, 2$, be numbers of class j cases in $\mathscr{L}_2$ classified as class i; R can be expressed

$$R = \alpha[N^{(2)}_{21}/N^{(2)}_1] + (1 - \alpha)[N^{(2)}_{12}/N^{(2)}_2], \tag{5.1}$$

where $0/0$ is taken to be 0.

We evaluate the mean-squared error of R as the sum of its variance and its squared bias. Let

$$A = \{N^{(1)}_1 = 0\} \cup \{N^{(1)}_1 = N/2\} \cup \{N^{(1)}_1 = N_1\}.$$

It is straightforward to compute that $N^{(1)}_1 \sim B(N/2, \alpha)$ and that $P(A) = O(\alpha^{N/2} + (1 - \alpha)^{N/2})$. On A^c, the conditional distribution of R given $\{N^{(k)}_j\}$ is that of

$$\alpha(W/N^{(2)}_1) + (1 - \alpha)(Z/N^{(2)}_2), \tag{5.2}$$

where W and Z are independent: $W \sim B(N^{(2)}_1, N^{(1)}_2/(N/2))$, $Z \sim B(N^{(2)}_2, N^{(1)}_1/(N/2))$. From these observations it follows that on A^c, $E(R|\{N^{(k)}_j\}) = 2\alpha[N^{(1)}_2/N] + 2(1 - \alpha)[N^{(1)}_1/N]$. Since $E\{2N^{(1)}_1/N\} = \alpha$ and $E\{2N^{(1)}_2/N\} = 1 - \alpha$, $E\{R\} = E\{E(R|\{N^{(k)}_j\})\} = 2\alpha(1 - \alpha) + O(\alpha^{N/2} + (1 - \alpha)^{N/2})$; and the squared bias of R is $O(\alpha^{N/2} + (1 - \alpha)^{N/2})$. Now

$$\text{Var}\{R\} = \text{Var}\{E(R|\{N^{(k)}_j\})\} + E\{\text{Var}(R|\{N^{(k)}_j\})\}; \tag{5.3}$$

and on A^c, $E(R|\{N^{(k)}_j\})$ can be rewritten $(4\alpha - 2)(N^{(1)}_2/N) + (1 - \alpha)$. Therefore, to within $O(\alpha^{N/2} + (1 - \alpha)^{N/2})$, the first term of (5.3) is $(4\alpha - 2)^2\text{Var}\{N^{(1)}_2\}/N^2 = (4\alpha - 2)^2 N\alpha(1 - \alpha)/2N^2 = 2\alpha(1 - \alpha)(2\alpha - 1)^2/N$. In particular,

$$\lim_{N \to \infty} N \text{Var}\{E(R|\{N^{(k)}_j\})\} = 2\alpha(1 - \alpha)(2\alpha - 1)^2. \tag{5.4}$$

On A^c, $\text{Var}(R|\{N^{(k)}_j\})$ can be computed from (5.2). If the resulting expression is multiplied by N it can be expressed as

$$V(N) = \frac{4\alpha^2}{[N^{(2)}_1/N]}\left[\frac{N^{(1)}_2}{N}\right]\left[\frac{N^{(1)}_1}{N}\right] + \frac{4(1 - \alpha)^2}{[N^{(2)}_2/N]}\left[\frac{N^{(1)}_2}{N}\right]\left[\frac{N^{(1)}_1}{N}\right].$$

So, with indicator variable notation,

$$NE\{\text{Var}(R|\{N^{(k)}_j\})\} = E\{V(N)I_{A^c}\} + O(N\alpha^{N/2} + N(1 - \alpha)^{N/2}).$$

Let

$$B = \left[\left|\frac{N^{(2)}_1}{N} - \frac{\alpha}{2}\right| > \frac{\alpha}{4}\right] \cup \left[\left|\frac{N^{(2)}_1}{N} - \frac{1 - \alpha}{2}\right| > \frac{1 - \alpha}{4}\right].$$

On $A^c \cap B$, $V(N) \leqslant N$; also, $P(A^c \cap B) = 0(N^{-2})$. On $A^c \cap B^c$, $V(N)$ is bounded uniformly in α by 16. From what has been shown, the law of large numbers, and the dominated convergence theorem it follows that

$$\lim_{N \to \infty} NE\{\mathrm{Var}(R|\{N_j^{(k)}\})\} = 2\alpha(1 - \alpha). \tag{5.5}$$

In view of (5.3), (5.4), and the remark that precedes (5.3), when test sample estimation is not stratified,

$$\lim_{N \to \infty} NE\{(R - 2\alpha(1 - \alpha))^2\} = \alpha(1 - \alpha)(8\alpha^2 - 8\alpha + 4). \tag{5.6}$$

In the stratified case, R can be expressed as in (5.1); but as was mentioned, the conditional distribution of $N_1^{(2)}$ given N_1 is degenerate. Arguments like those that led to (5.6) show that in the stratified case

$$\lim_{N \to \infty} NE\{(R - 2\alpha(1 - \alpha))\}^2 = \alpha(1 - \alpha)(4\alpha^2 - 4\alpha + 3). \tag{5.7}$$

From (5.6) and (5.7) it follows that the required

$$\text{efficiency} = (4\alpha^2 - 4\alpha + 3)/(8\alpha^2 - 8\alpha + 4). \tag{5.8}$$

Note that (5.8) is, as it clearly must be, symmetric about 1/2; it *never exceeds* 1, which is its value at 1/2; it is unimodal; and it has limiting value 3/4 at $\alpha = 0$ and $\alpha = 1$. Efficiency is discontinuous at $\alpha = 0$ and $\alpha = 1$, because its value at each is 1.

REFERENCES

Anderson, T. W. 1966. Some nonparametric multivariate procedures based on statistically equivalent blocks. In *Multivariate Analysis*, ed. P. R. Krishnaiah, pp. 5–27. New York: Academic Press.

Battler, A., Karliner, J. S., Higgins, C. D., Slutsky, R., Gilpin, E. A., Froelicher, V. F., and Ross, J., Jr. 1980. The initial chest x-ray in acute myocardial infarction: Prediction of early and late mortality and survival. *Circulation 61*, 1004–1009.

Bigger, J. T., Jr., Fleiss, J. L., Kleiger, R., Miller, J. P., Rolnitzky, L. M., and Multicenter Post-Infarction Research Group, 1984. The relationships among ventricular arrhythmias, left ventricular dysfunction, and mortality in the 2 years after myocardial infarction. *Circulation 69*, 250–258.

Breiman, L., Friedman, J. H., Olshen, R. A., and Stone, C. J. 1984. *Classification and Regression Trees*. Belmont, Calif.: Wadsworth International.

DuMouchel, W. H. 1981. Documentation for DREG. Technical Report. Cambridge: Massachusetts Institute of Technology.

Gilpin, E., Olshen, R., Henning, H., and Ross, J., Jr. 1983. Risk prediction after myocardial infarction: Comparison of three multivariate methodologies. *Cardiology 70*, 73–84.

Goldman, L., Weinberg, M., Weisberg, M., Olshen, R., Cook, F., Sargent, R. K., Lamas, G. A., Dennis, C., Deckelbaum, L., Fineberg, H., Stiratelli, R., and the Medical Housestaffs at Yale-New Haven Hospital and Brigham and Women's Hospital. 1982. A computer-derived protocol to aid in the diagnosis of emergency room patients with acute chest pain. *New England Journal of Medicine 307*, 588–596.

Helmers, C. 1973. Short and long term prognostic indices in acute myocardial infarction: A study of 606 patients initially treated in a coronary care unit. *Acta Medica Scandinavia (Suppl.) 555*, 7–26.

Henning, H., Gilpin, E. A., Covell, J. W., Swan, E. A., O'Rourke, R. A., and Ross, J., Jr. 1979. Prognosis after acute myocardial infarction: A multivariate analysis of mortality and survival. *Circulation 59*, 1134–1136.

Jennrich, R., and Sampson, P. 1981. Stepwise discriminant analysis. In *BMDP Statistical Software 1981*, ed. W. J. Dixon. Berkeley: University of California Press.

Lachenbruch, P. A. 1982. Robustness of discriminant functions. *SUGI-SAS Group Proceedings 7*, 626–632.

Luria, M. H., Knoke, J. D., Wachs, J. S., and Luria, M. A. 1979. Survival after recovery from acute myocardial infarction: Two and five year prognostic indices. *American Journal of Medicine 67*, 7–14.

Madsen, E. B., Gilpin, E., and Henning, H. 1983. Evaluation of prognosis one year after myocardial infarction. *Journal of the American College of Cardiology 1*, 985–993.

Madsen, E. B., Gilpin, E. A., and Henning H. 1984. Short-term prognosis in acute myocardial infarction: Evaluation by different prediction methods. *American Heart Journal 107*(6), 1241–1251.

Madsen, E. B., Gilpin, E. A., Ahnve, S., Henning, H., LeWinter, M., Ceretto, W., Joswig, W., Collins, D., Pitt, W., and Ross, J., Jr., 1984. Prediction of later mortality after myocardial infarction from variables obtained at different times during hospitalization. *American Journal of Cardiology 53*, 47–54.

Multicenter Postinfarction Research Group. 1983. Risk stratification and survival after myocardial infarction. *The New England Journal of Medicine 309*, 331–336.

Norris, R. M., Brandt, P. W. T., Caughey, D. E., Lee, A. J., and Scott, P. J. 1969. A new coronary prognostic index. *Lancet 1*, 274–278.

Peel, A. A. F., Semple, T., Wang, I., Lancaster, W. M., and Dall, J. L. G. 1962. A coronary prognostic index for grading the severity of infarction. *British Heart Journal 24*, 745–760.

Ryan, W., Karliner, J. S., Gilpin, E. A., Covell, J. W., De Luca, M., and Ross, J., Jr. 1981. The creatine kinase (CK) curve area and peak CK after acute myocardial infarction: Usefulness and limitations. *American Heart Journal 101*, 162–168.

IDENTIFICATION OF AGGREGATED MARKOVIAN MODELS: APPLICATION TO THE NICOTINIC ACETYLCHOLINE RECEPTOR

DONALD R. FREDKIN, MAURICIO MONTAL, and JOHN A. RICE
University of California, San Diego

1. INTRODUCTION

We consider a finite-state Markov process that is aggregated into two classes of states—*open* states and *closed* states, say—and discuss what can be learned about the original underlying process from the aggregated process. In particular, we are interested in determining conditions under which the underlying process is or is not identifiable and under which our observation of the aggregated process allows discrimination between different models. Such questions arise in the analysis of experimental data gathered in the study of the nicotinic acetylcholine receptor.

In Section 2 we give a brief introduction to biochemical and biophysical knowledge about the nicotinic acetylcholine receptor.

We acknowledge the collaboration of Dr. Pedro Labarca in the experimental work and thank him for many fruitful discussions. This research was supported by grants from the National Institutes of Health (EY-02084) and the Department of the Army Medical Research (17-82-C221) to M. Montal. J. Rice was partially supported by grants NSF MCS-7901800 and NIH 5ROICA16666-05.

In Section 3 we relate the one-dimensional distributions of the aggregated process to the parameters of the underlying process and consider what can be learned by varying the concentration of acetylcholine. In Section 4 we consider the higher dimensional distributions of the aggregated process and show that they are determined by two two-dimensional distributions. Consequently, questions of identifiability can be reduced to the study of the two-dimensional distributions only. Covariance functions for the aggregated process are derived in Section 5 and are shown to give information about the degree of complexity of the underlying process. Section 6 contains a summary of experimental implications and some final remarks.

2. BIOCHEMICAL AND BIOPHYSICAL BACKGROUND

One of the most studied synapses in neurobiology is the chemical cholinergic synapse. This synapse uses acetylcholine as the neurotransmitter. Transmission of the nerve impulse proceeds from the arrival of the impulse at the presynaptic cell and leads to the release of acetylcholine into a gap between two connected cells. Acetylcholine diffuses to the postsynaptic membrane and binds to specific receptor molecules. Binding of acetylcholine to its receptor produces a depolarization of the postsynaptic membrane. The impulse is then propagated along the electrically excitable membrane of the second nerve cell. This depolarization of the postsynaptic membrane arises from the transient opening of cation selective channels (for recent reviews see Karlin, 1980; Changeux, 1981; Ahnolt et al., 1983).

The acetylcholine receptor (AChR) is the best characterized neurotransmitter receptor, both electrophysiologically and biochemically, and can therefore serve as a model for understanding the structure and function of other neurotransmitter receptors. The AChR is a membrane protein that spans the membrane, protruding about 50 Å from the extracellular surface and about 15 Å from the intracellular surface; it has a diameter of approximately 85 Å. Electron microscopy suggests that the receptor molecule is shaped like a mushroom, and the ion channel is thought to be approximately 6.5 Å in diameter (Dwyer et al., 1980; Kistler et al., 1982; Zingsheim et al., 1982).

The availability of specific toxins that act on the AChR and a rich tissue source for its biochemical isolation and purification have facilitated the detailed characterization of its structure. Biochemical studies of the purified protein indicate that the receptor is a noncovalent complex of four distinct integral membrane glyco-protein subunits with subunit stoichiometry of $\alpha_2\beta\gamma\delta$ and a molecular weight of approximately 270,000 (Reynolds and Karlin, 1978; Lind-strom et al., 1979; Raftery et al., 1980; Noda et al., 1982; Claudio et al., 1983). The two smallest subunits (40,000 daltons) are respon-sible for the binding of acetylcholine (cf. Karlin, 1980). The cation channel is an integral component of the receptor, but it is not known which subunits form the channel. Recently, a number of laboratories have succeeded in cloning the genes that code for the various receptor subunits, and the entire amino acid sequence of the receptor subunits has been elucidated (Noda et al., 1982, 1983; Claudio et al., 1983; Devilliers-Theiry et al., 1983). Figure 1 illustrates some structural features of the AChR molecule.

This detailed knowledge of the structure of the AChR has been accompanied by a detailed characterization of the biophysical events associated with the opening and closing of individual receptor chan-nels. The dynamics of opening and closing have now been character-ized through patch recording in the native membrane (Neher and Sakmann, 1986; Sakmann et al., 1980; Colquhoun and Sakmann, 1981; Jackson et al., 1983); and more recently, the purified protein has been reconstituted into lipid bilayers in which the current flowing through individual channels can be recorded (Nelson et al., 1980; Boheim et al., 1981; Montal et al., 1983; Tank et al., 1983; see also Schindler and Quast, 1980). These studies demonstrated that the $\alpha_2\beta\gamma\delta$ subunit structure of the receptor contains both the agonist binding sites and the cation channel that they regulate. Therefore, the purified AChR that is reconstituted in planar lipid bilayers provides a vehicle for combining the biophysical analysis of the

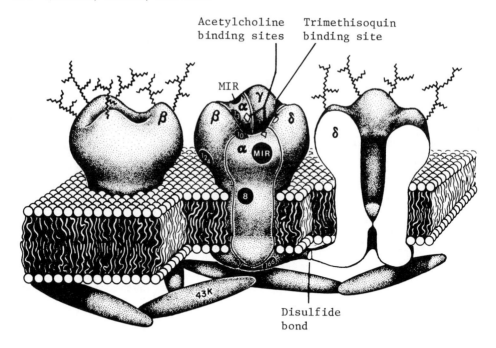

Acetylcholine   Trimethisoquin
binding sites   binding site

MIR

Disulfide
bond

Figure 1. A structural model of the acetylcholine receptor. (From Anholt et al., 1983; modified from Kistler et al., 1962). All the subunits are glycopeptides (arborizations protruding from the protein mass) and traverse the membrane. Aht acetylcholine binding sites are exposed at the extracellular surface. The width of the lipid bilayer membrane is approximately 50 Å. The section through the receptor molecule in the right-hand end of the scheme depicts the narrowing of the channel that accounts for its known ion selectivity. (Not drawn to scale.) (From Anholt et al., 1984. Reprinted with permission from Plenum Press, NY.)

receptor channel with the biochemical knowledge about the receptor marcomolecule and permits, for the first time, the study of structure-function correlates at the molecular level (Nelson et al., 1980; Labarca et al., 1983; Montal et al., 1983).

As described in detail elsewhere (Anholt et al., 1980, 1981, 1982, 1983), the AChR is purified from the electric organ of *Torpedo californica* and then reconstituted into phospholipid vesicles that mimic the lipid composition of the natural membrane. The reconstituted receptors in vesicles are in turn used to generate monolayers at the air-water interface that are composed of the purified receptor and

phospholipids. A planar lipid bilayer is then assembled by the hydro-
phobic apposition of two monolayers, either across a hole (approximately 200 μM in diameter) in a thin, teflon partition separating two
aqueous chambers (Montal, 1974; Nelson et al., 1980; Schindler and
Quast, 1980; Labarca et al., 1983; Montal et al., 1983), or more
recently, in the tip of a glass microelectrode pipet (Suarez-Isla et
al., 1983). Since the resistance of these bilayers is high (more
than 1 GΩ), the activity of the single channels can be resolved as
transient increases in the current that flows across the high resistance membrane when a constant voltage is applied (voltage clamping).
Figure 2 shows a single-channel recording from the purified ACh
receptor reconstituted in a planar lipid bilayer. The currents
fluctuate between two levels, which are associated with closed and
open states of the channel. The channel remains open for a few milli-
seconds before closing. The lower panel is a smoothed version of the
raw data in the upper panel.

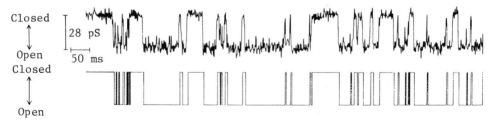

Figure 2. Single-channel currents activated by the agonist suberyl-
dicholine (0.1 μM) in planar lipid bilayers containing purified AChR.
Membranes were formed as described in detail elsewhere (Nelson et al.,
1980; Labarca et al., 1983; Montal et al., 1983) in 0.3 M NaCl, 5 mM
CaCl$_2$, 2.5 mM HEPES, pH 7.4. The applied voltage was -100 mV, and
the single-channel conductance was 28 pS.

We shall not discuss these data-processing techniques here but
point out that some interesting problems in classification, or signal
extraction, arise. Currently used techniques are fairly crude.

A simple model of the action of the receptor is shown in Figure
3a (Katz and Miledi, 1972). In this model, there are two closed
states—a resting state and a state with ACh (agonist) bound—and one
open state. For reasons that will become clear, this model is now

considered to be too simple; in particular, it is believed that there are at least two open states. Some other hypothetical models are given in Figures 3b-e. The scheme shown in Figure 3f includes, in addition a desensitized closed state and resembles the cyclic model of Katz and Thesleff (1957). The implicit dynamical assumption is that the process is Markovian. What is observed is not the underlying lying process but an aggregated version; from the data one knows only that the system is in an open or closed state, but not which one.

$$C + L \rightleftharpoons CL \rightleftharpoons OL \qquad (3a)$$

$$\begin{array}{ccccc} C + 2L & \rightleftharpoons & CL + L & \rightleftharpoons & CL_2 \\ & & \uparrow\downarrow & & \uparrow\downarrow \\ & & OL + L & & OL_2 \end{array} \qquad (3b)$$

$$\begin{array}{ccccc} C + 2L & \rightleftharpoons & CL + L & \rightleftharpoons & CL_2 \\ & & \uparrow\downarrow & & \uparrow\downarrow \\ & & OL + L & \rightleftharpoons & OL_2 \end{array} \qquad (3c)$$

$$\begin{array}{ccccc} C + 2L & \rightleftharpoons & CL + L & \rightleftharpoons & OL + L \\ & & \searrow\nwarrow & & \swarrow\nearrow \\ & & & OL_2 & \end{array} \qquad (3d)$$

$$C + 2L \rightleftharpoons CL + L \rightleftharpoons OL + L \rightleftharpoons OL_2 \qquad (3e)$$

$$\begin{array}{ccccccc} C + 2L & \rightleftharpoons & CL + L & \rightleftharpoons & CL_2 & \rightleftharpoons & OL_2 \\ \uparrow\downarrow & & \uparrow\downarrow & & \uparrow\downarrow & & \\ D + 2L & \rightleftharpoons & DL + L & \rightleftharpoons & DL_2 & & \end{array} \qquad (3f)$$

Figure 3. Several hypothetical models. L denotes ligand (acetylcholine, for example), C denotes a closed state, O denotes an open state, and D a desensitized state.

3. ONE-DIMENSIONAL DISTRIBUTIONS

Throughout, we shall assume that the process is stationary. There are two relevant one-dimensional distributions, the dwell times in the open state and closed state aggregates.

First we give some standard notation:

$$P_{ij}(t) = P\{\text{state } j \text{ at time } t \mid \text{state } i \text{ at time } 0\}. \tag{3.1}$$

The forward equation is

$$P'(t) = P(t)Q; \quad Q = \lim_{t \to 0} \frac{P(t) - I}{t}, \tag{3.2}$$

which has the solution

$$P(t) = e^{Qt}. \tag{3.3}$$

We wish to find the distribution of T, the open dwell time. We first calculate the distribution function F_{ij} and the corresponding density function f_{ij}, where

$$F_{ij} = P\{T \leqslant t \text{ and exits to state } j \mid \text{open state } i \text{ at } t = 0\}.$$

We follow a technique used by Colquhoun and Hawkes (1977, 1981) and described in Cox and Miller (1965). A new process is defined in which the closed states are absorbing:

$$\bar{P} = \begin{bmatrix} \bar{P}_{oo} & \bar{P}_{oc} \\ 0 & I \end{bmatrix}; \quad \bar{Q} = \begin{bmatrix} Q_{oo} & Q_{oc} \\ 0 & 0 \end{bmatrix}. \tag{3.4}$$

Then

$$F_{ij}(t) = [\bar{P}_{oc}(t)]_{ij}, \tag{3.5}$$

$$f_{ij}(t) = [\bar{P}'_{oc}(t)]_{ij}. \tag{3.6}$$

Now

$$\bar{P}'(t) = \bar{P}(t)\bar{Q} = \begin{bmatrix} \bar{P}_{oo} Q_{oo} & \bar{P}_{oo} Q_{oc} \\ 0 & 0 \end{bmatrix} \tag{3.7}$$

and

$$\bar{P}_{oo}(t) = e^{Q_{oo}t} \tag{3.8}$$

so that

$$f_{ij}(t) = [\bar{P}_{oo}(t)Q_{oc}]_{ij} = [e^{Q_{oo}t}Q_{oc}]_{ij}. \tag{3.9}$$

Let π_o be the row vector of equilibrium probabilities for the open entry states

$$\pi_{oi} = P\{\text{open aggregate is entered via open state } i\}, \tag{3.10}$$

and let u_c be a vector of 1's. (The determination of π_o and the similarly defined vector π_c is taken up in the next section.) Then the density of open dwell times is

$$f_o(t) = \pi_o e^{Q_{oo}t}Q_{oc}u_c. \tag{3.11}$$

We show below that Q_{oo} is diagonalizable; its spectral representation yields

$$f_o(t) = \sum_{i=1}^{N_o} \alpha_i e^{\lambda_i t}; \quad N_o = \text{number of open states.} \tag{3.12}$$

Similarly, the density of the closed dwell times is of the form

$$f_c(s) = \sum_{j=1}^{N_o} \beta_j e^{\omega_j s}; \quad N_c = \text{number of closed states.} \tag{3.13}$$

From these results we see that the number of closed and open states can be deduced from the marginal distribution of closed and open dwell times (if all α_i and β_j are nonzero). As those who have tried are aware, fitting sums of exponentials with a variable number of summands is tricky. Such analysis has indicated that there are at least two open states and three closed states in the underlying system (Colquhoun and Sakmann, 1981; Jackson et al., 1983; Labarca, 1983; Montal et al., 1983).

For a general aggregated Markov process the matrices Q_{oo} and Q_{cc} need not be diagonalizable, and (3.12) and (3.13) could contain polynomials arising from Jordan blocks. In this physical context a consequence of the law of detailed balance is that there is a set w_i of nonzero probabilities, corresponding to thermodynamic equilibrium, with the property that $w_i Q_{ij} = w_j Q_{ji}$ for all i and j. This implies

Lemma 3.1: Q, Q_{oo}, and Q_{cc} are diagonalizable, and

$$(W_o^{1/2} Q_{oc} W_c^{-1/2}) = (W_c^{1/2} Q_{co} W_o^{-1/2})^T,$$

where W_o and W_c are submatrices for the open and closed states of the diagonal matrix W with elements w_i.

Proof: The law of detailed balance is

$$WQ = Q^T W$$

or

$$W^{1/2} Q W^{-1/2} = W^{-1/2} Q \, W^{1/2}$$
$$= (W^{1/2} Q W^{-1/2})^T.$$

Q is thus similar to a symmetric, hence diagonalizable, matrix and therefore is itself diagonalizable. Expanding the relation above in partitioned form shows that Q_{oo} and Q_{cc} are also similar to diagonal matrices and gives the result involving Q_{oc} and Q_{co}.

Aspects of the system may be deducible from studying the way that λ and ω change as functions of concentration of agonist. The λ, for example, are the eigenvalues of Q_{oo} and can be conveniently related to that matrix via their elementary symmetric functions (Mirsky, 1955). The rth elementary symmetric function of λ_1, λ_2, ..., λ_n is $\sum \lambda_{i_1} \lambda_{i_2} \cdots \lambda_{i_r}$, where the sum is over all subsets of size r, and is equal to the sum of the r-rowed principal minors of Q_{oo}. The first elementary symmetric function is the sum of the eigenvalues and equals the trace of Q_{oo}, and the nth is their product, which is the determinant of Q_{oo}.

As an example we consider a hypothetical system (Figure 4). Denoting the agonist concentration by c, the transition matrix for this system is

$$
\begin{bmatrix}
-ck_{12} & ck_{12} & 0 & 0 & 0 \\
k_{21} & -(k_{21}+ck_{23}+k_{24}) & ck_{23} & k_{24} & 0 \\
0 & k_{32} & -(k_{32}+k_{35}) & 0 & k_{35} \\
0 & k_{42} & 0 & -(k_{42}+ck_{45}) & ck_{45} \\
0 & 0 & k_{53} & k_{54} & -(k_{53}+k_{54})
\end{bmatrix}
$$

$$= \begin{bmatrix} Q_{cc} & Q_{co} \\ Q_{oc} & Q_{oo} \end{bmatrix}.$$

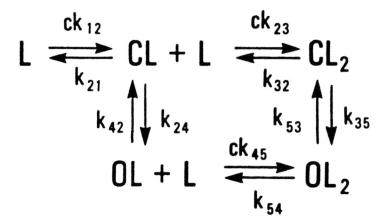

Figure 4. A hypothetical model discussed in the text. States 1, 2, and 3 are closed, and 4 and 5 are open; c denotes the concentration of acetylcholine.

For this scheme the sum of the product of λ_1 and λ_2 should be linear functions of concentration. From their observed slopes and intercepts, we could determine the values of four combinations of k_{42}, k_{45}, k_{53}, and k_{54}, so these rates would be established. The sum of all ω is a linear function of concentration, and their product and the second elementary symmetric function are quadratic functions of concentration. These functions contain eight coefficients, which are more than sufficient to determine the remaining rates.

Experimental agreement with the predicted form of the dependence of the elementary symmetric functions on agonist concentration would provide partial confirmation of the model.

4. HIGHER DIMENSIONAL DISTRIBUTIONS

In this section we consider what can be learned about the underlying process from the higher dimensional distributions of the observed process.

We first consider the two-dimensional distributions of (a) an open dwell time and the following closed dwell time and (b) a closed dwell time and the following open dwell time. We denote the corresponding densities by f_{oc} and f_{co}. By arguing in the same manner as was done to determine the one-dimensional distributions, we find

$$f_{oc}(t, s) = \pi_o e^{Q_{oo}t} Q_{oc} e^{Q_{cc}t} Q_{co} u_o = \sum_{i=1}^{N_o} \sum_{j=1}^{N_o} \alpha_{ij} e^{\lambda_i t + \omega_j s}, \qquad (4.1)$$

$$f_{co}(s, t) = \pi_c e^{Q_{cc}t} Q_{co} e^{Q_{oo}t} Q_{oc} u_c = \sum_{i=1}^{N_o} \sum_{j=1}^{N_o} \beta_{ji} e^{\omega_j s + \lambda_i t}. \qquad (4.2)$$

It is interesting that the same exponential parameters occur in the two-dimensional distributions as in the one-dimensional distributions. A comparison of the one- and two-dimensional distributions thus gives a check on the Markov assumption, which would be interesting to apply to experimental data.

The same argument gives the joint density of any finite succession of open and closed dwell times. For example, the joint density of a sequence starting with an open, ending with a closed, and containing n open and n closed dwell times is

$$f(t_1, \ldots, t_n, s_1, \ldots, s_n) = \pi_o e^{Q_{oo}t_1} Q_{oc} e^{Q_{cc}s_1} Q_{co} \cdots$$
$$Q_{oc} e^{Q_{cc}s_n} Q_{co} u_0. \qquad (4.3)$$

The vectors π_o and π_c can be expressed in terms of Q by constructing appropriate Markov chains. For example, consider the Markov chain whose states are open states in which an open dwell time begins; there are N_o states in all, some of which may be trivial (never entered). From a two-dimensional analog of (3.9), we find the transition matrix

$$\int e^{Q_{oo}t} Q_{oc} e^{Q_{cc}s} Q_{co} \, dt \, ds = Q_{oo}^{-1} Q_{oc} Q_{cc}^{-1} Q_{co}.$$

A similar expression holds for the transition matrix of the Markov chain of closed entry states. The vectors π_o and π_c are the equilibrium probabilities for these matrices, i.e., the eigenvectors corresponding to the eigenvalue 1.

Since the distribution of any finite stretch of the observed process can thus be expressed explicitly in terms of Q, the method of maximum likelihood could in principle be used to estimate Q, providing that Q is identifiable. Maximum likelihood estimation has been considered by Horn and Lange (1983) from a slightly different point of view.

In some cases the higher dimensional distributions are degenerate, and the observed process is an alternating renewal process, that is, the successive dwell times are independent of each other. This is the case for the process schematized in Figures 3d and e; there is only one way to leave the closed state aggregate so that the open dwell time is independent of the preceding closed dwell time and of the preceding open dwell time, and similarly for closed dwell times.

To generalize this scheme we consider the graph corresponding to a finite state Markov process, nodes of which correspond to the states and in which two nodes are joined by an edge if transitions between the corresponding states are possible. Suppose that we have a model with the property that there is a particular state, which we shall call the *gateway*, such that upon removal of its node, the graph is disconnected into two components, one of which has the closed states as its nodes and the other the open states. In this case the successive dwell times are independent; to see this, note that the assumption of a single gateway state implies that Q_{oc} and Q_{co} have rank one and therefore that (4.3) factors.

This result allows us to reject immediately a whole class of possible kinetic models if it can be demonstrated experimentally that successive dwell times are dependent. There is evidence that this is the case: Jackson et al., 1983; Labarca et al., 1983.

A further consequence is that if the one-dimensional distributions do not suffice to identify such a model, then the model is not identifiable (the reaction rates cannot be uniquely determined by observation of the aggregated process). The one-dimensional open dwell time distribution is a function of $2N_o - 1$ independent parameters (N_o exponentials λ_i and $N_o - 1$ linear constants α_i); there is a constraint that the density integrates to one. The same reasoning applies to the N_o closed states, giving $2(N_o + N_o - 1)$ parameters in all. This is exactly the number of rate constants if the graph contains no cycles. A model containing cycles, as in Figure 3d, is not identifiable. Moreover, the models of Figures 3d and e cannot be distinguished.

Although considerations of space prevent us from discussing in detail the identifiability of acyclic models of this type, we can show that, by analogy with certain compartmental models (Corbelli et al., 1979), they are identifiable up to a known ambiguity.

In the more general case there is additional information in the higher dimensional distributions, but not as much as one might suppose. The following theorem shows that there is no more information in the higher dimensional distribution than that contained in the two two-dimensional distributions (4.1) and (4.2).

Theorem 4.1: Under the assumption that all α_i and β_j of (3.12) and (3.13) are nonzero, the parameters of all higher dimensional distributions can be determined from the parameters of the two two-dimensional distributions (4.1) and (4.2).

Proof: We first introduce some notation. As before, π_o and π_c denote the stationary distribution of entry states for the open and closed processed. From Lemma 3.1, Q_{oo} and Q_{cc} have the spectral representations

$$Q_{oo} = \sum_{i=1}^{N_o} \lambda_i x_i y_i^T,$$

$$Q_{cc} = \sum_{j=1}^{N_o} \omega_j z_j w_j^T.$$

Let

$$\gamma_i = \pi_o x_i \qquad \rho_j = \pi_c z_j$$
$$\delta_i = y_i^T Q_{oc} u_c \qquad \mu_j = w_j^T Q_{co} u_o$$
$$\xi_{ij} = y_i^T Q_{oc} z_j \qquad \eta_{ji} = w_j^T Q_{co} x_i$$

The one-dimensional open dwell time density is

$$f_o(t) = \pi_o e^{Q_{oo}t} Q_{oc} u_c$$

$$= \pi_o \sum e^{\lambda_i t} x_i y_i Q_{oc} u_c$$

$$= \sum e^{\lambda_i t} \alpha_i,$$

where $\alpha_i = \gamma_i \delta_i$. Similarly, the closed dwell time density is

$$f_c(s) = \sum e^{\omega_j s} \beta_j,$$

where $\beta_j = \rho_j \mu_j$. Now consider the two two-dimensional distributions;

$$f_{oc}(t, s) = \pi_o e^{Q_{oo}t} Q_{oc} e^{Q_{cc}s} Q_{co} u_o$$

$$= \sum \sum e^{\lambda_i t + \omega_j s} \pi_o x_i y_i^T Q_{oc} z_j w_j^T Q_{co} u_o$$

$$= \sum \sum e^{\lambda_i t + \omega_j s} \alpha_{ij},$$

where $\alpha_{ij} = \gamma_i \xi_{ij} \mu_j$; and

$$f_{co}(s, t) = \sum \sum e^{\omega_j s + \lambda_i t} \beta_{ji},$$

and where $\beta_{ji} = \rho_j \eta_{ji} \delta_i$. We now establish the assertion for the two three-dimensional distributions:

$$f_{oco}(t_1, s, t_2) = \pi_o e^{Q_{oo}t_1} Q_{oc} e^{Q_{cc}s} Q_{co} e^{Q_{oo}t_2} Q_{oc} u_c$$

$$= \sum \sum \sum e^{\lambda_{i_1} t_1 + \omega_j s + \lambda_{i_2} t_2} \gamma_{i_1} \xi_{i_1 j} \eta_{ji_2} \delta_{i_2}$$

$$= \sum \sum \sum e^{\lambda_{i_1} t_1 + \omega_j s + \lambda_{i_1} t_2} \alpha_{i_1 j i_2},$$

where $\alpha_{i_1 j i_2} = \gamma_{i_1} \xi_{i_1 j} \eta_{ji_2} \delta_{i_2}$. Similarly,

$$f_{coc}(s_1, t, s_2) = \sum \sum \sum e^{\omega_{j_1} s_1 + \lambda_i t + \omega_{j_2} s_2} \beta_{j_1 i j_2},$$

where $\beta_{j_1 i j_2} = \rho_{j_1} \eta_{j_1 i} \xi_{i j_2} \mu_{j_2}$. Now observe that

$$\alpha_{i_1 j i_2} = \frac{\alpha_{i_1 j}}{\mu_j} \times \frac{\beta_{ji_2}}{\rho_j} = \frac{\alpha_{i_1 j} \beta_{ji_2}}{\beta_j}.$$

A similar expression holds for $\beta_{j_1 i j_2}$. The assertion thus holds for the three-dimensional distributions.

The same kind of argument works for higher dimensions. For example,

$$\alpha_{i_1 j_1 i_2 j_2} = \gamma_{i_1} \xi_{i_1 j_1} \eta_{j_1 i_2} \xi_{i_2 j_2} \mu_{j_2}$$

$$= \frac{\alpha_{i_1 j_1 i_2}}{\delta_{i_2}} \frac{\alpha_{i_2 j_2}}{\gamma_{i_2}}$$

$$= \frac{\alpha_{i_1 j_1 i_2} \alpha_{i_2 j_2}}{\alpha_{i_2}}.$$

The general case follows by induction.

Corollary 4.1: Under the assumption of Theorem 4.1, the underlying process is not identifiable if it depends on more than $2N_o N_c$ independent parameters.

Proof: The two two-dimensional distributions are given in the proof of the previous theorem. There are $N_c + N_o$ exponential parameters. There are $N_o N_c - 1$ free α_{ij} because the f_{oc} density must integrate to 1. There are $N_o N_c$ β_{ij}, but the marginal densities must agree, giving the equations

$$\sum_j \frac{\beta_{ji}}{\omega_j} = \sum_j \frac{\alpha_{ij}}{\omega_j}, \quad i = 1, \ldots, N_o,$$

$$\sum_i \frac{\beta_{ji}}{\lambda_i} = \sum_i \frac{\alpha_{ij}}{\lambda_i}, \quad j = 1, \ldots, N_c.$$

The constraint that f_{co} must integrate to 1 is implied by these constraints and the fact that f_{oc} integrates to 1. The β_{ji} constitute a two-way table, the (weighted) margins of which are fixed, and there are $(N_o - 1)(N_c - 1)$ degrees of freedom in such a table. The total number of free parameters in the two two-dimensional distributions is thus $N_o + N_c + (N_o N_c - 1) + (N_o - 1)(N_c - 1) = 2N_o N_c$.

Therefore, to ascertain whether a model is identifiable it is sufficient to ascertain whether it is identifiable from the parameters of the two-dimensional distributions, a considerable reduction of complexity. Unfortunately, we have not yet developed nice, checkable, sufficient conditions. That there are no more than $2N_o N_c$ transition rates is clearly not sufficient; consider Figure 3d, which we known is not identifiable. On the other hand, we conjecture, but have not been able to prove, that Figures 3b and c are identifiable.

5. COVARIANCE FUNCTIONS

Since the observed process is stationary, it is natural to consider what information can be obtained from covariance functions. Letting $S_0, S_1, \ldots$ denote the sequence of closed dwell times and $T_0, T_1, \ldots$ the sequence of open dwell times, we consider

$$\Gamma_c(k) = \text{Cov}(S_i, S_{i+k}), \tag{5.1}$$

$$\Gamma_o(k) = \text{Cov}(T_i, T_{i+k}). \tag{5.2}$$

Let M denote the minimum of the number of open entry states, open exit states, closed entry states, and closed exit states.* M is a measure of the complexity of the process of transition from open to closed and closed to open. The covariance functions give us information about M:

Theorem 5.1: The covariance functions are of the form

$$\Gamma_c(k) = \sum_{i=1}^{M-1} u_i \kappa_i^{|k|},$$

$$\Gamma_o(k) = \sum_{i=1}^{M-1} v_i \sigma_i^{|k|},$$

where $0 \leqslant \kappa_i < 1$, $0 \leqslant \sigma_i < 1$, and $k \neq 0$.

Proof:

$$f(t_0, t_k) = \pi_0 e^{Q_{oo} t_0} [Q_{oc} Q_{cc}^{-1} Q_{co} Q_{oo}^{-1}]^{k-1} Q_{oc} (-Q_{cc}^{-1}) Q_{co} e^{Q_{oo} t_k} Q_{oc} u_c$$

from integrating (4.3).

$$ET_0 T_k = -\pi_0 Q_{oo}^{-2} [Q_{oc} Q_{cc}^{-1} Q_{co} Q_{oo}^{-1}]^{k-1} Q_{oc} Q_{cc}^{-1} Q_{co} Q_{oo}^{-2} Q_{oc} u_c$$

$$= \pi_0 Q_{oo}^{-2} [Q_{oc} Q_{cc}^{-1} Q_{co} Q_{oo}^{-1}]^k u_o,$$

since $\sum_j Q_{ij} = 0$ implies $Q_{oc} u_c + Q_{oo} u_o = 0$, or $-Q_{oo}^{-1} Q_{oc} u_c = u_o$.
Continuing,

$$ET_0 T_k = \pi_0 Q_{oo}^{-3/2} [Q_{oo}^{-1/2} Q_{oc} Q_{cc}^{-1} Q_{co} Q_{oo}^{-1/2}]^k Q_{oo}^{-1/2} u_o$$

$$= \pi_0 Q_{oo}^{-3/2} W_o^{-1/2} [(W_o^{1/2} Q_{oo} W_o^{-1/2})^{-1/2} (W_o^{1/2} Q_{oc} W_c^{-1/2})$$
$$(W_c^{1/2} Q_{cc} W_c^{-1/2})^{-1} (W_c^{1/2} Q_{co} W_o^{-1/2})$$
$$(W_o^{1/2} Q_{oo} W_o^{-1/2})^{-1/2}]^k W_o^{1/2} Q_{oo}^{-1/2} u_o$$

$$= \pi_0 Q_{oo}^{-3/2} W_o^{-1/2} [A^T A]^k W_o^{1/2} Q_{oo}^{-1/2} u_o,$$

where

$$A = (W_c^{1/2} Q_{cc} W_c^{-1/2})^{-1/2} (W_c^{1/2} Q_{co} W_o^{-1/2})(W_o^{1/2} Q_{oo} W_o^{-1/2})^{-1/2}.$$

Here we have used Lemma 3.1. $A^T A$ is symmetric, positive semidefinite with nonnegative real eigenvalues σ_i. Now, from the expression for A and Lemma 3.1,

---

*The number of open (closed) entry states is, in fact, equal to the number of open (closed) exit states because of detailed balance.

rank (A) = rank (Q_{co}) = rank (Q_{oc}),

which is less than or equal to M. (Generally, we would expect the rank to be equal to M, but for certain values of the elements of Q_{cc}, the rank could be less than M.)

It can be further argued that one of the eigenvalues is one and the remainder are less than one. (The eigenvalues of $A^T A$ can be related to those of a transition matrix for a particular Markov chain.)

Expanding the spectral decomposition of $(A^T A)^k$ and identifying the term corresponding to the engenvalue one as $E(T_0)E(T_k)$ proves the theorem for $\Gamma_o(k)$. The proof for $\Gamma_c(k)$ is analogous.

The implications of the theorem are that the underlying Markovian model dictates that the covariance functions be of a certain form, and that by counting the number of geometrically decaying components a lower bound on M can be obtained.

6. DISCUSSION

The kinetics of the nicotinic acetylcholine receptor and of many other gated channels are conventionally described by an aggregated Markovian model. Experimental methods now permit the collection of large quantities of high-quality data on these systems that are fundamental to our lives (and to our ability to think about them). What can we learn from these data? If a model is proposed, based on biological intuition, can we, *in principle* (neglecting real life laboratory limitations of instrumental noise, finite bandwidth, etc.), use the data to accept or reject the model or to choose among different models? If a model is accepted, the parameters have direct physico-chemical significance and could be compared with the results of ab initio calculations; can we determine the parameters of the model? We have seen that the answer to these questions can be *no* for examples that are intuitively plausible.

Some of the results have direct experimental implications: In Section 3 it was shown that the behavior of the elementary symmetric functions of λ and ω (3.12, 3.13) as functions of agonist concentration can yield information about the binding steps in the underlying process and in some cases makes possible the determination of all the rate constants (Figure 4). In Section 4 it was shown that the two- and higher-dimensional distributions, which can in principle be estimated from experimental data, have the form of sums of exponentials with the same decay rates (λ and ω, 4.1 and 4.2) that occur in the one-dimensional distributions (3.12, 3.13), thus providing a possible check on the Markov assumption. A closed form for the likelihood function, under the simplifying assumption of one channel, was also developed. It was shown in Section 4 that dependence of successive dwell times rules out a large class of possible models. Covariance functions, which can easily be estimated from experimental data, were shown in Section 5 to contain information about the number of transition routes between open and closed states.

The present work is but a first step toward development of hypothesis testing and estimation procedures for aggregated Markovian models. Application to the nicotinic acetylcholine receptor and other gated channels is immediate. More broadly, it is an article of faith in physical science and engineering that any system can be modeled as an aggregated Markov process if the set of states is sufficiently rich. In the future, therefore, we shall generalize our results to Markov processes that are aggregated into more than two states, and apply them to a broad range of problems.

REFERENCES

Anholt, R., J. Lindstrom, and M. Montal (1980). Functional equivalence of monomeric and dimeric forms of purified acetylcholine receptor from *Torpedo californica* in reconstituted lipid vesicles. *Eur. J. Biochem. 109*, 481–487.

Anholt, R., J. Lindstrom, and M. Montal (1981). Stabilization of acetylcholine receptor channels by lipids in cholate solution and during reconstitution in vesicles. *J. Biol. Chem. 256*, 4377–4387.

Anholt, R., D. R. Fredkin, T. Deerinck, M. Ellisman, M. Montal, and J. Lindstrom (1982). Incorporation of acetylcholine receptors into liposomes: Vesicle structure and acetylcholine receptor function. *J. Biol. Chme. 257*, 7122–7134.

Anholt, R., J. Lindstrom, and M. Montal (1983). The molecular basis of neurotransmission: Structure and function of the nicotinic acetylcholine receptor. In *The Enzymes of Biological Membranes*. (A. Martonosi, ed.) Plenum Press, New York (in press).

Boheim, G., W. Hanke, F. J. Barrantes, H. Eibl, B. Sakmann, G. Fels, and A. Maelicke (1981). Agonist-activated ionic channels in acetylcholine receptor reconstituted into planar lipid bilayers. *Proc. Nat'l. Acad. Sci. USA 78*, 3586-3590.

Changeux, J.-P. (1981). The acetylcholine receptor: An allosteric membrane protein. *Harvey Lect. 75*, 85-254.

Claudio, T., M. Ballivet, J. Patrick, and S. Heinemann (1983). Nucleotide and deduced amino acid sequences of *Torpedo californica* acetylcholine receptor γ subunit. *Proc. Nat'l. Acad. Sci. USA 80*, 1111-1115.

Colquhoun, D., and A. G. Hawkes (1977). Relaxation and fluctuations of membrane currents that flow through drug-operated channels. *Proc. R. Soc. London B. 199*, 231-262.

Colquhoun, D., and A. G. Hawkes (1981). On the stochastic properties of single ion channels. *Proc. R. Soc. London B. 211*, 205-235.

Colquhoun, D., and B. Sakmann (1981). Fluctuations in the microsecond time range of the current through single acetylcholine receptor ion channels. *Nature 294*, 464-466.

Corbelli, C., A. Lepschy, and G. R. Jarur (1979). Identifiability results on some constrained compartmental models. *Mathematical Biosciences 47*, 173-195.

Cox, D. R., and H. D. Miller (1965). *The Theory of Stochastic Processes*. Methuen.

Devilliers-Thiery, A., J. Diraudat, M. Bentaboulet, and J. P. Changeux (1963). Complete mRNA coding sequence of the acetylcholine binding α-subunit of *Torpedo marmorata* acetylcholine receptor: A model for the transmembrane organization of the polypeptide chain. *Proc. Nat'l. Acad. Sci. USA 80*, 2067-2071.

Dwyer, T. M., D. Adams, and B. Hille (1980). The permeability of the end-plate channel to organic cations in frog muscle. *J. Gen. Physiol. 75*, 469-492.

Horn, R., and K. Lange (1983). Estimating kinetic constants from single channel data. *Biophys. J. 43*, 207-223.

Jackson, M. B., B. S. Wong, C. E. Morris, H. Lecar, and C. N. Christian (1983). Successive openings of the same acetylcholine receptor-channel are correlated in their open times. *Biophys. J. 42*, 109-114.

Karlin, A. (1980). Molecular properties of nicotinic acetylcholine receptors. In *The Cell Surface and Neuronal Function*, pp. 191-260. (G. Poste, G. Nicholson, and C. Cotman, eds.) Elsevier/North Holland Biomedical Press, New York.

Katz, B., and S. Thesleff (1957). A study of the "desensitization" produced by acetylcholine at the motor end-plate. *J. Physiol. 138*, 63-80.

Katz, B., and R. Miledi (1972). The statistical nature of the acetylcholine potential and its molecular components. *J. Physiol. 224*, 665-699.

Kistler, J., R. M. Stroud, M. W. Klymkowsky, R. A. Lalancette, and R. H. Fairchoigh (1982). Structure and function of an acetyl-choline receptor. *Biophys. J. 37*, 371-383.

Labarca, P., J. Lindstrom, and M. Montal (1983). The acetylcholine receptor from *Torpedo californica* has two open states. *J. Neuroscience* (in press).

Lindstrom, J., J. Merlie, and G. Yogeeswaran (1979). Biochemical properties of acetylcholine receptor subunits from *Torpedo californica. Biochemistry 18*, 4465-4470.

Mirsky, L. (1955). *An Introduction to Linear Algebra.* Oxford at the Clarendon Press.

Montal, M. (1974). Formation of bimolecular membranes from lipid monolayers. *Meth. Enzymol. 32*, 545-556.

Montal, M., P. Labarca, D. R. Fredkin, B. A. Suarez-Isla, and J. Lindstrom (1983). Channel properties of the purified acetylcholine receptor from *Torpedo californica*, reconstituted in planar lipid bilayer membranes. *Biophys. J. 41* (in press).

Neher, E., and B. Sakmann (1976). Single channel currents recorded from membrane of denervated frog muscle fibers. *Nature 260*, 799-802.

Nelson, N., R. Anholt, J. Lindstrom, and M. Montal (1980). Reconsti-tution of purified acetylcholine receptors with functional ion channels in planar lipid bilayers. *Proc. Nat'l. Acad. Sci. USA 77*, 3057-3061.

Noda, M., H. Takahashi, T. Tanabe, M. Toyosato, Y. Fusutani, T. Hirose, M. Asai, S. Inayama, T. Miyata, and S. Numa (1982). Primary structure of α-subunit precursor of *Torpedo californica* acetyl-choline receptor deduced from cDNA sequence. *Nature 299*, 793-797.

Noda, M., H. Takahashi, T. Tanabe, M. Toyosato, S. Kikyotani, T. Hirose, M. Asai, H. Takashima, S. Inayama, T. Miyata, and S. Numa (1983). Primary structure of β-subunit precursors of *Torpedo californica* acetylcholine receptor deduced from cDNA sequences. *Nature 301*, 251-255.

Raftery, M. A., M. W. Hunkapiller, C. D. Strader, and L. E. Hood (1980). Acetylcholine receptor: Complex of homologous subunits. *Science 208*, 1454-1457.

Reynolds, J. A., and A. Karlin (1978). Molecular weight in detergent solution of acetylcholine receptor from *Torpedo californica. Biochemistry 17*, 2035-2038.

Sakmann, B., J. Patlak, and E. Heher (1980). Single acetylcholine-activated channels show burst-kinetics in presence of desensitizing concentrations of agonist. *Nature 286*, 71-79.

Schindler, H., and U. Quast (1980). Functional acetylcholine receptor from *Torpedo marmorata* in planar membranes. *Proc. Nat'l. Acad. Sci. USA 77*, 3052-3056.

Suarez-Isla, B. A., K. Wan, J. Lindstrom, and M. Montal (1983). Single-channel recordings from purified acetylcholine receptors reconstituted in bilayers formed at the tip of patch pipets. *Biochemistry 22*, 2319-2323.

Tank, D. W., R. L. Juganir, P. Greengard, and W. W. Webb (1983). Patch-recorded single-channel currents of the purified and reconstituted *Torpedo* acetylcholine receptor. *Proc. Nat'l. Acad. Sci. USA 80*, 5129-5133.

Zingsheim, H. P., F. J. Barrantes, J. Frank, W. Haenicke, and D.-Ch. Neugebauer (1982). Direct structural localization of two toxin-recognition sites on an Ach receptor protein. *Nature 299*, 81-84.

MONOCLONAL ANTIBODY 791T/36 FOR TUMOR DETECTION AND DRUG TARGETING

R. W. BALDWIN University of Nottingham
V. S. BYERS Xoma Corp., Berkeley

INTRODUCTION

The notion that tumor cell binding antibodies might be used for tumor detection and for targeting therapeutic agents has been under investigation for many years. Antibodies are generated as one component of the immune response following recognition of antigens. In the present context, these tumor-associated antigens are either specific to, or preferentially expressed in, tumors. In the normal course of events, antigen stimulation begins the generation of multiple populations of sensitized cells (B lymphocytes), each of which is programmed to produce antibody to a single determinant. Because a single protein may contain many hundreds of such determinants (*epitopes*), the outcome of this type of immunization is the production of *polyclonal* antisera that contain many species of antibodies

These investigations were supported by the Cancer Research Campaign, U.K.

that react with the many different epitopes on the immunizing antigen. This makes the problem of isolating a specific antibody very
difficult.

Developments in hybridoma technology make it possible to overcome this problem because in this technique, hybrid cells (hybridomas)
are formed in culture by fusing a single antibody-producing B lymphocyte with a myeloma cell [1]. Each sensitized B cell possesses the
genetic information to produce antibody against a single determinant,
but is not capable of indefinite survival in culture and so cannot be
used as a continuous source of antibody-producing cells. The other
partner in the hybridoma is a myeloma cell that can be maintained
indefinitely in culture and retains the capacity to produce immunoglobulins, which are the building blocks for antibody. When these
two cells are fused, the resultant hybridoma may have acquired the
genetic information of the B lymphocyte partner. As a result it
produces the antibody that this B cell is programmed to produce.
By the selection of single hybridoma cells and their expansion and
growth in tissue culture, it is possible to obtain a continuous
supply of monoclonal antibodies that recognize specific tumor cell-
associated substances.

It was originally anticipated that the fusion of spleen cells
from donors immunized against human tumor cells with myeloma cells of
the appropriate species (rat/mouse) would yield hybridomas that
secrete antibodies that would react specifically with tumors.
This ideal requirement has not often been achieved, because hybridomas have produced antibodies showing restricted, but not necessarily tumor specific, reactivity [2, 3]. Nevertheless, some of the
monoclonal antibodies that are produced against human tumor products
are potentially useful for targeting diagnostic or therapeutic
agents [4, 5]. These developments are considered here with respect
to the application of an anti-human-tumor monoclonal antibody
designated 791T/36 [6]. This monoclonal antibody is produced by a
hybridoma obtained by the fusion of spleen cells from a mouse that
has been immunized against cells of a human osteogenic sarcoma 791T
and a mouse myeloma P3NS1. Originally, it was found to react in vitro

with cells from a high proportion of human osteogenic sarcoma lines
[6]. The antigen detected by 791T/36 antibody is expressed in many
human sarcomas, this having been shown by antibody reaction with
extracts from surgically derived tumor specimens [7]. Furthermore,
the 791T/36 antibody-defined antigen is not restricted to sarcomas,
because the antibody was early found to bind to the products of other
tumors [6]. This increases its potential for targeting diagnostic
and therapeutic agents to a range of human tumors, rather than being
restricted to osteogenic sarcoma.

RADIOIMMUNOLOCALIZATION OF TUMORS

The idea of using antibodies to target diagnostic agents to tumors
has been under investigation for some years following the successful gamma camera imaging of colorectal carcinoma patients who have
received injections of ^{131}I-labeled sheep antibodies which react
with carcinoembryonic antigen, CEA [8]. Attention is now being
directed to the application of radioisotope-labeled monoclonal
antibodies for tumor imaging, especially because several of the
antibody preparations localize in human tumors that develop as
xenografts in immunodeprived mice [9]. These approaches can be
illustrated by a series of clinical trials in which monoclonal
antibody 791T/36 has been used for radioimmunodetection (immunoscintigraphy) of human tumors.

Osteogenic Sarcoma

Immunoscintigraphy with ^{131}I-labeled 791T/36 antibody has been used
to detect a primary osteogenic sarcoma in the right knee of a young
girl [10]. One, two, and three days after intravenous administration
of ^{131}I-labeled antibody (200 µg protein, 70 MBq ^{131}I) anterior and
posterior images of the legs, abdomen, and chest were taken with a
gamma camera and images recorded by the computer. Image enhancement
was then achieved by subtraction of blood pool images acquired
following in vivo labeling of the patient's red blood cells with

^{99m}Tc. The results of these images (Figure 1a) show the uptake
of radioactivity into the tumor in the right leg, following infusion
of ^{131}I-labeled antibody. Figure 1b shows the blood pool image
obtained with ^{99m}Tc indicating abnormal uptake into the right leg,
but after subtraction (Figure 1c) obvious reactivity, representing
^{131}I-labeled antibody is seen in the area of the tumor.

^{131}I Image ^{99m}Tc Blood Pool   Subtracted Image

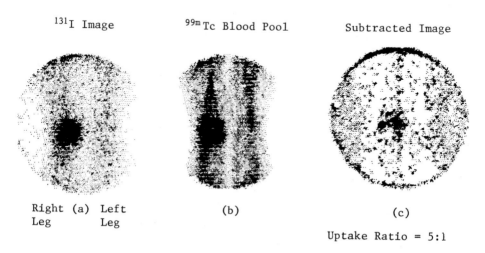

Right (a) Left   (b)   (c)
Leg   Leg

Uptake Ratio = 5:1

Figure 1. Gamma camera views of an osteogenic sarcoma patient
infused 24 h earlier with ^{131}I-labeled 791T/36 monoclonal antibody
(200 µg protein 70 MBq ^{131}I): The patient's red blood cells were
labeled in vivo with ^{99m}Tc. After normalization to equalize the
counts recorded for each image, the ^{99m}Tc image of the blood pool
was subtracted from the ^{131}I-labeled antibody image [10].
(a) ^{131}I-labeled antibody image. (b) ^{99m}Tc blood pool image
showing abnormal concentration of blood in tumor. (c) Subtracted
image showing tumor localization.

In extending this approach, positive images have been obtained
in twelve other patients showing primary osteogenic sarcoma following
injection of ^{131}I-labeled 791T/36 monoclonal antibody (Armitage et
al., unpublished findings). This trial is continuing to assess the
potential of the technique for the detection of osteogenic sarcoma
metastases. These clinical tests also provide evidence that the
level of localization of 791T/36 monoclonal antibody in tumors is
sufficiently different from that in normal tissues to make it feasible to use the antibody for drug targeting. For example, computer

analysis of the images acquired after blood pool subtraction in the patient in Figure 1 indicated that the tumor-to-nontumor uptake of ^{131}I-labeled antibody was 5:1 [10].

Colorectal Carcinoma

Clinical studies have been developed using the imaging procedures outlined with the osteogenic sarcoma patient to examine the potential of 791T/36 monoclonal antibody for the radioimmunodetection of primary and metastatic colorectal carcinomas. In the first trial [11], eleven patients with primary and metastatic tumors were studied, and in nine of them accurate tumor localization was observed (Table 1). The patients in the first trial included five with primary colorectal cancers and six with extensive disseminated disease histologically confirmed at a previous laparotomy. As summarized in Table 1, positive images were obtained with metastases. In case 10, for example, a brain metastasis was clearly identified (Figure 2). This is the smallest tumor identified in this series, measuring 2 × 2 × 1 cm [11].

This trial has been extended to study the localization of ^{131}I-labeled 791T/36 monoclonal antibody in 59 other patients with colorectal cancer, noncolonic gastrointestinal cancers, and benign colonic disease [12, 13]. No antibody localization was detected in patients with benign colorectal disease [6]. In comparison, all but one (13 of 14) of the patients with metastatic colorectal cancer showed positive antibody localization in the tumor.

Ovarian Carcinoma

Monoclonal antibody 791T/36 also localizes in ovarian tumors and, in the initial trial, 10 out of 11 patients gave positive scans following injection of ^{131}I-labeled antibody [14]. In five patients, uptake of antibody could be visualized on the ^{131}I images alone. However, this scan is not considered satisfactory in itself, because localization may reflect blood-borne radioactivity in highly vascular tumors, and subtraction of the blood pool image is routinely

Table 1. External Imaging of Colorectal Carcinomas with ^{131}I-Labeled 791T/36 Monoclonal Antibody

Patient Number	Macroscopic Tumor	Site of Image	Tumor-to-Non-tumor Ratio*
1	Primary carcinoma	Primary in pelvis	8.0:1
2	Primary carcinoma	Primary in pelvis	2.1:1
3	Primary carcinoma	Negative (behind bladder)	
4	Two primary carcinomas	Both primaries	2.1:1
5	Primary carcinoma	Primary in pelvis	1.5:1
6	Disseminated carcinoma, omental metastasis	Primary pelvis mass and secondary	
7	Disseminated carcinoma, liver metastasis	Liver metastasis	5.1:1
8	Disseminated carcinoma, liver metastasis	Liver metastasis	4.4:1
9	Inoperable carcinoma, treated by radio-therapy (30 Gy)	Negative	
10	Disseminated carcinoma, liver and brain metastases	Liver and brain metastases	4.0:1
11	Disseminated carcinoma, pelvic recurrence	Recurrent tumor	4.3:1

*Tumor-to-nontumor ratio is the ratio of radioactivity concentrated over the area of the macroscopic tumor divided by the mean radio-activity in adjacent areas.

performed in all patients. In this series of patients, the target-to-nontarget ratio of radioactivity ranged from 25:1 to 31:1 after blood pool subtraction. Interestingly, one of the patients in which [131]I 791T/36 antibody failed to localize was proven to have a benign lesion.

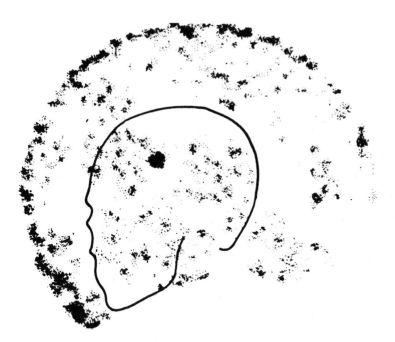

Figure 2. Brain metastasis detected in a colorectal patient (case 10, Table 1) following imaging with ^{131}I-labeled 791T/36 monoclonal antibody and blood pool subtraction with ^{99m}Tc labeled red blood cells [11].

MONOCLONAL ANTIBODY TARGETING OF ANTITUMOR AGENTS

A fundamental objective in cancer therapy is to destroy malignant cells while restricting damage to normal tissues. In this respect, the administration of agents linked to monoclonal antibodies that react with human tumors is a new approach to selective therapy, which should be particularly applicable for the treatment of small deposits of tumor cells (metastases) that have spread from a primary tumor [4, 9, 13].

It should be possible to link highly toxic substances to monoclonal antibodies, so that after binding specifically to tumor cells, conjugates internalize to produce a cytotoxic response. One approach under investigation is to link plant and bacterial toxins

(e.g., ricin and diphtheria toxin) to antibody to produce immuno-
toxins [5]. These toxins consist of two polypeptide chains, a B
chain which serves to bind toxin to cells and an A chain which, on
entry into cells, catalyzes lethal biochemical lesions. Toxins
have thus been fragmented and the separated A chains conjugated to
antibody so that target cell killing is dictated by the binding of
antibody to tumor cells. This approach to the design of immuno-
toxins is well advanced, and several monoclonal antibodies including
791T/36 have been linked to ricin A chain to yield conjugates that
are specifically cytotoxic for cells expressing the antigen recognized by the monoclonal antibody [5].

There are difficult problems with the clinical application of
toxin-antibody conjugates including their potential toxicity for
normal tissues after nonspecific binding. The targeting of conventional cytotoxic drugs is, however, a more acceptable proposal because
the toxic properties of these drugs are already well known. This is
the most common approach that has been developed using monoclonal
antibody 791T/36.

CYTOTOXIC DRUGS

The potential of anti-osteogenic sarcoma monoclonal antibody 791T/36
for targeting cytotoxic drugs is illustrated by investigations with
conjugates containing the vinca alkaloid analogue *vindesine* VDS
[15]. These conjugates, containing approximately 6 mole VDS/mole
antibody, retained a high degree of antibody reactivity and, as
illustrated in Figure 3, are cytotoxic for tumor cells expressing
the 791T/36 antibody-defined antigen. In this representative
experiment, exposure of osteogenic sarcoma 791T cells to 791T/36
antibody-VDS conjugate for 15 min, followed by extensive washing of
the cells before plating, markedly inhibited protein synthesis when
assayed 24 h later by ^{75}Se-selenomethionine uptake [15]. Similar
treatment of ovarian carcinoma PA1-cells, which do not express the
791T/36 antibody-defined antigen, did not affect survival of these

cells. That this is not due to inherent resistance of PA1 cells to VDS in comparison with 791T cells is indicated by the finding that both cell lines showed comparable susceptibility to free drug. These in vitro studies establish that conjugation of VDS to 791T/36 antibody produces conjugates that retain antibody and drug cytotoxic activities [15].

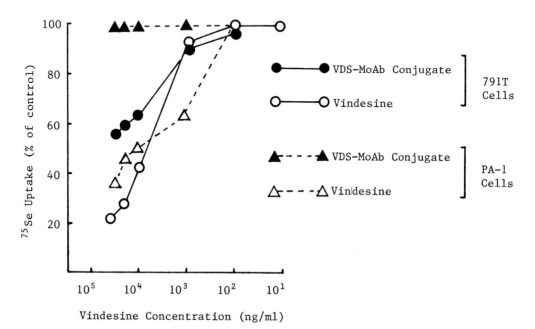

Vindesine Concentration (ng/ml)

Figure 3. Effect of vindesine (VDS) and VDS-791T/36 monoclonal antibody conjugate on survival of tumor cells in culture. Cells were treated with reagents for 15 min and then washed and cultured for 24 h. Cells were then labeled for 16 h with ^{75}Se-selenomethionine and washed three times. Uptake of ^{75}Se by tumor cells is expressed as a percentage of that in controls treated with medium. 791T:osteogenic sarcoma; PA1:ovarian carcinoma. [15]

From the in vitro studies, trials to evaluate the in vivo effectiveness of VDS-antibody conjugates have been developed and carried out [16, 17]. In the initial experiments, osteogenic sarcoma 791T cells were inoculated subcutaneously into immunodeprived mice which then received a course of treatment with either VDS or VDS-791T/36 conjugates (Figure 4). The agents were administered intraperitoneally in seven equal doses after tumor challenge, the total doses being equivalent to 19.2 mg VDS and 500 mg antibody/kg of body weight. Treatment with VDS-791T/36 antibody conjugate markedly suppressed tumor growth compared with untreated controls. Also, the conjugates were nontoxic when administered by this regime of treatment. Although free VDS produced a more marked inhibition of tumor growth, the dose was above that which was tolerated resulting in 2/6 deaths in the VDS group. Treatment with unconjugated 791T/36 antibody does not influence growth of 791T xenografts [18].

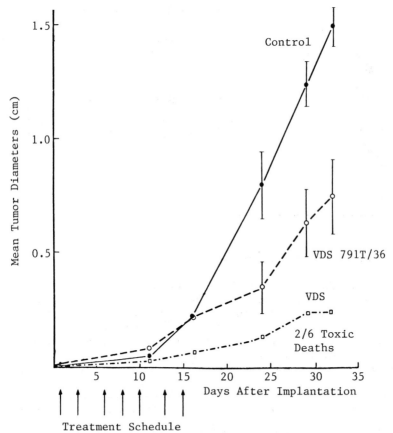

Figure 4. Effect of vindesine (VDS) and VDS-791T/36 monoclonal antibody conjugate on growth of osteogenic sarcoma xenografts in immunodeprived CBA mice. Each agent was given intraperitoneally in seven divided doses according to the treatment schedule shown. Total dose per mouse was 19.2 mg/kg body weight of VDS alone of conjugated to 791T/36 antibody (500 mg/kg).

Conjugates of 791T/36 monoclonal antibody with methotrexate (MTX) have also been synthesized [19]. In this case, MTX was first conjugated to human serum albumin (HSA); then the MTX-HSA product was coupled to 791T/36 antibody to yield products with a probable empirical formula $(MTX_{32} - HSA)_{1-3} - 791T/36$. These conjugates were specifically cytotoxic for osteogenic sarcoma target cells when assayed in vitro using the ^{75}Se-selenomethionine uptake assay (Table 2). Target cells that do not express the 791T/36 antibody-defined antigen, such as bladder carcinoma T24, were not affected by the methotrexate antibody conjugate.

Table 2. Inhibition of Cell Survival Following Treatment with 791T/36-Human Serum Albumin-Methotrexate Conjugate

Cell Line	Antibody Binding*	Percent inhibition of ^{75}Se-methionine uptake at following concentrations[†]		
		40 µg/ml	20 µg/ml	10 µg/ml
791T	++	79	82	79
T278	+	73	79	81
T24	−	7	0	0
RPMl 5966	−	−5	29	0

*Determined by radioimmunoassay and flow cytometry.

[†]Concentration expressed as µg/ml methotrexate. The percent inhibition of ^{75}Se-methionine uptake is relative to that in controls treated with phosphate-buffered saline.

BIOLOGICAL RESPONSE MODIFIERS

The use of monoclonal antibodies as carriers for biological response modifiers has received little attention so far but may be potentially valuable in immunotherapy as a method for focusing host immune responses in tumors, particularly metastatic deposits. This requirement is well established in numerous studies on *regional immunotherapy* in which it has been shown that localization of immunomodulating agents in tumor deposits provokes powerful antitumor responses [20].

Interferons (IFN), which can act both directly as antiprolifera-
tive agents and indirectly via augmentation of preexisting cellular
immunity, represent an important class of biological response modifying agents. The potential of antibody targeting systems has been
examined, therefore, by constructing conjugates of 791T/36 monoclonal
antibody with interferon α derived from Namalwa cells [21]. These
conjugates retained antibody reactivity when assayed in various ways
and the IFN moiety, like free IFN, augmented the natural killer cell
activity of human peripheral blood lymphocytes [22, 23]. Finally,
IFN-791T/36 antibody conjugates localized specifically in xenografts
of human osteogenic sarcomas in immunodeprived mice. These investigations establish the validity of using monoclonal antibodies for
targeting immunomodulating agents to tumors.

DISCUSSION

One of the long-term objectives in developing tumor-localizing monoclonal antibodies is to use them therapeutically, especially for the
treatment of occult or established metastases. In some cases,
antibody alone can exert a tumor-inhibitory effect through the
mediation of host responses, although a more general approach
involves targeting of anti-tumor agents following their conjugation
to antibody. The potential of this approach is emphasized by the
experiments on the conjugation of vindesine (Figure 3) and methotrexate (Table 2) to 791T/36 monoclonal antibody to produce conjugates that are cytotoxic in vitro for tumor cells expressing the
791T/36 antibody-defined antigen.

There are major problems to be resolved in this approach,
however, both with respect to the capacity of drug-antibody conjugates to localize in tumors [9] and then for the drug moieties to
exert their cytotoxic effects. The in vivo tumor-localizing
capacity of monoclonal antibodies (or drug conjugates) is influenced
by multiple factors that are difficult to control. Thus, this will
probably have to be determined empirically for each antibody. It is

now possible, however, to consider processes involved in antibody binding to tumor cells that should be applicable to antibody conjugates. In this respect, a general cell surface antigen-bivalent, antibody-binding theory has been developed [24] to interpret experimental results derived from analyses of the association and dissociation of fluorescein isothiocyanate (FITC) linked 791T/36 antibody to 791T tumor cells, this being determined by flow cytofluorimetry [25]. A basic assumption in this model is that the initial association between antibody and cell-sufrace antigen is monovalent and only afterward does bivalent attachment take place. Conversely, dissociation of bivalently attached antibody proceeds first to monovalent attachment and only then may complete dissociation from cell surface occur.

The mode of entry of drug moieties into tumor cells following cell binding membrane of antibody conjugates is more complicated, but a number of pathways can be proposed. Internalization of antibody conjugates may occur following their binding to tumor cell surface membranes, and a number of pathways for intracellular membrane traffic may be involved, including the lysosomal pathway and the transcellular route [26]. Alternatively, drugs may be linked to antibodies by biodegradable bonds so that the agents are liberated within the milieu of the tumor following conjugate binding to tumors. In this case, antibody-targeted drugs will function essentially as free drugs and be subject to the normal pathways of drug metabolism.

Antibody conjugation of agents that dc not require metabolic conversion following binding to tumor cells provides an alternative approach to therapy. Monoclonal antibodies have been shown to localize in tumors by gamma camera scanning of patients injected with radioisotope preparations; so an obvious approach is to link therapeutic doses of radioisotopes to antibodies. This approach is appealing because conventional antitumor antisera have already been tried for this purpose [27]. Another possibility, as already outlined with studies on interferon-791T/36 antibody conjugates [21], is to target immunomodulating agents so as to provoke local immune responses in tumors. The advantage of this approach is that activation of host cells in tumors will lead to an escalation of antitumor immunity.

ACKNOWLEDGMENTS

The authors gratefully acknowledge the permission of their many colleagues in the Cancer Research Campaign Laboratories and the Queens Medical Centre, University of Nottingham, England, to incorporate their findings in this review.

REFERENCES

1. McMichael, A. J., and Fabre, J. W. (eds.). *Monoclonal Antibodies in Clinical Medicine*, p. 661. New York: Academic Press, 1982.

2. Hellström, K. E., Hellström, I., and Brown, J. P. Human tumor-associated antigens identified by monoclonal antibodies. *Springer Seminars in Immunopathology 5*, no. 2 (1982):127.

3. Baldwin, R. W., Embleton, M. J., and Price, M. R. Monoclonal antibody-defined antigens on tumor cells. In *Biomembranes*, Vol. 11, p. 285. Ed. by Alois Nowotny. New York: Plenum Press, 1983.

4. Baldwin, R. W. Monoclonal antibodies for drug-targeting in cancer therapy. *Pharmacy International 4*, no. 6 (1983):137.

5. Thorpe, P. E., and Ross, W. C. J. The preparation and cytotoxic properties of antibody-toxin conjugates. *Immunological Reviews 62* (1982):119.

6. Embleton, M. J., Gunn, B., Byers, V. S., and Baldwin, R. W. Antitumour reactions of monoclonal antibody against a human osteogenic sarcoma cell line. *Brit. J. Cancer 43*, no. 5 (1981):582.

7. Roth, J. A., Restropo, C., Scuderi, P., Baldwin, R. W., Reichert, C. M., and Hosoi, S. Analysis of antigenic expression by primary and autologous metastatic sarcoma using monoclonal antibodies. *Cancer Res. 44* (1985):5320-5325.

8. Goldenberg, D. M., Deland, F., Kim, E., Bennett, S., Primus, F. J., Nagel, J. R., Esles, N., Desimore, P., and Rayburn, P. Use of radiolabelled antibodies to carcinoembryonic antigen for the detection of localization of diverse cancers by external photoscanning. *N. Engl. J. Med. 298*, no. 25 (1978):1384.

9. Baldwin, R. W., and Pimm, M. V. Antitumor monoclonal antibodies
    for radioimmunodetection of tumors and drug targeting. *Cancer
    Metastasis Reviews*, Vol. 2, p. 89. Ed. by I. J. Fidler.
    Boston: Martinus Nijhoff Publishers, 1983.

10. Farrands, P. A., Perkins, A., Sully, L., Hopkins, J. S., Pimm,
    M. V., Baldwin, R. W., and Hardcastle, J. D. Localisation of
    human osteosarcoma by anti-tumour monoclonal antibody (791T/36).
    J. Bone Joint Surgery 65 (1983):638.

11. Farrands, P. A., Perkins, A. C., Pimm, M. V., Embleton, M. J.,
    Hardy, J. D., Baldwin, R. W., and Hardcastle, J. D. Radioimmuno-
    detection of human colorectal cancers by an anti-tumour monoclonal
    antibody. *Lancet 2*, no. 8295 (1982):397.

12. Armitage, N. C., Perkins, A. C., Pimm, M. V., Farrands, P. A.,
    Baldwin, R. W., and Hardcastle, J. D. The localisation of an
    anti-tumour monoclonal antibody (791T/36) in gastrointestinal
    tumours. *Brit. J. Surgery 71* (1984):407.

13. Baldwin, R. W., Pimm, M. V., Embleton, M. J., Armitage, N. A.,
    Farrands, P. A., Hardcastle, J. D., and Perkins, A. Monoclonal
    antibody 791T/36 for tumor detection and therapy of metastases.
    In *Cancer Invasion and Metastasis: Biologic and Therapeutic
    Aspects.* Ed. by Garth L. Nicholason and Luka Milas. New York:
    Raven Press, 1984, pp. 437-456.

14. Symonds, E. M., Perkins, A. C., Pimm, M. V., Baldwin, R. W.,
    Hardy, J. G., and Williams, D. A. Clinical implications for
    immunoscintigraphy in patients with ovarian malignancy: A pre-
    liminary study using monoclonal antibody 791T/36. *Br. J. Obst.
    & Gynae. 92* (1985):270.

15. Embleton, M. J., Rowland, G. F., Simmonds, R. G., Jacobs, E.,
    Marsden, C. H., and Baldwin, R. W. Selective cytotoxicity
    against human tumour cells by a vindesine-monoclonal antibody
    conjugate. *Brit. J. Cancer 47*, no. 1 (1983):43.

16. Baldwin, R. W., Embleton, M. J., Garnett, M. C., Pimm, M. V.,
    Price, M. R., Armitage, N. A., Farrands, P. A., Hardcastle,
    J. D., Perkins, A., and Rowland, G. F. Application of monoclonal
    antibody 791T/36 for radioimmunodetection of human tumours and
    for targeting cytotoxic drugs. *Protides of the Biological
    Fluids 31* (1984):775

17. Rowland, G. F., Axton, C. A., Baldwin, R. W., Brown, J. P.,
    Corvalan, J. R. F., Embleton, M. J., Gore, V. A., Hellstrom, I.,
    Hellstrom, K. E., Jacobs, E., Marsden, C. H., Pimm, M. V.,
    Simmonds, R. G., and Smith, W. Anti-tumor properties of
    vindesine-monoclonal antibody conjugates. *Cancer Immunol.,
    Immunother. 19* (1985):1.

18. Price, M. R., Pimm, M. V., and Baldwin, R. W. Complement-
    dependent cytotoxicity of anti-human osteogenic sarcoma monoclonal
    antibodies. *Brit. J. Cancer 46*, no. 4 (1982):601.

19. Garnett, M. C., Embleton, M. J., Jacobs, E., and Baldwin, R. W.
    Preparation and properties of a drug-carrier-antibody conjugate
    showing selective antibody directed cytotoxicity in vitro. *Int.
    J. Cancer 31*, no. 5 (1983):661.

20. Baldwin, R. W., and Byers, V. S. Immunoregulation by bacterial
    organisms and their role in the immunotherapy of cancer. In
    Immunostimulation, p. 73. Ed. by L. Chedid, P. A. Miescher,
    and H. J. Mueller-Eberhard. New York: Springer-Verlag, 1980.

21. Pelham, Julie M., Gray, J. D., Flannery, G. R., Pimm, M. V.,
    and Baldwin, R. W. Interferon conjugation to human osteogenic
    sarcoma monoclonal antibody 791T/36. *Cancer Immunol. Immunother.
    15* (1983):210.

22. Baldwin, R. W., Flannery, G. R., Pelham, J. M., and Gray, J. D.
    Immunomodulation by IFN-conjugated monoclonal antibody to human
    osteogenic sarcoma. *Proc. Amer. Assoc. Cancer Res. 23* (1982):
    254.

23. Flannery, G. R., Pelham, J. M., Gray, J. D., and Baldwin, R. W.
    Immunomodulation: NK cells activated by interferon-conjugated
    monoclonal antibody against human osteosarcoma. *Europ. J.
    Cancer Clin. Oncol. 20* (1984):791.

24. Laxton, R. R., Roe, R., Robins, R. A., Baldwin, R. W. A mathe-
    matical theory for bivalent monoclonal antibody interactions
    with cell surface antigens. In preparation, 1983.

25. Roe, R., Robins, R. A., Laxton, R. R., and Baldwin, R. W.
    Kinetics of divalent monoclonal antibody binding to tumour cell
    surface antigens using flow cytometry: Standardization and
    mathematical analysis. *Molecular Immunol. 22* (1985):11.

26. Farquhar, M. G. Multiple pathways of exocytosis, endocytosis,
    and membrane recycling: Validation of a Golgi route. *Federa-
    tion Proceedings 42*, no. 8 (1983):2406.

27. Ettinger, D. S., Order, S. E., Wharam, M. D., Parker, M. K.,
    Klein, J. L., and Leichner, K. Phase-I-II study of isotopic
    immunoglobulin therapy for primary liver cancer. *Cancer Treat-
    ment Reports 66*, no. 2 (1982):289.

MATHEMATICAL MODELING FOR MONOCLONAL ANTIBODY THERAPY OF LEUKEMIA

ROBERT O. DILLMAN AND **JAMES A. KOZIOL**   University of California, San Diego; and Veterans Administration Medical Center

ABSTRACT: We propose and develop a pharmacokinetic model for the quantitative analysis of dose-time-cell-survival curves devolving from our clinical investigations of infusions of the murine monoclonal antibody T101 into patients with advanced refractory chronic lymphocytic leukemia (CLL) and cutaneous T cell lymphoma (CTCL). Our model offers tentative explanations for the dynamics of monoclonal antibody therapy in CLL and CTCL.

1. INTRODUCTION

The advent of hybridoma technology has made readily available highly specific monoclonal antibodies that may be of significant value for cancer detection and therapy (Kohler and Milstein, 1975; Dillman, 1984). Already, monoclonal antibodies (MoAbs) serve as immunologic reagents in assays that screen for the presence of tumor antigens shed into the circulatory system and, linked to radioactive tracers, are being tested for tumor imaging in both murine and human tumor systems (Halpern et al., 1983; Goldenberg and DeLand, 1982). Trials of MoAb

From *Proceedings of the Berkeley Conference in Honor of Jerzy Neyman and Jack Kiefer*, Volume I, Lucien M. Le Cam and Richard A. Olshen, eds., copyright © 1985 by Wadsworth, Inc. All rights reserved.

serotherapy have produced promising results in animal tumor model
systems (Foon et al., 1982; Badger and Bernstein, 1983). We thus
have embarked on preliminary trials to evaluate the use of MoAbs as
passive therapy in man. We have limited our treatments to individuals
with advanced chronic lymphocytic leukemia (CLL) and cutaneous T cell
lymphoma (CTCL) who have been refractory to standard therapy. We
have utilized T101, a murine monoclonal antibody that detects a
65,000 dalton antigen (T65) on the surface of CLL cells, normal and
malignant T cells, and thymocytes, but which has yet to be detected in
nonlymphoid tissues (Royston et al., 1980). The rationale for the
doses and schedules of such therapy would be enhanced were it based
upon knowledge of the kinetics of the injected antibody, including
rates of clearance from the blood and diffusion into extravascular
areas, and of the mechanisms of antibody-induced leukemic cell kill
via the humoral and cell-mediated pathways of the immune system.
These mechanisms may in fact include complement-mediated cytotoxicity,
antibody-dependent effector cell-mediated cytotoxicity, and phagocy-
tosis of opsonized circulating tumor cells by the recticuloendothelial
(RE) system. We address these issues herein and provide a kinetic
analysis of our preliminary MoAb studies.

2. THE MODELS AND COMPARISON WITH CLINICAL RESULTS

A number of aspects of T101 MoAb therapy are of clinical relevance and,
indeed, have been reported in detail elsewhere (Dillman et al., 1982a-
1982c, 1984). However, because optimal implementation of immunotherapeutic regimens will involve the consideration of kinetic phenomena, we
focus here on the interplay between cell population kinetics and MoAb
levels in the body. Our approach is deterministic, using compartmental models suggested by our experimental data for the quantitation
and prediction of the cell-killing behavior associated with MoAb
therapy. The representation of the body as a system of compartments
is a commonly employed technique leading to the pharmacokinetic characterization of drugs, and it is not our purpose to survey this

procedure anew; we instead refer the interested reader to Jacquez (1972), and Wagner (1975) for mathematical developments of compartmental analyses in biology and medicine.

Let us begin with cell kinetics of CLL. In Figure 1 we depict a steady-state preinfusion model of CLL. Stem or mother cells, denoted by M, reside in the marrow and lymph nodes and give rise to offspring daughter cells (D) at some low rate (rate constants k_+, k_-). Cell surface antigen that is reactive with T101 is, relatively speaking, either present (+) or absent (-) on the surface membranes of mother cells and their daughters. The daughter cells are eventually eliminated or removed from the central circulatory compartment at some rate k_e.

Preinfusion

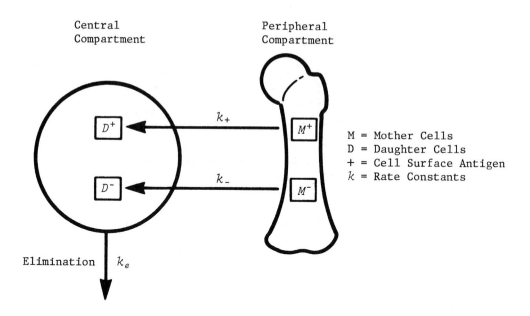

Figure 1. Dynamics of CLL prior to first infusion of MoAbs

Typically in CLL there may be with time a proliferation of leukemic cells in the marrow and in the central compartment; it has been suggested that this accumulation of cells is due not to an increased rate of proliferation over that of normal stem cells, but to a decreased rate of cell death. Thus, in terms of our model, we have $k_+ + k_- \geq k_e$, and the population of circulating leukemic cells ($D^+ + D^-$) may increase with time.

Consider next the dynamics at time of first infusion, as depicted in Figure 2. We have a continuous infusion of MoAb into the central compartment at rate $i(\cdot)$. These infusions typically last from 1 to 24 hours, a period that is usually too short to alter the transformation rate constants k_+, k_-, and k_e. We assume instantaneous and uniform mixing of MoAbs throughout the central compartment and first-order transfer of MoAb between the central and the peripheral compartments.

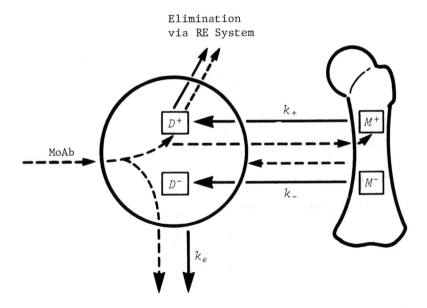

Figure 2. Dynamics at time of first infusion

In a patient previously unexposed to T101, free antibody either is irreversibly bound to D^+ or M^+ cells or is eliminated via usual protein degradative networks. Besides the usual elimination of daughter cells, D^+ cells whose surface anitgen attracts T101 antibody are selectively eliminated through the RE system (Dillman et al., 1982c).

The dynamics of the system subsequent to infusion are shown in Figure 3. Stem cells may give rise to daughter cells at an augmented rate to fill the depleted reservoir in the central compartment. Another possibility is that daughter cells residing outside the central compartment enter that compartment in response to a leukostatic feedback pathway. In addition, a process called *modulation* occurs, whereby the presence of antibody results in disappearance of reactive surface antigen in M⁺ cells. That is, M⁺ cells are biotransformed to M⁻ cells. Further discussion of this phenomenon and its implications in our clinical studies is reported elsewhere (Dillman, 1984).

Postinfusion

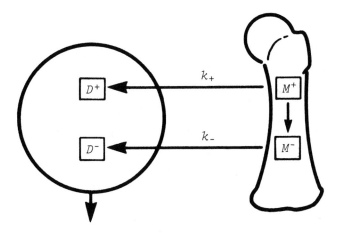

Figure 3. Dynamics subsequent to first infusion

Since our experimental data were obtained primarily during periods of infusion, we shall focus on the schema of Figure 2 and examine the pharmacokinetics for this period in the following manner. Let t denote time, with start of infusion at $t = 0$, and let

$A_f(t)$ = concentration of free (unbound) T101 MoAb in the central compartment at time t;

$M_f(t)$ = concentration of free T101 MoAb in the peripheral compartment;

$L(t)$ = concentration of D⁺ cells in the central compartment;

$R_f(t)$ = number of free (unbound) receptors on a D^+ cell;

$F(t) = R_f(t) \cdot L(t)$ = concentration of free receptors in the central compartment;

k_i, k_{ij} = generic rate constants.

(Note in particular that we are assuming homogeneity of D^+ cells in terms of numbers of cell surface receptors for T101 antibody.) We now consider the model described by the following system of differential equations:

$$dA_f(t)/dt = k_0 i(t) - k_1 A_f - k_{12} A_f + k_{21} M_f - k_3 A_f F; \qquad (2.1)$$

$$dL(t)/dt = \begin{cases} -k_4 L(t), & L(t) \leqslant \ell_0, \; R_f(t) \leqslant r_0, \\ -k_5, & L(t) \geqslant \ell_0, \; R_f(t) \leqslant r_0; \end{cases} \qquad (2.2)$$

$$dF(t)/dt = -k_6 F A_f - k_7 F. \qquad (2.3)$$

In equation (2.1), k_0 is a normalizing constant; $-k_1 A_f$ corresponds to first-order elimination of MoAb via protein metabolic and degradative networks; the terms $-k_{12} A_f$ and $k_{21} M_f$ correspond to first-order transfer of free MoAb between the central and peripheral compartments; the last term $-k_3 A_f F$, is reflective of an irreversible biomolecular reaction between MoAb and antigen with forward rate constant k_3. Although we consider the antibody-receptor binding to be a first-order process, we recognize that a Michaelis-Menten relation might be preferable (though differences would be minimal at low-concentration values).

Equation (2.2) characterizes the elimination of D^+ cells via the RE network during infusion. Three points merit attention: typically, generation of new D^+ cells at the preinfusion rate k_+ is inconsequential relative to the active elimination of D cells during infusion, and hence we assume $k_+ = 0$; we postulate a threshold level of bound receptors per cell before elimination (see Dillman et al., 1982c, 1983a); and saturation of the RE system can theoretically occur. With regard to this last point, note that a constant blood flow is maintained through the RE system, regardless of the varying concentration of D^+ cells, and that the RE system likely has a fixed capacity for eliminating antibody-cell complexes. Thus, once threshold is achieved, elimination should be zero order or first order, depending on whether the system is or is not saturated. The RE system might preferentially remove cells with increased numbers of bound antibodies, but this should not alter the elimination rate.

Equation (2.3) incorporates two dynamic processes: the biomolecular reaction between circulating MoAb and free receptors, and the elimination of D^+ cells. Since $F = R_f \cdot L$, further development of equation (2.3) is feasible and leads to useful formulations for numerical analysis.

We emphasize that our models are not merely artificial constructs but are designed to provide biological mechanisms for experimental data. Our compartments are distinct anatomic or physiologic entities, and the equations (2.1)-(2.3) are meant to reflect the time course of measurable parameters during periods of infusion. We are thus building models to explain our observations; the suitability of our models and derived equations can therefore appropriately be adjudged by their degree of consonance with experimental data.

Unfortunately, equations (2.1)-(2.3) do not describe a system of first-order, linear, ordinary differential equations; so closed-form solutions will in general not be available. Instead, we rely upon Runge-Kutta methods and nonlinear least squares for numerical solution of these equations.

We note first that equation (2.2) describes either zero-order or first-order elimination of D^+ cells, beyond a threshold level of MoAb binding. Regrettably, our experimental data are not sufficiently informative to discriminate between the two alternatives. In Figure 4 we plot (the logs of) surviving fractions of D^+ cells over time of infusion for one of our patients. Although first-order elimination, as evinced by a linear fit, seems an adequate description of the observed kinetics ($R^2 = .82$), this does not preclude the theoretical possibility of saturation of the RE system. In fact, a capacity-limited or Michaelis-Menten type of elimination process would seem satisfactory to us, but at the cost of undue complications in our numerical calculations.

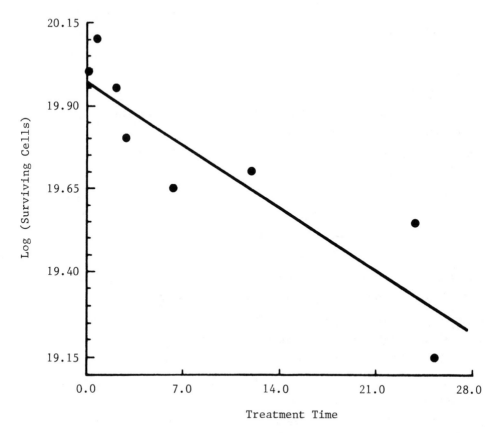

Figure 4. D⁺ cells remaining in the peripheral circulation over the period of initial treatment of a patient with CLL, who received a 24-h infusion of 10 mg of T101 monoclonal antibody

We turn next to joint consideration of free MoAb levels and receptor sites. Our modeling here is hampered by insufficient data: We have at best only indirect evidence, by immunofluorescent techniques, pertaining to numbers and proportions of unbound receptors, and we have measured levels of free MoAb only in the central compartment. Nevertheless, with appropriate choice of parameter values, our system of equations seems to model observed MoAb levels in our patients in a quite satisfactory manner, as can be seen in Figure 5.

Let us briefly consider the kinetic phenomena depicted in Figure 3, We have little available data pertaining to the repopulation of hematopoietic cells in the central compartment following discontinuance

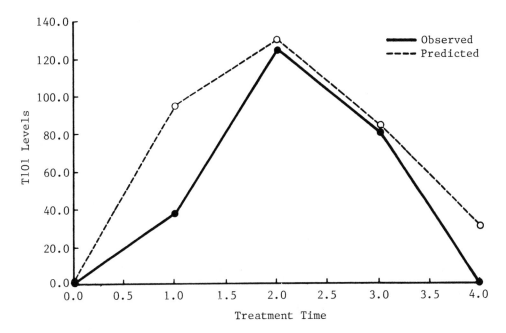

Figure 5. Observed and predicted levels of free T101 monoclonal antibody in the peripheral circulation over the period of initial treatment of a patient with CTCL, who received a 2-h infusion of 2 mg of T101

of infusion and removal of circulating antibody, though we have little reason to doubt that repopulation is likely at an early exponential growth rate that gradually tapers off as a steady-state number of cells, about equal to the pretreatment level, is attained. (In fact, we occasionally observe an overshoot, suggestive of stimulation, followed by a decline to the pretreatment level.) The Gompertz relative growth curve has been successfully used in situations such as this (Laird, 1964; Skipper and Perry, 1970). The occurrence of modulation merits attention, however, as its presence has significant impact on the optimal design of immunotherapeutic regimens. Given our inability to sample the peripheral compartment systematically, modeling of this phenomenon is, to a certain extent, moot; nevertheless, in some patients, repeated courses of immunotherapy were of decreasing efficacy. Figure 6 summarizes some findings bearing on this matter.

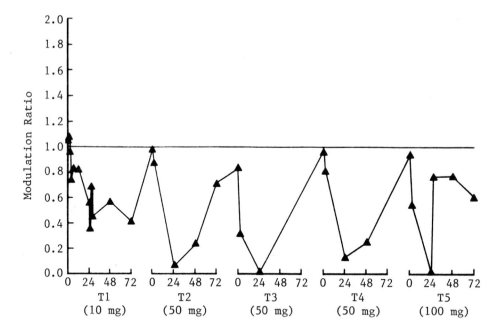

Figure 6. Modulation ratio (ratio of number of cells expressing T65 antigen reactive with T101 MoAb, to number of cells expressing Ia antigen, which is found on CLL cells) in a patient with CLL, over five courses of T101 infusion

3. DISCUSSION

We have tried to incorporate the salient features of MoAb therapy in CLL and CTCL into our mathematical models; nevertheless, we must recognize certain limitations in the models. The models of the underlying kinetics are inherently nonlinear, leading, in particular, to formidable problems with numerical investigations; unfortunately, we do not think this difficulty can be circumvented. Undoubtedly, the models can be refined as more knowledge of such aspects as the interplay between MoAbs and receptors, the mechanism of cell kill in the RE system, or the phenomenon of modulation becomes available. This should be balanced by the paucity of experimental data in the clinical setting (arising in part from obvious limitations on peripheral blood drawing

and our inability to identify, let alone sample from, the malignant
hemotopoietic stem cell pool), which precludes exploratory or confirmatory model fitting to all but the more elementary models. Model
identifiability and in vivo applications of sophisticated models will
thus remain significant problems.

The phenomenon of antigenic modulation merits additional comment.
In vitro incubation of excess monoclonal antibody with target cells
results in binding of antibody to the surface antigen. In the presence
of appropriate effector cells, or complement, the target cells are
destroyed. In the absence of these, antibody and antigen are internal-
ized, and antigen is ultimately re-expressed; additional antibody-
antigen complexes internalize until no antibody is left in the media.
At any time, if cells are removed and examined for presence of surface
antigen, only a small amount will be found. However, if the cells
are allowed to incubate in media, free of the antibody, the antigen
will be re-expressed at usual levels.

This same process is evident in vivo. Prolonged serum levels of
T101 are eventually associated with appearance in the circulation of
cells that have very low levels of T65 on their surface but otherwise are phenotypically the same target cells. Once the T101 dis-
appears completely, cells in the circulation exhibit the same pheno-
type and the same density and intensity of T65 antigen expression as
before treatment. The same phenomenon has been observed in the skin
of patients with CTCL and in the bone marrow of patients with CLL.
Our data suggest that antibody-coated target cells in the circulation
are efficiently removed by the RE system but that modulation is occurring in those cells that are outside the circulation. Those cells
that enter the circulation have an insufficient level of antibody on
their surface to induce removal in the RE system. Thus, therapy beyond
a certain dose level, or beyond a certain time, is apparently wasted in
terms of potential tumor cell elimination.

We should emphasize that equations (2.1)-(2.3) most accurately
describe the system kinetics during *initial* infusion of MoAb. In our
clinical experiences, with subsequent courses of MoAb infusion, certain individuals seem to evince increased sensitivity in their RE

systems to MoAb-leukemic cell complexes, leading to heightened elimination; yet others produce endogenous antibodies to the murine T101 MoAb,
which substantially alters the bioavailability and thereby the potency
of these subsequent infusions. Such factors need to be accounted for
in our models; but more importantly, the combination of host immune
response to murine MoAb and the occurrence of modulation have significant implications for the frequency and duration of MoAb therapeutic
dosage.

What, then, is the potential of MoAb therapy of leukemia? Note
that the mechanism of antibody-induced cell kill in our model is solely
via the RE system. We have no experimental evidence of (complement-
mediated) toxicity to leukemic stem cells residing in the peripheral
compartment (although animal models suggest this can occur). To the
contrary: MoAb is delivered to the peripheral compartment, where its
presence leads perversely to modulation. In other words, MoAb infusion
can be eminently successful in temporarily reducing the circulating
pool of leukemic cells; but this therapy is not curative (because it
does not appear to eliminate the malignant stem cell population in
tissues), and with time it will become increasingly ineffective because
of modulation and host immune rejection of foreign antibody. (To be
sure, this last factor is not evident in all cases and can be eliminated by the development of human MoAb lines, but the other criticisms
remain.) On the other hand, we believe a two-pronged approach to
immunotherapy could be potentially curative: Initially, a bolus injection would be given to clear the circulating compartment of leukemic
cells; then an infusion of MoAb linked to a radioisotope (or, perhaps,
a toxin or chemotherapeutic agent) could be given with the hope of
eliminating the malignant stem cells at the tissue level. (This same
principle might similarly apply in colon cancer and prostate cancer
therapy.) Details would need to be worked out. For example,
our clinical experience suggests that large bolus injections of MoAb
can produce undesirable reactions in certain leukemic patients; consideration might be given to the infusion of a "cocktail" of MoAbs to
avoid difficulties with modulation. Also, the integrity of the MoAb-
radioisotope complex and its deliverability to the peripheral

compartment must be assured. Still, we believe this method holds promise and certainly merits further evaluation. Our clinical results should be construed not as contraindicative of the efficacy of MoAb therapy but as providing requisite information preliminary to this curative approach. In fact, dramatic clinical responses to monoclonal antibodies have already been reported (Miller et al., 1981a, 1981b). Further evaluation of passive serotherapy logically would include investigation of other antibodies to different antigens, cocktails of monoclonal antibodies, and use of a variety of doses and schedules of administration. If the problems of modulation and endogenous antimouse production are insurmountable, then the only in vivo therapeutic application of monoclonal antibodies may indeed be as carriers of cytotoxic substances. In that situation, modulation— that is, internalization of antigen-antibody complexes—could be advantageous for the elimination of tumor cells, and repeated treatments over a short period of time would presumably not be complicated by the production of antimouse antibodies.

REFERENCES

Badger, C. C., and Bernstein, I. D. Therapy of murine leukemia with monoclonal antibody against a normal differentiation antigen. *J. Exp. Med.* 1983; 157: 828-842.

Dillman, R. O. Monoclonal antibodies in the treatment of cancer. *CRC Crit. Rev. Hematol./Oncol.* 1 (1984):357-385.

Dillman, R. O., Sobol, R. E., Collins, H., Beauregard, J., and Royston, I. T101 monoclonal antibody therapy in chronic lymphocytic leukemia. In *Hybridomas in the Diagnosis and Treatment of Cancer*, Mitchell, M. S., and Oettgen, H. F., eds. Raven Press, New York, Prog. Cancer Research Ther., 1982a; 21: 151-172.

Dillman, R. O., Sobol, R. E., and Toyston, I. Preliminary experiences with murine monoclonal antibody infusions in cancer patients. In *Protides of the Biological Fluids*, Peeters, H., ed. Pergamon Press, New York, 1982b; 29: 915-920.

Dillman, R. O., Beauregard, J. C., Shawler, D. L., Sobol, R. E., and Royston, I. Results of early trials using murine monoclonal antibodies as anti-cancer therapy. In *Protides of the Biological Fluids*, Peeters, H., ed. Pergamon Press, New York, 1983a; 30: 353-358.

Dillman, R. O., Shawler, D. L., Dillman, J. B, and Toyston, I. R. Therapy of chronic lymphocytic leukemia and cutaneous T-cell lymphoma with T101 monoclonal antibody. *J. Clin. Oncol. 2* (1984):881–891.

Dillman, R. O., Shawler, D. L., Sobol, R. E., Collins, H. A., Beauregard, J. C., Wormsley, S. B., and Toyston, I. Murine monoclonal antibody therapy in two patients with chronic lymphocytic leukemia. *Blood* 1982c; 59: 1036–1045.

Foon, K. A., Bernhard, M. I., and Oldham, R. K. Monoclonal antibody therapy: assessment by animal tumor models. *J. Biol. Resp. Mod.* 1982; 1: 227–304.

Goldenberg, D. M., and De land, F. H. History and status of tumor imaging with radiolabeled antibodies. *J. Biol. Resp. Mod.* 1982; 1: 121–136.

Halpern, S. E., Dillman, R. O., and Hagan, P. L. The problems and promise of monoclonal antitumor antibodies. *Diag. Imag.* 1983; 5 (No. 6): 40–47.

Jacquez, J. A. *Compartmental Analysis in Biology and Medicine*. Elsevier, New York, 1972.

Kohler, G., and Milstein, C. Continuous cultures of fused cells secreting antibody of predetermined specificity. *Nature* 1975; 256: 495–597.

Laird, A. K. Dynamics of tumour growth. *Brit. J. Cancer* 1964; 18: 841–902.

Miller, R. A., and Levy, R. Response of cutaneous T cell lymphoma to therapy with hybridoma monoclonal antibody. *Lancet* 1981; 2: 226–230.

Miller, R. A., Maloney, D. B., Warnke, R., et al. Treatment of B-cell lymphoma with monoclonal anti-idiotype antibody. *N. Engl. J. Med.* 1981; 306: 517–522.

Ritz, J., Pesando, J. M., Sallan, S. E., et al. Serotherapy of acute lymphoblastic leukemia with monoclonal antibody. *Blood* 1981; 58: 141–152.

Royston, I., Majda, J. A. Baird, S. M., Meserve, B. L., and Griffiths, J. C. Human T cell antigens defined by monoclonal antibodies: the 65,000-dalton antigen of T cells (T65) is also found on chronic lymphocytic leukemic cells bearing surface immunoglobulin. *J. Immunol.* 1980; 125: 725–731.

Rubinow, S. E. *Mathematical Problems in the Biological Sciences*. SIAM, Philadelphia, 1973.

Skipper, H. E., and Perry, S. Kinetics of normal and leukemic leukocyte populations and relevance to chemotherapy. *Cancer Res.* 1970; 30: 1883–1892.

Wagner, J. G. *Fundamentals of Clinical Pharmacokinetics*. Drug Intelligence Publications, Inc., Hamilton, Illinois, 1975.

THE SIGNIFICANCE OF CELL VARIATION
IN UNDERSTANDING CANCER

H. RUBIN University of California, Berkeley

It seems self-evident that cancer is a heritable state of cells.
If it were not so, how would a self-perpetuating tumor arise? As
obvious as it may seem, even this minimal description of the cancer
cell must be qualified. There are *conditional* cancers in which the
expression of the malignant state is dependent on the hormonal state
of the host. Mouse mammary tumors, for example, can arise and regress
repeatedly depending on the hormonal state of the female during preg-
nancy [1]. Eventually, however, the tumor becomes autonomous, that
is, no longer regulated by hormonal state of the host. In the early
stages, therefore, the cancerous condition of the tumor cells cannot
be called heritable without qualification because this condition is
expressed only in certain restricted environments.

There is, however, a more subtle aspect to the heritability of
the cancerous state. Cellular heritability usually implies a precise

The research reported here was supported by NIH Public Health Service
Grant CA 15744 and Department of Energy Contract 03-79-EV10277 and
was done with the technical assistance of Berbie M. Chu.

replication of the parental cell in its progeny. The idea accords
with the concept of DNA as the sole transmitter of heritable information. At this point, complications arise because cells change a
very large number of characteristics during differentiation and,
with the sole exception of immunogenic cells, do so with no evidence
of change in the primary structure of DNA. These changes are
inheritable: Liver cells reproduce as liver cells, and skin cells
likewise. Unlike mutations, which are rare events, differentiation
occurs in all cells of highly developed organisms and therefore must
be considered the most common source of heritable change in such
organisms. They share at least one character with conditional tumors
in that they require a particular environment to be induced (*determination*) and another one to be expressed (*differentiation*).

It is less well appreciated that fully differentiated cells lose
their identifying characteristics when placed in a strange environment. This is most dramatically shown when cells in a tissue are
dispersed into a suspension of single cells and grown in culture.
Cells of most epithelial tissues quickly lose their identifying
traits, and this loss soon becomes irreversible [2]. The major exception to this rule is the connective tissue cell known as the *fibro-
blast*, which exists in a dispersed state in tissue and maintains many
of its characteristics in cell culture. Even these, however, undergo
many changes in culture. For example, normal human fibroblasts
gradually lose the capacity to multiply and finally die out [3].
Rodent cells do the same at first, but then abnormal variants appear
that can grow indefinitely in culture [4]. Almost all these established lines can produce tumors when inoculated into animals [5].
So it seems that cells try to adapt to their new environment in culture. Most of them fail, but those that do adapt usually become
neoplastic in the process. Certainly the initial gradual loss of
reproductive capacity, because it involves the entire population as
well as many cellular properties [4], is a change that has more in
common with differentiation than mutation. Establishment is a later
step in the evolution of cell populations in culture, but there is
nothing to indicate it differs in kind from the preceding changes.

Given the neoplastic state of most established lines, it would not appear to be amiss to consider the malignant change a misguided form of differentiation.

When fibroblasts are put in culture, another form of change occurs that is even more subtle than the loss of differentiation or the loss of growth capacity. Indeed, it is so subtle that it was just recognized a few years ago and is still rarely acknowledged. When fibroblasts from the same small explant of human foreskin are cloned (i.e., single cells are grown into large populations), the enzyme contents of the individual clones differ widely from one another [6]. What is at least as surprising is that when the descendants of a single cloned cell are themselves subcloned, the subclones differ from one another. Subclones that have a high content of one enzyme may have a low or high content of another: There is no correlation between them. When more than a few enzymes are considered, the number of possible combinations becomes immense, and it is not farfetched to consider each cell unique. This extreme variegation comes under the heading of *cellular heterogeneity*. It seems that when the ordering forces that operate in normal organized tissues are disrupted in cell culture, each cell goes its own way.

The foregoing discussion of heterogeneity has been restricted to normal cells whose tissue organization has been disrupted. Heterogeneity is much better known in tumor cells, but here it is found in the undisrupted tumor [7, 8]. In fact, a generalized mark of the malignancy of a tumor is the term *pleiomorphic*, which means differences in the appearance of the individual cells constituting a tumor. Nonetheless, the predominant classical view of tumors was that they consisted of homogeneous populations of malignant cells.

It was not until the early 1950s that tumor heterogeneity took on functional significance, and the classical view began to change. This occurred when Leslie Foulds, the British pathologist, described the phenomenon of *progression*, in which he recognized that a tumor can go through many stages, involving many combinations of characteristics, as it progresses to autonomous growth and death of the host [1]. The implication of this sequential development was that a tumor may consist of cells in various stages of progression from atypical but relatively normal cells to fully malignant ones.

The advent of tissue culture techniques allowed a more quantitative analysis of tumor heterogeneity. It was then clearly established that cells from the same tumor can be heterogeneous for a wide spectrum of characteristics including chromosome constitution (karyotype), antigenic composition, drug sensitivity, morphology, growth rate, and transplantability [10]. The use of cloning techniques allowed a detailed analysis of the metastatic capacity of tumor cells, which some consider the definitive attribute of the malignant state [11]. This analysis showed that metastatic variants occur at high frequency in tumors and in individual metastases arising from the tumors. As in the case of dispersed normal fibroblasts in culture, if we consider the large ensemble of characteristics already known to be expressed heterogeneously in tumors, we can easily imagine each tumor cell to be unique.

There is one crucial difference between the normal and the tumor cells, namely, that the former have to be dispersed in cell culture to express their capacity for heterogeneity whereas the latter have already done so in the intact tumor. This suggests that tumors might arise in a tissue because it has lost the capacity to maintain the organized state of its constituent cells. If such were the case, one would have to look for ultimate causes in cancer at the level of the morphogenetic field rather than the individual cell. This suggests a parallel with the advent of Maxwell's field concepts in physics, which were required to explain Faraday's electromagnetic phenomena. It may be instructive that the field concepts themselves had to undergo further revolutionary development in relativity and quantum mechanics to deal with otherwise intractable problems. Biology may have to develop equally revolutionary concepts before it can cope with the complexities of cancer.

After this excursion into basic theory, we can now return to our original question of whether the malignant state of a cell is heritable in its progeny. The answer in a broad sense is *yes*, but in a more detailed and precise sense is *no*. I have encountered this almost paradoxical situation in my own work at several levels. My examples come from the study of anchorage-independent multiplication by tumor cells. These cells, unlike normal cells, have no need for attachment to a solid substratum to multiply. Individual cells will multiply to form colonies when suspended in a semisolid medium such as an agar gel [12]. The number of such colonies can be counted and is expressed, in percentage, as the *colony-forming efficiency in agar* (CFEag). We have found that different subpopulations derived from the same clone of cells have different CFEag [13].

One can take a population with a low CFEag, seed it in agar, pick one of the rare colonies that arises, grow it out attached to a plastic dish, and then reassay it in agar. By this procedure one might expect to select a colony of cells that has a high CFEag. In fact, we end up with a population that has the same low CEFag as the original population [12]. This is strange indeed because the colony we picked contained several hundred cells at least, which means that many cell divisions occurred in its formation from a single cell. Yet most of these cells and their progeny do not initiate a colony when reseeded into agar. The implication is that at any given time some cells in a given tumor cell population have the capacity to multiply in suspension and pass that capacity on to enough of their progeny to maintain the increase in cell number required to produce a grossly visible colony. Yet, when the colony is dispersed, the probability of initiating a new colony is the same as that of the original population.

The other side of the coin is to clone the cells with low CFEag on plastic dishes, where all the cells produce colonies. In a random selection of colonies, one might expect some that can produce colonies in agar and some that cannot. What we find instead is that all the colonies grown on plastic have cells that can produce colonies in agar, but with the same low CFEag as the original population. At this point, it seems that the problem is as much one for the probability-oriented mathematician as it is for the empirically-oriented biologist.

I said that the colonies grown on plastic have the same low CFEag as the original population. Actually, the truth is that the CFEag varies from colony to colony, and the average is like that of the original population [12]. However, if one continues to grow the clones, and repeatedly assays them in agar at 1- or 2-week intervals, one finds that ehe CFEag of some individual clones fluctuate considerably from week to week. If we average the CFEag for the same clone after many assays, we find it is similar to the CFEag of the original uncloned population. Occasionally we encounter a clone that has an average CFEag in many assays that is significantly different from the others, but such clones are a distinct minority. So, there is heterogeneity from clone to clone in any given assay, and within a clone from assay to assay. We do not have the foggiest idea of what the source of the variation is, except that it is certainly not in the assay technique itself, which is highly reproducible with replicates of the same cell population.

There is still another type of unexplained variation. When our transformed mouse cell lines are inoculated into nude mice, they produce tumors. If a tumor grows very rapidly, it indicates that the cells are well adapted to growth in the animal. Such cells, when returned to culture are indistinguishable, so far, from the cultured cells used to initiate the tumor in appearance, growth rate, and CFEag. If, however, there is a delay in appearance of the tumor, and if once it appears, it grows slowly, the tumor cells have a much lower CFEag when returned to culture than the cells used to initiate the tumor. This is so although they look like the parental cells and multiply as rapidly on plastic. Clearly the population has changed on being introduced to a new environment. In several cases that we have studied, the tumor cells eventually return to the higher CFEag of the source cells after a month or two of growth on plastic. (However, we have recently found that cultures derived from some tumors retain their low CFEag indefinitely in culture.) The question that arises here is whether the change in CFEag that occurs when cells are switched from culture to the animal and vice versa is an induced, adaptive type of change or is the result

of a process of mutation and selection. We have some evidence
for the former, but the evidence is far from conclusive. Since the
ability to induce tumors in the animal, knowledge about the type of
change that controls the former should give some hint about the type
of change that determines tumor production.

It should be apparent by now that the cellular heritability of
the malignant state is a complex problem, as is the heritability
of the differentiated state. Heritability in both these cases is
likely to have a basis that is very different from that which determines the color or shape of a pea, which served as the starting point
for modern genetics. The basis for the latter, of course, lies in
the nucleotide sequence of DNA as eventually translated into the
structure of proteins. In the malignant and the differentiated
states, we have a form of control very much dependent on the devel-
opmental history and environment of the cell. When the environment
of the differentiated cell is changed, particularly when it is placed
in dispersed cell culture, the state of the cell is changed in an
unpredictable way, and each cell may differ from the other [6]. As
noted above, cells in a tumor within an animal seem to behave like
dispersed normal cells in culture. This observation directs our
attention to the tissue environment surrounding the cancer cells.
Could it be that the tissue as a whole undergoes some yet undetected
form of disorganization before a tumor appears? We know that tumors
undergo a prolonged, complex development known as *progression* before
they become highly malignant. It has been estimated that the average
mammary tumor in women has been developing through various silent
stages for ten years before it becomes detectable [1]. The exact
path that these stages take is never predictable, but it seems that
each step in any sequence is a low probability event, that is,
only one or a very few of the cells that have reached a certain
stage progress to the next stage. The implication of this line of
reasoning is that the immediate environment of the frankly malignant
tumor cells is not made up of fully normal cells. Indeed, Dr. Smith
presented evidence in this symposium that supposedly nonmalignant
cells adjacent to mammary tumors respond to irradiation in an abnormal

manner. This supports the notion of the tumor arising in an abnormal microenvironment and suggests the possibility that the altered environment plays a role in initiating the tumor in the first place.

This discussion has been a background to considering the possible molecular mechanism of the malignant transformation. The widely shared assumption of those in cancer research is that the heritable malignant state of cells must involve a change in the primary structure of DNA, because this is the only material basis we know for inheritance. The problem with jumping to this conclusion is that there are quasi-inheritable states, such as differentiation, which quite clearly do not involve changes in DNA structure; and the malignant state appears to be closer in kind to these states than to conventional mutations. Unfortunately, we do not know how the process of differentiation is controlled. I emphasize the term *process*, because there is no reason to believe that it can be explained by a single molecular event or even a few sequential events. An astute analysis has been made by Elsasser of the basis of heritability in organisms in which he concludes that two levels of control are involved [14, 15]. The first is precise replication of DNA, and the second is a less precise process which is involved in determining form and organization. He argues that the latter is of such complexity that it is not susceptible to reduction to its component parts. In this he carries forward the idea of Niels Bohr that those biological phenomena that characterize the living state cannot be exhaustively analyzed into their components without profoundly disturbing or destroying that state [16, 17]. There is an obvious similarity between the reasoning used here and that which occurred in physics in the development of Heisenberg's uncertainty principle. Bohr, in a general sense, and Elsasser, in a much more detailed way, propose that the defining characteristics of life must be studied without destroying life—an approach "from above" rather than "below"—and that a new epistemology is required in biology, just as it was in modern physics when it took the form of quantum mechanics. Indeed, Elsasser has developed just such a radical new epistemology for biology, which, having heterogeneity and indeterminacy as its basic assumptions, actually anticipated the more recent fundings on cellular variation described above. It is my feeling that a fundamental development in biological theory of this magnitude is a prerequisite for deeper understanding of differentiation and cancer.

REFERENCES

1. Foulds, L. (1969). *Neoplastic Development*, Vols. I and II. New York: Academic Press.

2. Harris, M. (1964). *Cell Culture and Somatic Variation*. New York: Holt, Rinehart and Winston.

3. Hayflick, L., and Moorhead, P. (1961). The serial cultivation of human diploid cell strains. *Exp. Cell Res. 25*, 585-621.

4. Todaro, G., and Green, H. (1963). Quantitative studies of the growth of mouse embryo cells in culture and their development into established lines. *J. Cell Biol. 17*, 299-313.

5. Boone, C., Takeichi, N., Paranjpij, M., and Gilden, R. (1976). Vasoformative sarcomas arising from Balb/3T3 cells attached to solid substrates. *Cancer Res. 36*, 1626-1633.

6. Griffin, J. E., Allman, D. R., Durrant, J. L., and Wilson, J. D. (1981). Variation in steroid 5α-reductase activity in cloned human skin fibroblasts. *J. Biol. Chem. 256*, 3662-3666.

7. Dexter, D. L., Kowalski, H. M., Blazar, B. A., Fligiel, Z., Vogel, R., and Heppner, G. H. (1978). Heterogeneity of tumor cells from a single mouse mammary tumor. *Cancer Res. 38*, 3174-3181.

8. Henderson, J. S., and Rous, P. (1962). The plating of tumor components on the subcutaneous expanses of young mice: Findings with benign and malignant epidermal growths and with mammary carcinomas. *J. Exp. Med. 115*, 1211.

9. Dunn, T. B. (1959). Morphology of mammary tumors in mice. In *Physiopathology of Cancer*. Ed. by F. Homburger. 2nd ed. New York: Hoeber-Harper.

10. Dexter, D. L., and Calabresi, P. (1982). Intraneoplastic diversity. *Biochim Biophys. Acta 695*, 97-112.

11. Fidler, I. J., and Hart, I. R. (1982). Biological diversity in metastatic neoplasms: Origins and implications. *Science 217*, 998-1003.

12. Rubin, H., Romerdahl, C. A., and Chu, B. M. (1983). Colony morphology and heritability of anchorage-independent growth among spontaneously transformed Balb/c3T3 cells. *JNCI 70*, 1087-1096.

13. Rubin, H., Chu, B. M., and Arnstein, P. (1983). Heritable variations in growth potential and morphology within a clone of Balb/c3T3 cells and their relation to tumor formation. *JNCI* (in press).

14. Elsasser, W. M. (1981). Principles of a new biological theory: A summary. *J. Theor. Biol. 89*, 131-150.

15. Elsasser, W. M. (1982). *Biological Theory on a Holistic Basis*. Privately printed; to be published in 1983.

16. Bohr, N. (19233). Light and life. *Nature 131*, 421-423, 457-459.

17. Bohr, N. (1954). *Unity of KNowledge in Atomic Physics and Human Knowledge*. New York: John Wiley & Sons.

THE REQUIREMENT FOR SURVIVAL MEASUREMENTS AT LOW DOSES

B. PALCIC, J. W. BROSING, G. Y. K. LAM, AND L. D. SKARSGARD
B.C. Cancer Foundation

ABSTRACT: Several mathematical models describing cell survival as a function of dose D were used to calculate the *oxygen enhancement ratio* (*OER*) as a function of radiation dose in oxygen. Four models among the many available in the literature were examined: multitarget, multitarget with initial slope, linear-quadratic, and repair-misrepair models. These are the models most often used in cellular radiobiology.

It was shown that though one cannot distinguish among the different models on the basis of their goodness of fit to experimental data at high doses, they predict quite different values for the OER when they are used to extrapolate the data at high doses (at which measurements are generally made) to low doses (at which the values are often needed). It has been concluded that, at present, meaningful OER values in the low-dose region can be ontained only from experimental measurements. It is likely that this conclusion also applies to other radiobiological parameters.

This work was supported by the Medical Research Council of Canada, the National Institute of Canada, and the B.C. Cancer Foundation.

INTRODUCTION

A significant proportion of all radiobiological knowledge is derived
from experiments with mammalian cells grown *in vitro* using cell survival after exposure to ionizing radiation as the end point. This
assay involves exposing cells to radiation under controlled conditions
and then testing their proliferative capacity. This is done by plating
cells into petri dishes and incubating for such time intervals that
cell colonies of 50 cells or more emerge from the individual surviving
cells after 10-15 generations. The colonies are then counted and the
relative surviving fraction S is established. This assay has changed
little since its introduction by Puck and Marcus in 1956 [1].

Relatively high doses of ionizing radiation (e.g., x-rays) are
required to produce effects large enough for accurate measurements
with this assay. With x-rays, for example, the dose used is generally
between 3 and 30 Gy (300 and 3000 rads) for mammalian cells *in vitro*.
This inactivates the cells to surviving fractions between 0.8 and
0.0001. Much of our radiobiological knowledge has been obtained by
studying the behavior of a very few surviving cells among many killed
cells.

Yet, for practical and fundamental reasons, we require radiobiological information at much lower dose levels, 2 Gy or less, at
which cell survival is high ($S > 0.8$). For example, radiotherapy of
patients delivers small doses of 2 Gy or less in many fractions over
several weeks. Questions of radiological health require knowledge at
even lower doses. The mechanism(s) by which radiation interacts with
the biological system also could be much better understood if data at
low doses were available.

Our knowledge of radiobiology in the low-dose region is
derived predominantly from extrapolation of data obtained at high
doses, for which one needs a mathematical model. Various models have
been used to describe cell survival as a function of dose. The most
commonly used model for fitting of data has been the multitarget model
[2]. A two-component model, which includes an exponential term in the
multitarget equation, is sometimes employed to introduce a nonzero

slope in the low-dose region [3]. The linear-quadratic (LQ) model has also been used extensively. This equation was derived from models that assume widely different initial postulates [4, 5, 6, 7]. More recently another model, the repair-misrepair (RMR) model developed by Dr. Tobias [8], has been used to fit data from several laboratories.

Many other mathematical models can be found in the literature, and new ones emerge regularly. The greatest problem with many of these models is that they contain more than two independent parameters. At best, our present experimental techniques allow us to define only two parameters with some precision. It can be argued that only the ratio of the two parameters can be determined accurately, with an error estimate of only a few percent. Three or more parameters usually make a reasonable error estimate impossible.

We shall limit this discussion to the following models:

$$S = 1 - (1 - e^{-D/D_0})^n \qquad \text{multitarget model (MT)} \qquad (1)$$

$$S = e^{-D/D_1}(1 - (1 - e^{-D/D_2})2)^n \quad \text{multitarget with initial slope (MTS)} \qquad (2)$$

$$S = e^{-\alpha D - \beta D^2} \qquad \text{linear-quadratic model (LQ)} \qquad (3)$$

$$S = e^{\alpha D}(1 + \alpha D/\varepsilon)^{\varepsilon\phi} \qquad \text{repair-misrepair (RMR) model} \qquad (4)$$

where S is the surviving fraction, D the dose, and D_0, n, D_1, D_2, α, β, ε, and ϕ are the independent parameters by which these models are described. These parameters usually have physical or biological meanings or implications.

Which model should be chosen to fit the data best? If all of them fit the data equally well, do they predict the same radiobiological values in the low-dose region? We shall use the oxygen enhancement ratio (OER) as an example in addressing these two questions.

It is well established that molecular oxygen radiosensitizes cells when it is present in cells during irradiation, and that in hypoxia (lack of oxygen), cells are much more radioresistant than are fully oxygenated cells. The effect of the presence or absence of molecular oxygen in cells at the time of irradiation is of great importance to radiotherapy. It is postulated that a large proportion of local failure of radiotherapy treatment of tumors is due to the

presence of radioresistant hypoxic cells in tumors.   The OER is
defined as the ratio of the doses needed to yield the same biological
effect under hypoxic and oxic conditions.

$$\text{OER} = \frac{D(\text{hypoxic})}{D(\text{oxic})} \tag{5}$$

The OER is usually calculated from full survival curves that are
measured under the two conditions.   Figure 1 shows curves for hypoxic
CHO cells in purified nitrogen (curve N_2) and oxic CHO cells in air
(curve O_2).   The cells were then irradiated to graded doses of x-rays,
and survival of cells was measured as described.   The shaded area in
the figure is the area where data are most needed but are very hard to
obtain with sufficient precision [9].

If one employs a mathematical model, the OER can be readily calculated as a function of dose.   Figures 2, 3, and 4 show such calculations, in which the OER was computed as a function of dose in oxygen
for different values of the independent parameters in equations (1),
(3), and (4), respectively.   By choosing the right combination of the
independent parameters, one can obtain a family of widely diverging
curves.   The data have one constraint at high doses; namely, that the
OER is approximately 3.   But even with this constraint, all models
allow the OER to assume any desired value at low doses.

Millar et al. [10] performed an analysis of their survival data
from experiments using Chinese hamster cells (V-79 line).   Using equations (1), (2), and (3) to fit the data, they showed by χ^2 analysis
that at a significance level of $p < 0.0005$ these models cannot be
distinguished; all models fit the data equally well.   Their analysis
also gave the best estimates of the independent parameters for the
models and their associated uncertainties.   Using the best estimates
of the independent parameters from the data of Millar et al., we have
calculated the OER as a function of dose for equations (1), (2), and
(3), the models that they used.   The results are shown in Figure 5.
Again, in the high-dose region all of the models converge; but at low
doses, predicted OER values range from 1 to 4, depending on the model
used.   Thus, if one is asked by a radiotherapist to predict the OER

value at doses below 2 Gy, one would have some misgivings about doing so on the basis of any of these mathematical models, even though they fit the cell survival data.

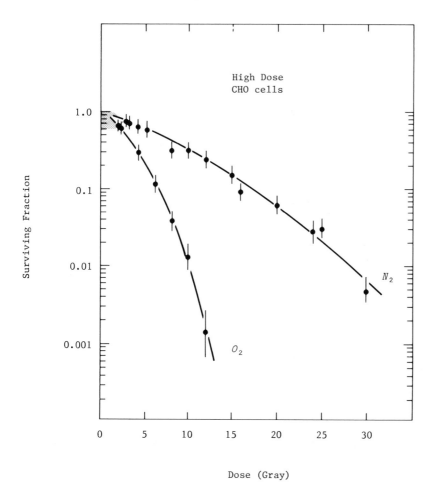

Figure 1. CHO cells were made hypoxic (N_2) or oxic (O_2) and were irradiated at 4°C to graded doses of x-rays. Survival measurements were made with the conventional assay. The solid lines represent best fits of the data to the equation $S = e^{-\alpha D - \beta D^2}$. The shaded area indicates where data are most needed.

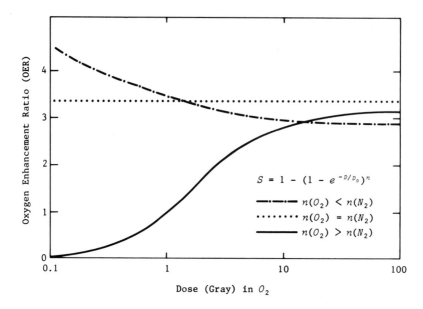

Figure 2. Equation (1) was used to calculate the OER as a function of dose in oxygen for different sets of values of D_0 and n.

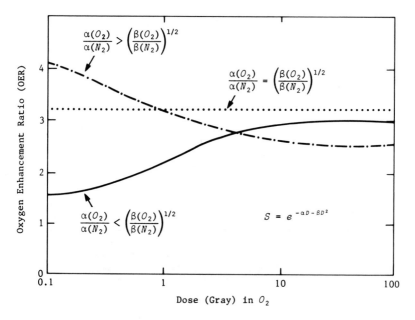

Figure 3. Equation (3) was used to calculate the OER as a function of dose in oxygen for three different ratios of α and β.

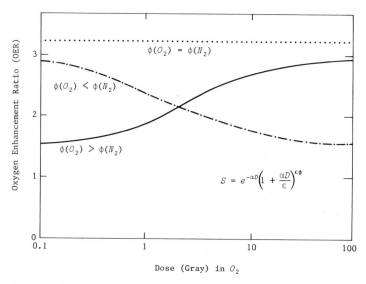

Figure 4. Equation (4) was used to calculate the OER as a function of dose in oxygen for different values of the parameter ϕ.

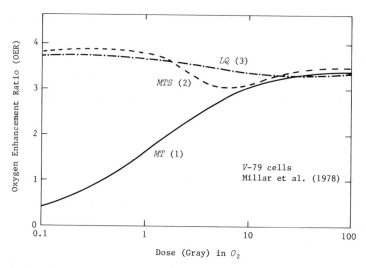

Figure 5. The best estimates of the independent parameters in equations (1), (2), and (3) were obtained from the data analysis of Millar et al. [10]. The OER as a function of dose was then computed with these parameters. Whereas in the high-dose region, $D > 10$ Gy, all functions converge, in the low-dose region they diverge and thus predict different OER values. χ^2 analysis showed that one cannot predict which of the above models is the best, because they all fit the experimental data at high doses equally well.

We believe, therefore, that it is imperative to acquire experimental data in the low-dose region if meaningful values of radiobiological parameters such as the OER are to be obtained. We have used a method whereby one can derive better survival data in the low-dose region by counting both the killed cells and the surviving cells [11, 9]. With this technique, we have obtained complete survival curves in the shaded area of Figure 1. The results are shown in Figure 6 and have been reported elsewhere [9]. We employed an iterative procedure [12, 13] to compute best estimates for the OER as the function of dose and to estimate the associated uncertainties [9]. These values are plotted in Figures 7 and 8. An attempt was made to fit the data by equations (1) and (3), in which only two independent parameters are needed. We searched for the best parameters to accommodate the data at high and low doses. It can be seen that equation (3) is better able to accommodate the data than equation (1). In the multitarget equation (1), if data are fitted for the low-dose region, then they do not fit the data at high doses, and vice versa. The linear-quadratic equation (3) fits the data rather well.

In addition to obtaining actual experimental values of the OER at low doses, these data impose other constraints on the mathematical models. For example, they suggest that in the multitarget model, equation (1), the extrapolation numbers $n(O_2)$ and $n(N_2)$ are not equal for oxic and hypoxic cells, respectively, but that $n(N_2)$ is less than $n(O_2)$. This observation is in agreement with those of Rèvèsz and colleagues [14] and others who have argued that the data from hypoxic and oxic cells cannot be fitted by forcing both curves through one and the same extrapolation number. Hence, if the multitarget model is correct, these data do not support the postulate that the OER is independent of dose, i.e., that it acts as a constant dose-modifying factor throughout the dose range [15].

In conclusion, we have shown that extrapolation of data from high doses into the low-dose region can be model dependent. Until one finds out which model is the "correct" one, one is unable to predict the values of radiobiological parameters at low doses. In particular, at least at present, determination of the oxygen enhancement ratios

(OER) in the low dose region requires actual measurement of them. Such measurements will put additional constraints on the independent parameters for various models and may, in fact, help to eliminate some of them.

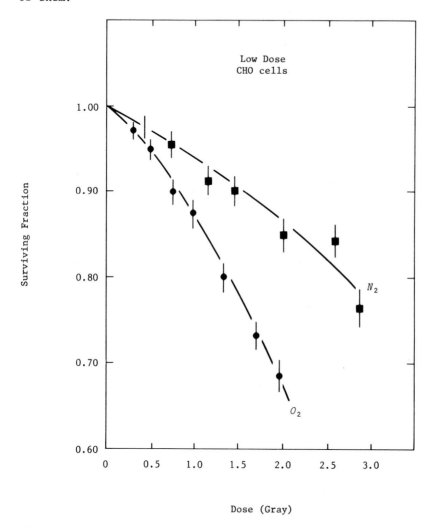

Dose (Gray)

Figure 6. CHO cells were irradiated with low doses of x-rays, and the survival was measured using the low-dose assay. The solid lines are best fits to the equation $S = e^{-\alpha D - \beta D^2}$.

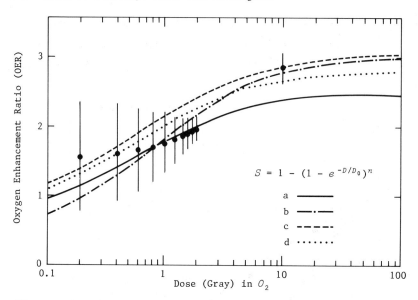

Figure 7. The best estimates for the OER and the associated uncertainties were computed by an iterative procedure. The data were then fitted by different combinations of the parameters n and D_0, using equation (1): (a) $n(O_2) = 8$, $n(N_2) = 6$, $D_0(N_2) = 4.0$ Gy; (b) $n(O_2) = 12$, $n(N_2) = 8$, $D_0(N_2) = 4.8$ Gy; (c) $n(O_2) = 8$, $n(N_2) = 6$, $D_0(N_2) = 4.8$ Gy; (d) $n(O_2) = 8$, $n(N_2) = 6$, $D_0(N_2) = 4.5$ Gy.

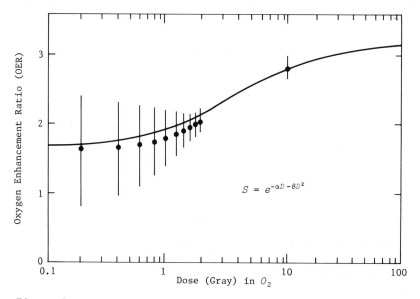

Figure 8. Best estimates for α and β values in equation (3) were obtained from the combined data of Figs. 1 and 6. The fitted values ($\bullet$) are the same as those in Fig. 7. The solid line is the OER as a function of dose calculated using the best estimates of α and β as described above.

REFERENCES

1. Puck, T. T., and Marcus, P. I. Action of x-rays on mammalian cells. *J. Exp. Med. 103* (1956), 653-666.

2. Elkind, M. M., and Whitmore, G. F. *The Radiobiology of Cultured Mammalian Cells.* Gordon and Breach, New York, 1967.

3. Bender, M. A., and Gooch, P. C. The kinetics of x-ray survival of mammalian cells *in vitro. Int. J. Radiat. Biol. 5* (1962), 133-145.

4. Sinclair, W. K. The shape of radiation survival curves of mammalian cells cultured *in vitro.* In *Biophysical Aspects of Radiation Quality*, pp. 21-48. IAEA, Vienna, 1966.

5. Kellerer, A. M., and Rossi, H. H. The theory of dual radiation action. In *Current Topics of Radiation Research 8* (1972), 85-158.

6. Chadwick, K. H., and Leenhouts, H. P. A molecular theory of cell survival. *Physics Med. Biol. 18* (1973), 78-87.

7. Douglas, B. G., and Fowler, J. F. The effect of multiple small doses of x-rays on skin reactions in the mouse and a basic interpretation. *Radiat. Res. 66* (1976), 401-426.

8. Tobias, C. A., Blakely, E. A., Ngo, F. Q. H., and Yang, T. C. H. The repair-misrepair model of cell survival. In *Radiation Biology in Cancer Research* (R. E. Meyn and H. R. Withers, eds.), pp. 195-230. Raven Press, New York, 1979.

9. Palcic, B., Brosing, J. W., and Skarsgard, L. D. Survival measurements at low doses: oxygen enhancement ratio. *Br. J. Cancer 46* (1982), 980-984.

10. Millar, B. C., Fielden, E. M., and Millar, J. L. Interpretation of survival curve data for Chinese hamster cells, line V-79 using multi-target, multi-target with initial slope, and α, β equations. *Int. J. Radiat. Biol. 33* (1978), 599-603.

11. Bedford, J. S., and Griggs, H. G. The estimation of survival at low doses and the limits of resolution of the single-cell-plating technique. In *Cell Survival at Low Doses of Radiation: Theoretical and Clinical Implications* (T. Alper, ed.), pp. 34-39. John Wiley & Sons, London, 1975.

12. Lam, G. K. Y., Henkelman, R. M., Douglas, B. G., and Eaves, C. J. Method of analysis to derive cell survival from observation of tissue damage following fractionated irradiation. *Radiat. Res. 77* (1979), 440-452.

13. Lam, G. K. Y., Henkelman, R. M., Douglas, B. G., and Eaves, C. J. Dose dependence of pion RBE values for mouse foot skin reactions. *Int. J. Radiat. Oncol. Biol. Phys. 7* (1981), 1689-1694.

14. Rèvèsz, L., Littbrand, B., Midander, J., and Scott, O. C. A. Oxygen effects in the shoulder region of cell survival curves. In *Cell Survival at Low Doses: Theoretical and Clinical Implications* (T. Alper, ed.), pp. 141-149. John Wiley & Sons, London, 1975.

15. Koch, C. J. Measurement of very low oxygen tensions in liquids: does the extrapolation number for mammalizn survival curves decrease after x-irradiation under anoxic conditions? *Ibidus* (1975), pp. 167-173.

MECHANISMS OF RADIATION-INDUCED NEOPLASTIC CELL TRANSFORMATION

TRACY CHUI-HSU YANG and CORNELIUS A. TOBIAS
University of California, Berkeley

I. INTRODUCTION

In 1896, one year after the discovery of x rays, Clarance Dally, assistant to Thomas Alva Edison, may have been the first individual to develop a tumor from overexposure to x rays. He was demonstrating Edison's new invention, x-ray fluoroscopy, by exposing his arm many hundreds of times for the benefit of visitors to the Chicago World's Fiar. Although the deleterious effects of x rays became known quite early during the first forty years of diagnostic radiology, quite a few physicians suffered deformities and tumors on their hands because of repeated exposures. During the last sixty years there have been extensive studies on animals, mostly small rodents, on the long-term effects of ionizing radiation, and we now know how to shield and protect ourselves.

These studies were supported by NASA, the National Cancer Institute under grant CA-15184, and the Department of Energy under contract De-Ac03-76SF00098.

It is generally believed that ionizing radiation can cause various types of tumors in vivo and that within a relatively low dose
range the frequency of cancer generally increases with dose (Walburg,
1974). Among radiation-induced tumors, leukemia appears to have a
viral etiology in certain strains of mice (Gross, 1958; Kaplan, 1967),
and certain forms of leukemia in cattle also seem to have viral
etiology. Hoever, in man there is no evidence that radiation-induced
leukemia has viral origins, although Burkitt's lymphoma in humans may
relate to Epstein-Barr virus (Rapp, 1978).

During the last fifteen years, using mammalian cell cultures
as experimental models, investigators in different laboratories
have demonstrated that radiation can induce neoplastic cell transformation both directly and indirectly (Borek and Sachs, 1968; Terzaghi
and Little, 1975; Han and Elkind, 1979; Yang and Tobias, 1980, 1982;
Stoker, 1963; Pollock and Todaro, 1968).

How radiation can induce neoplastic cell transformation both
directly and indirectly is an intriguing question. Although as yet
there is no clear answer, double-strand DNA breaks and cellular repair
processes appear to play important roles in both radiation carcinogenesis and cocarcinogenesis. An analysis of available experimental
data indicates that radiation can produce certain DNA damage (but not
necessarily a single specific gene mutation) and can initiate the
cell transformation directly, but that some unknown epigenetic changes
may also be involved in promoting the progression and the expression
of transforming properties of cells.

II. COCARCINOGENESIS: RADIATION ENHANCEMENT
OF VIRAL TRANSFORMATION

The effect of radiation on the sensitivity of mammalian cells to
transformation by oncogenic DNA virus has been well studied, and
enhancement of viral transformation has been commonly found (Stoker,

1963; Pollock and Todaro, 1968; Lytle et al., 1970).* Several investigators suggested that DNA breaks that result from x irradiation
directly or from repair of UV-induced base damage, may aid viral
transformation. They did not, however, specify which type(s) of
DNA break (e.g.. single or double strand) is the important one for
the radiation-enhancement effect. It has been reported that DNA
double-strand breaks are formed in UV-irradiated human cells during
repair incubation (Bradley, 1981).

The oncogene theory of carcinogenesis assumes that specific
base sequences, or oncogenes, exist that can cause cancer and cell
transformation. The DNA viruses and the RNA retroviruses carry some
of the oncogenes. Indeed, it has been shown that in the course of
cell transformation by viruses, the oncogenes become integrated in
the mammalian cell genome. For simian virus SV40, the oncogene
sequence has been identified. It has also been shown that many
copies of the oncogene can integrate into the same cell and that
most of these copies do not express the transforming property. We
also know that normal mammalian cells often carry several oncogenes;
their presence has been proved by hybridization of DNA sequences
against known oncogene DNA. Cell transformation, therefore, has
genetic components; however, equally important are the gene regulatory components, which determine whether or not an oncogene is
expressed.

The role of ionizing radiation is inducing oncogenic transformation is poorly understood at present. Radiation has been
applied to many mammalian and human cell strains in culture. In
most cases it has not produced a significant number of transformed
colonies. An exception is the case of a special mouse fibroblastic

---

Enhancement is the ratio of the yield of transformed cells after
viruses and radiation are applied to the yield produced by virus
alone. We use an operational definition for "enhancement." The
enhancement ratio (E.R.), is

$$E.R. = \frac{\text{colonies transformed by virus + radiation}}{\text{colonies transformed by virus alone}}.$$

strain derived from an embryo: the C3H10T1/2 strain. It is believed,
however, in this case that radiation may accomplish merely the last
step in a complex chain of processes that change a normal cell into
a cancer cell.

Radiation can also cause point mutations in genes that may turn
them into oncogenes or bring about chromosome deletions and translocations. Radiation might bring about transposition of genes also;
however, at this state there is very little evidence that these are
the pathways on which radiation acts.

Because the DNA of oncogenic viruses (SV40, for example) is
double-stranded and integrated into DNA of transformed cells, it
seems plausible to assume that radiation-produced double-strand breaks
in cellular DNA are among the main type of lesions involved in the
radiation enhancement of viral transformation. Consequently, we
hypothesized that radiation can produce double-strand DNA breaks in
the cell nucleus and that a misrepair of these DNA lesions enhances
the integration of viral genomes and thus cell transformation. To
test this hypothesis, we used a physical agent that can produce
double-strand breaks efficiently; heavy ions with high linear-energy
transfer (LET) have been found to induce double-strand breaks more
effectively than conventional x or gamma rays (Christensen et al.,
1972; Ritter et al., 1977).

At the Lawrence Berkeley Laboratory, the Bevalac can accelerate
heavy nuclei with atomic numbers up to 92 (uranium) to several hundred million electron volts per nucleon (MeV/u) (Alonso et al.,
1982). We carried out a series of cell transformation experiments
with heavy ions to test our hypothesis, and the experimental results
with x rays and argon ions (570 MeV/u) are shown in Figure 1 and in
the table. The number of transformants per survivor increases
rapidly with dose, and the increase of transformation frequency as
a function of radiation dose is curvilinear for both argon particles
and x rays. However, the initial slope of the transformation curves
are higher for heavy ions than for x rays, and it appears that at
low doses single heavy ions can cause transformation. The relative

biological effectiveness (RBE) for argon ions is about 1.6 to 2.1
at 300 rad and about 3.0 at 100 rad.

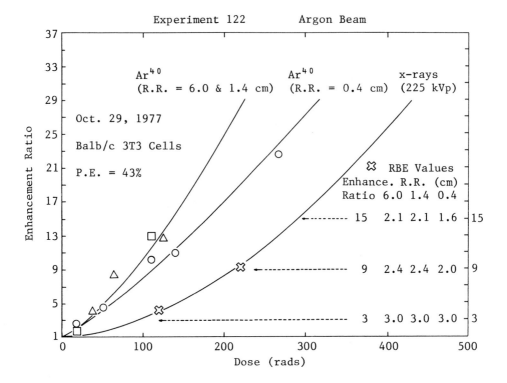

Figure 1. Enhancement effect of argon ions with various residual
ranges (R.R.) and x rays on SV40 transformation of embryonic mouse
cells in vitro.   (XBL 785-8408)

Similar results were found with energetic neon particles, as
reported earlier (Yang et al., 1980). Our results with other heavy
ions also consistently show that high-LET radiation can be more
effective than x rays in enhancing viral transformation. The RBE
calculated for a single particle also increases rapidly with LET
values up to 350 keV/μm, as shown in Figure 2. These data are,
therefore, in agreement with the hypothesis that double-strand DNA
breaks may be important lesions in enhancing cell transformation by
an oncogenic DNA virus.

Table. Effect of X-Rays or Argon Ions on the SV40 Viral Transformation of Cells In Vitro

Radiation	Residual Range in water (cm)	Dose (rad)	Cells plated per dish	Survival (%)	Survivals per dish	Average number of transformants per dish	Transformants per survival		Enhancement ratio
X ray (225 kVp)	—	0	4130	43	1776	0.66	3.72	10	1.00
		120	4630	40	1852	2.50	13.50	10	3.63
		220	6115	25	1528	5.25	34.36	10	9.24
		360	8590	14	1202	10.38	86.36	10	23.21
Argon ions (570 MeV/amu)	6.0	0	4898	51	2106	0.78	3.70	10	1.00
		40	5055	43	2174	2.88	13.25	10	3.56
		65	6150	34	2091	6.50	31.10	10	8.36
		125	8914	20	1783	8.50	47.67	10	12.81
	1.4	20	5163	45	2323	1.50	6.46	10	1.73
		52	5685	37	2103	3.20	15.22	10	4.09
		110	9506	24	2281	11.00	48.22	10	12.96
	0.4	20	4850	50	2425	2.37	9.77	10	2.63
		52	5784	41	2371	3.87	16.32	10	4.39
		110	9388	27	2535	9.62	37.95	10	10.20
		140	14265	20	2853	11.62	40.37	10	10.95
		267	48446	7	3294	27.62	83.85	10	22.54

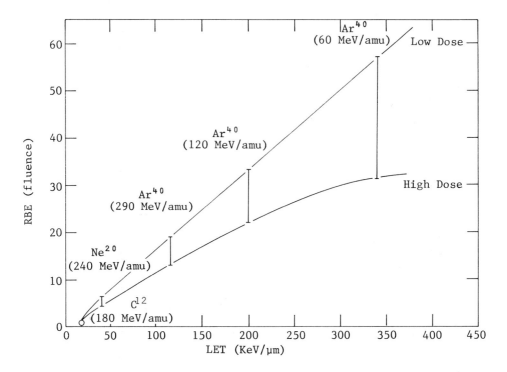

Figure 2. The relative biological effectiveness (RBE) for a single heavy particle as a function of LET in enhancing the SV40 viral transformation. (XBL 785-8399)

Some of the enhancement lesions are repairable. When the viral infection of confluent G_1 cells was delayed after x-irradiation, the transformation frequency decreased exponentially, as shown in Figure 3. An interference of repair processes with repair inhibitors (e.g., caffeine) increased the transformation frequency significantly (Figure 4). For repair inhibitor studies, cells were infected with SV40 12 h before being exposed to x rays; right after irradiation the cells were plated into culture medium with 2 mM caffeine. One day after plating, the drug medium was replaced with fresh medium containing no drug, and the cells were fixed and stained after a 10-day and a 16-day incubation for colony-forming ability and transformation-frequency determination, respectively. Caffeine appeared to enhance further the radiation effect on viral transformation.

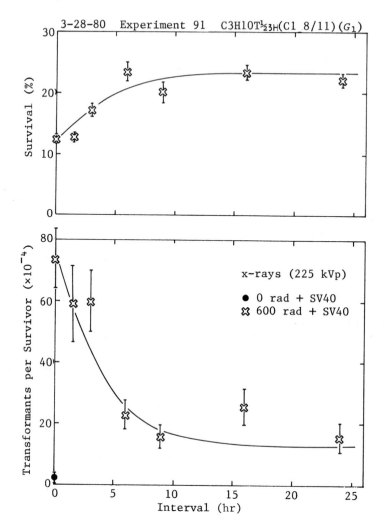

Figure 3. Kinetics of repair in confluent cells irradiated with too rad x rays. Top: potential lethal damage repair. Bottom: potential enhancement lesions repair. Cells were infected with SV40 at various repair intervals. (XBL 806-9830)

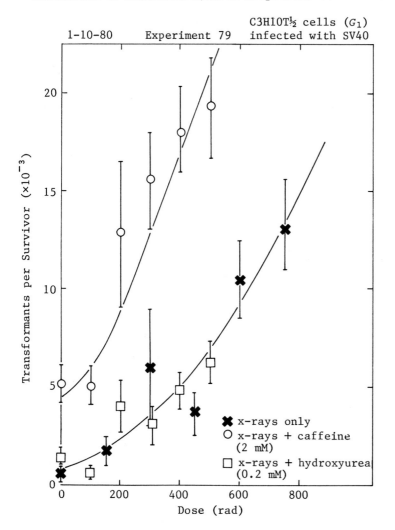

Figure 4. Effect of caffeine and hydroxyurea on the radiation enhancement of viral transformation. Cells were infected with SV40 before the irradiation and drug treatment. (XBL 806-9829)

To date our results of heavy-ion radiation experiments and repair studies appear to support the hypothesis that the enhancement of viral transformation may be due to the induction of double-strand breaks and a subsequent increase in the rate of integration of viral DNA into the DNA of the host cell. We believe that double-strand breaks may act as sites for integration and proceed via misrepair of these lesions. The exact molecular mechanisms for the radiation-enhancement effect is, however, still unknown and needs further study.

III. CARCINOGENIC EFFECT: NEOPLASTIC CELL TRANSFORMATION WITHOUT THE NEED FOR VIRAL INFECTION

Many investigators have studied the radiation-induced neoplastic cell transformation in vitro, because the cultured-cell system makes it possible to investigate malignant transformation directly at the cellular level without involving the complex biological properties of the whole animal. When a population of normal cells is irradiated in vitro and allowed to proliferate at an optimal growth condition for several weeks, a small fraction of cells becomes transformed. These transformed cells can form a tumor in syngeneic test animals. Radiation can transform not only established mouse cell lines, such as C3H10T1/2 and Balb/3T3 cells, but also primary cultures of hamster embryos (Borek and Sachs, 1968). In addition to rodent cells, human fibroblasts have been transformed by x rays (Borek and Hall, 1982; Borek, 1980) and UV radiation (Maher et al., 1982; Cleaver, 1969; Setlow et al., 1969). Results obtained to date clearly demonstrate that radiation can transform mammalian cells directly.

1.0 Heavy Ion Radiation and Cell Transformation

Because of our interest in the mechanisms of radiation carcinogenesis, we have approached this problem using quantitative studies on cell transformation by heavy ions. Heavy-ion radiation can provide a wide spectrum of LET, and a variety of biological effects (e.g., cell

killing, recovery kinetics, and DNA breaks have shown a strong LET
dependency). Consequently, a systematic study of the carcinogenesis
of heavy particles can yield quantitative information that can be
analysed and compared with results of other biological effects of
heavy ions to shed light on the relationships among mutation, malignant cell transformation, cell killing, repair, and DNA lesions.
More insight on the mechanisms of radiation carcinogenesis may thus
be gained.

Figure 5 shows the cell transformation results we obtained
with energetic argon ions (330 MeV/u; LET = 140 keV/μm) and x rays.
The general method we used for quantitative determination of neoplastic cell transformation in vitro is similar to that used by other
investigators, except only confluent G_1 cells were used for all
heavy-ion radiation experiments. The cell system and experimental
techniques for radiation transformation study have been reported
earlier (Yang and Tobias, 1982).

For both x rays and argon ions, the frequency of cell transformation per survivor increases curvilinearly with radiation dose.
Argon particles, however, show a linear increase at low doses and a
greater effect than x rays in inducing neoplastic cell transformation
for a given dose because the number of transformants per survivor
for all argon-ion doses used is consistently above the curve for
x rays.

The RBE value of argon particles varies with the dose level,
from about 2.5 at relatively high doses to about 5.5 for low doses,
below 300 rad of x rays. The effectiveness of heavy ions in inducing
neoplastic cell transformation is also LET-dependent. A plot of RBE
as a function of LET for the various heavy ions studied in our laboratory is shown in Figure 6. The RBE is determined at the transformation frequency per survival induced by x-ray dose that kills 50% of
the cells. Apparently, there is an increase of RBE with LET, at
least up to about 200 keV/μm. At present we do not have data for
heavy ions with LET values greater than 200 keV/μm, and we do not
know at what LET value the RBE will reach its maximum. The higher
RBE value for cells plated one day after irradiation is due to the

fact that high-LET heavy ions produce less repairable lethal and transformation lesions in the cell than x rays.

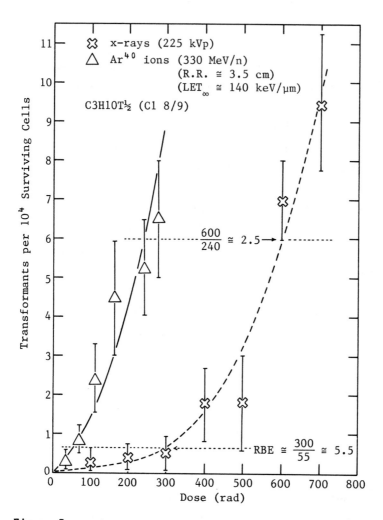

Figure 5. Induction of neoplastic cell transformation by energetic argon ions (330 MeV/u) and x rays.   (XBL 835-9915)

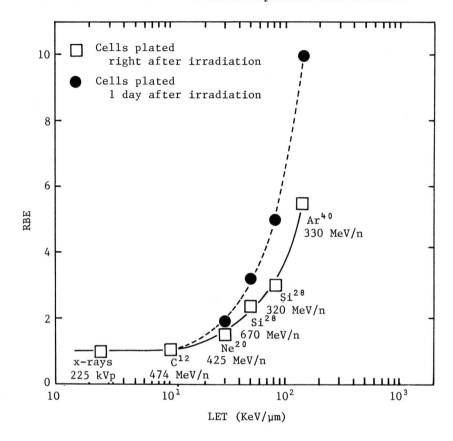

Figure 6. RBE as a function of LET of heavy ions in the production of neoplastic cell transformation. Well-confluent cells (G_1) were used for all experiments. (XBL 833-8741)

2.0 Repair of the Cell Transformation Lesion

Some of the radiation-induced transformation lesions can be repaired, and for a given dose that amount of repairable lesions formed in the cells depends on the radiation quality. When x-irradiated confluent C3H10T1/2 cells were kept at 37°C for one day before being plated into dishes at a low density, the transformation frequency decreased significantly, and the dose modifying factor (DMF) was found to be about 1.6 to 1.8 (Figure 7). This decrease of transformation frequency in cells plated one day after irradiation is most likely due to the proper repair (or eurepair) of potential cell transformation

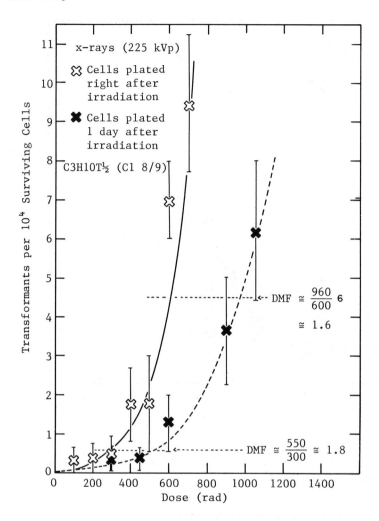

Figure 7. Repair of potential transformation lesions in x-irradiated confluent cells. (XBL 835-9914)

lesions. An inhibition of beta DNA polymerase, which is believed to be a repair enzyme, with β-araA (adenine-β-D-arabinofuranoside) greatly enhanced the ratiation transformation (Figure 8). Radion-induced neoplastic cell transformation may, therefore, be a result of misrepair of some DNA lesions. Unlike low-LET x rays, some high-LET radiation induces very few if any repairable transformation lesions. Confluent cells irradiated with high-LET argon ions (330 MeV/u) and allowed to repair for one day at 37°C, for example, showed the same transformation frequency as cells plated immediately after irradiation (Figure 9).

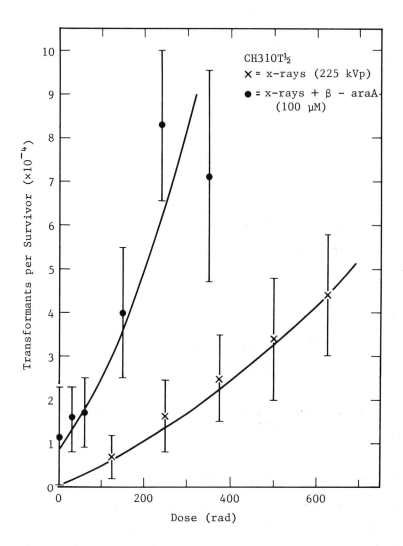

Figure 8. Enhancement of radiation-induced cell transformation by β-araA (a DNA polymerase inhibitor).   (XBL 8212-4294)

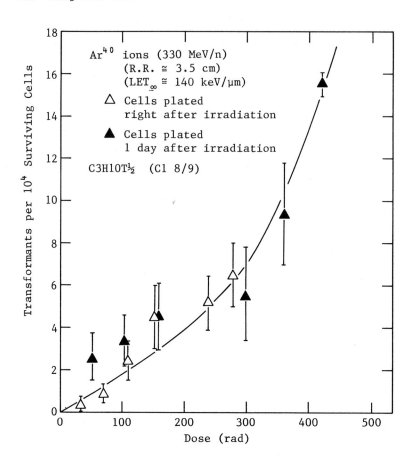

Figure 9. Effect of delayed plating on the transformation frequency of cells irradiated with energetic argon particles (330 MeV/u). (XBL 835-9951)

3.0 Target for Neoplastic Cell Transformation

For cell killing by low and moderately high doses of ionizing radiation, nuclear DNA is an important primary target. The primary target for radiation-induced neoplastic cell transformation appears also to be the DNA. There are experimental data to suggest that radiation damage in the DNA may be the cause of neoplastic cell transformation. The first clear evidence that damaged DNA may result in neoplastic transformation came from a carcinogenesis study with fish cells (hart and Setlow, 1975). When cells from the thyroid tissue of a fish

(*Poecilia formosa*) were irradiated with UV and subsequently injected into isogenic recipients, thyroid carcinomas developed in almost all the fish. If the cells were photoreactivated immediately after UV exposure, however, very few tumors arose. This decrease of tumor incidence is due to an activation of cellular DNA repair enzymes that can remove the UV-induced DNA lesions.

The suggestion that UV-induced DNA lesions are important in neoplastic cell transformation is also supported by studies with human cells. Repair-deficient cells from patients with xeroderma pigmentosum (XP) show a higher rate of transformation than normal cells for a given UV dose (Maher et al., 1982).

Perhaps the most direct evidence suggesting that DNA may be the primary target for neoplastic cell transformation is the study of malignant transformation by isotope-labeled nucleotides. Both $5-(^{125}I)$ iododeoxyuridine and (^3H) thymidine were found to induce neoplastic cell transformation effectively when incorporated into cellular DNA (LeMotte et al., 1982). At low doses, ^{125}I was over twenty-five times more effective than ^3H and x rays. This high efficiency in transforming mammalian cells in vitro might be due to the fact that ^{125}I radiation was highly localized to small regions of DNA at the site of each decay and produced DNA double-strand breaks. An additional evidence in favor of DNA as the target is the study with DNA analogs. Cells exposed to 5-bromodeoxyuridine at a noncarcinogenic concentration before x-irradiation gave a transformation frequency several times higher than cells treated with radiation alone (Little, 1977).

4.0 Molecular Nature of the Cell Transformation Lesion

Although experimental evidence shows that DNA is the possible primary target, we know very little about the molecular nature of the transformation lesion. The somatic cell mutation theory of cencer has been an attractive possibility, and the potential mutagenic effects of radiation have been investigated extensively. We have attempted to identify the molecular nature of the transformation lesion by comparing the experimental data for cell transformation with that for mutation; we have concluded that neither point mutation nor simple deletion mutation nor sister chromatide exchange can be the primary lesion for cell transformation.

UV radiation can induce both cell transformation and ouabain-resistant mutation (a point mutation), but no ouabain-resistant mutant in either x or heavy-ion irradiated cells has ever been found in our laboratory. Because ionizing radiation can transform cells effectively, the ouabain-resistant related point mutation cannot be the type of mutation lesion involved in the malignant cell transformation. Studies with XP cells suggest that sister chromatide exchange may not be the transformation lesion. XP cells, which are highly susceptible to UV radiation in neoplastic cell transformation, showed the same frequency of sister chromatid exchange as normal cells for a given UV dose (Wolff et al., 1975).

Another genetic marker used extensively for studying radiation mutagenesis in mammalian cells is the 6-thioguanine resistant (6-TG^r) mutation. The 6-TG^r mutation induced by radiation appears to be a result of DNA deletion (cox and Masson, 1978). Heavy ions induce 6-TG^r mutation very effectively in mammalian cells, and the RBE for this mutation is about 9 at a LET range of 100 to 200 keV/μm (Cox et al., 1977). In the same LET range, however, the RBE for cell transformation is only about 5, as shown in Figure 6. An analysis of the cross-section size also shows a discrepancy between 6-TG^r mutation and cell transformation. The cross-section for cell transformation is about 10 to 100 times larger than that for mutation (Goodhead, 1983). In addition, it has been observed that diethylstilbestrol, a synthetic estrogen, can induce morphological and neoplastic cell transformation but not 6-thioguanine-resistant mutation (Barrett et al., 1983). Therefore, it is difficult to equate the deletion type lesion with the transformation damage. In short, no experimental data available at present support the suggestion that cell transformation arises from a simple specific radiation-induced mutation.

5.0 The Radiogenic Cell Transformation Processes

Experimental observations on the modulation of radiation cell transformation indicate that the process of neoplastic cell transformation
is a complicated one and includes at least two different states:
induction and expression. The results of studies using internal
emitters and heavy ions suggest that some radiation-induced DNA
damage may initiate the transformation. A subsequent misrepair of
these DNA lesions may cause the fixation of transformation lesions,
because an inhibition of repair enzymes can enhance greatly the
transformation frequency.

The induction state is a relatively short one and may be completed after one cell division. The expression stage, however,
will usually take several weeks and can be modulated by various
physical and chemical agents. Phorbol ester derivative 12-0-tetra-
decanoyl-phorbol-13-acetate (TPA), for example, when applied to cells
after irradiation, promotes transformation significantly (Kennedy et
al., 1978, 1980; Miller et al., 1981; Han and Elkind, 1981). Other
agents reported to enhance radiogenic transformation include interferon and β-estradiol (Brouty-Boyce and Little, 1977; Borek, 1982).
The expression process can also be inhibited by various chemicals,
such as retinoids (Harisiadis et al., 1978; Miller et al., 1981).
In our laboratory we have observed that fungizone (amphotericin B),
an antibiotic for fungus, and dimethyl sulfoxide (DMSO) can decrease
the frequency of cell transformation by ionizing radiation.

Figure 10 shows the effect of fungizone on radiation-induced
cell transformation. Right after irradiation, the cells were incubated in growth medium with various concentrations of fungizone for
5.5 weeks at 37°C. Several well-transformed clones developed in
cells irradiated with 600 rad and kept in the medium with no fungizone, but only one transformant showed in irradiated cells treated
with 1.14 µg/ml fungizone. At a higher concentration of fungizone
(2.27 µg/ml), no transformant was observed, suggesting an efficient
inhibition of transformation. This suppression of radiation transformation by fungizone is not due to an inhibition of cell growth,

because the cells treated with this antibiotic give the same number
of survivors as the nontreated cells.

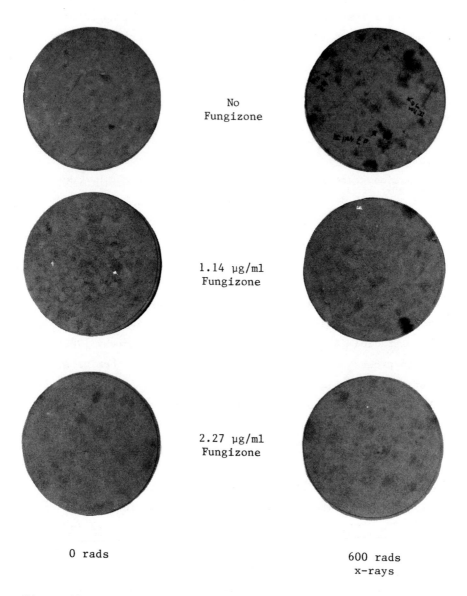

No
Fungizone

1.14 µg/ml
Fungizone

2.27 µg/ml
Fungizone

0 rads

600 rads
x-rays

Figure 10. An inhibitory effect of fungizone on the radiation-
induced cell transformation.   (BBC 809-10309)

The inhibitory effect of amphotericin B on radiation-induced cell transformation also depends on the time of addition. Terzaghi and Little (1976) showed that mouse embryo fibroblasts x-irradiated with 300 rad give $< 10^{-4}$, 1.2×10^{-4}, and 5×10^{-4} transformants per surviaor when the fungizone was added to cells at 0, 2, and 3 weeks after seeding, respectively.

We observed a similar result with DMSO, which is shown in Figure 11. X-irradiated cells were treated for 4 weeks with DMSO medium,

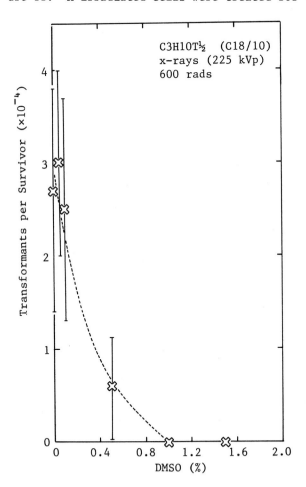

Figure 11. Effect of dimethylsulfoxide (DMSO) on the transformation frequency of cells irradiated with 600 rad x rays. Cells were treated with DMSO medium 10 days after irradiation and plating. (XBL 835-9899)

beginning 10 days after irradiation and plating. There is a decrease of transformation frequency with an increase in concentration of DMSO; 1% DMSO blocked the radiation transformation completely, without a significant effect on the cell growth (Figure 12).

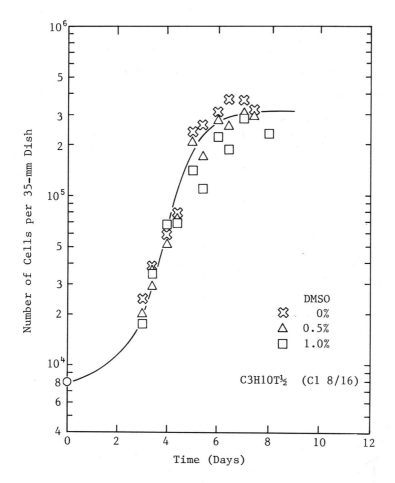

Figure 12. The growth rate of cells treated with various concentrations of DMSO. (XBL 835-9902)

The effect of fungizone and DMSO on the progression of cell transformation is interesting because neither chemical has been found to be mutagenic at the concentrations we used. Although the mechanism of progression and expression of transformation is far from understood and may be quite complicated, our studies with fungizone and DMSO indicate that some nonmutagenic changes in the cell can interfere with the expression process. Some epigenetic changes, therefore, may be important and may be involved in the progression and expression processes. A simplified diagram of the possible processes of radia- tion cell transformation is shown in Figure 13.

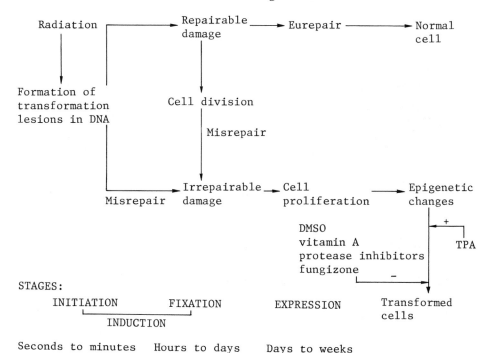

Figure 13. A schematic diagram of the possible mechanisms of radiation-induced neoplastic cell transformation. (XBL 8310-4024)

Other experimental evidence, indicating that epigenetic changes can be important in the expression of tumor cell properties, is the observation of the interaction between normal and transformed cells. Unlike normal fibroblasts, which show density inhibition of growth, transformed fibroblasts tend to grow randomly and form colonies

with multilayers of cells.   This uninhibited growth is a prominent
property of transformed fibroblasts.   Recently, we have studied the
effect of normal cells on the expression of transformed properties of
tumorigenic cells in vitro and found that normal cells can modulate
the growth of transformed cells.   When well-transformed cells
(10T1/2-6S), which were transformed by silicon ions and formed tumors
in syngeneic hosts, were seeded into tissue culture dishes containing
nontransformed cells (C3H10T1/2), the number of transformed foci
decreased as the number of normal cells increased, as shown in Figure 14.   We observed an average of about 130 transformed foci in the

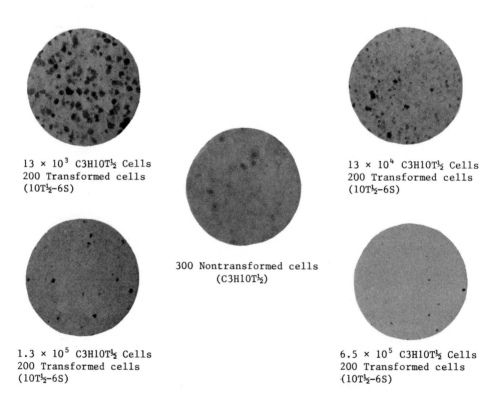

13 × 10^3 C3H10T½ Cells
200 Transformed cells
(10T½-6S)

13 × 10^4 C3H10T½ Cells
200 Transformed cells
(10T½-6S)

300 Nontransformed cells
(C3H10T½)

1.3 × 10^5 C3H10T½ Cells
200 Transformed cells
(10T½-6S)

6.5 × 10^5 C3H10T½ Cells
200 Transformed cells
(10T½-6S)

Figure 14.   Effect of normal cells on the expression of tumor-cell
growth property of transformed cells.   Two hundred transformed cells
(10T1/2-6S) were mixed with varying numbers of nontransformed cells
and plated into 60-mm tissue culture dishes.   Cells were incubated
for three weeks at 37°C.   A decrease in the number of observable
transformed foci with an increase of density of nontransformed cells
in the dish is evident.   (BBC 833-2652)

dish seeded with only 200 transformed cells (10T1/2-6S), which is a
plating efficiency of about 65%, and less than 10 transformed
colonies in the dish plated with 200 transformed and 6.5×10^5
nontransformed cells.

Figure 15 is a plot of the number of observable transformed foci
as a function of the number of nontransformed cells plated in a 60-mm
tissue culture dish. There is clearly a rapid but nonlinear decrease
of transformed foci with an increase of number of nontransformed cells
seeded. At present it in unknown how the normal cells modulate the
growth of tumor cells and whether this inhibition effect is irreversible. Further investigations are needed to elucidate the mechanisms
of this interesting cellular phenomenon.

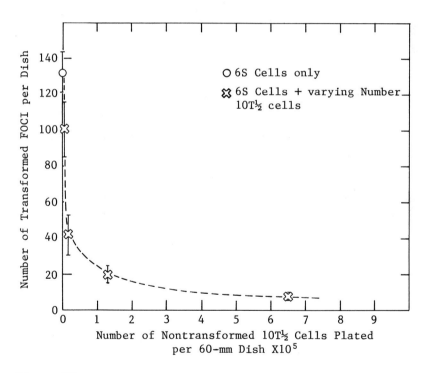

Figure 15. A plot of the number of transformed foci per dish as a
function of the density of nontransformed cells. (XBL 838-11286)

IV. SUMMARY

Studies with cultured mammalian cells demonstrated clearly that
radiation can transform cells directly and can enhance the cell
transformation by oncogenic DNA viruses. In general, high-LET heavy-
ion radiation can be more effective than x and gamma rays in inducing
neoplastic cell transformation. Various experimental results indicate that radiation-induced DNA damage, most likely doublt-strand
breaks, is important for both the initiation of cell transformation
and the enhancement of viral transformation.

   Some of the transformation and enhancement lesions can be
repaired properly in the cell, and the amount of irrepairable lesions
produced by a given dose depends on the quality of radiation. An
inhibition of repair processes with chemical agents can increase the
transformation frequency of cells exposed to radiation and/or oncogenic viruses, suggesting that repair mechanisms may play an important role in the radiation transformation.

   The progression of radiation-transformed cells appears to be a
long and complicated process that can be modulated by some nonmutagenic chemical agensts, such as DMSO. Normal cells can inhibit the
expression of transforming properties of tumorigenic cells through
an as yet unknown mechanism. The progression and expression of
transformation may involve some epigenetic changed in the irradiated
cells.

ACKNOWLEDGMENT

The excellent technical help from L. Craise and D. Tse is appreciated;
we thank the Bevalac crew for their consistent and valuable help.

REFERENCES

Alonso, J. R., R. T. Avery, T. Elioff, R. J. Force, H. A. Grunder, H. D. Lancaster, E. J. Lofgren, J. R. Meneghetti, F. B. Selph, R. R. Stevenson, and R. B. Yourd. 1982. Acceleration of uranium at the Bevalac. *Science 217*, 1135-1137.

Barrett, J. C., J. A. McLachlan, and E. Elmore. 1983. Inability of diethylstilbestrol to induce 6-thioquanine-resistant mutants and to inhibit metabolic cooperation of V79 Chinese hamster cells. *Mutat. Res. 107*, 427-432.

Borek, C. 1980. X-ray-induced transformation of human diploid cells. *Nature* (London) *283*, 776-778.

Borek, C. 1982. Radiation oncogenesis in cell culture. *Advan. In Cancer Res. 37*, 159-232.

Borek, C., and E. J. Hall. 1982. Oncogenic transformation produced by agents and modalities used in cancer therapy and its motulation. *Ann. Acad. Sci. 397*, 193-210.

Borek, C., and E. Sachs. 1968. In vitro cell transformation by X-irradiation. *Nature* (London) *210*, 276-278.

Bradley, M. O. 1981. Double-strand breaks in DNA caused by repair of damage due to ultraviolet light. *J. Supramolec. Struc. Cell. Biochem. 16*, 337-343.

Brouty-Boyce, D., and J. B. Little. 1977. Enhancement of X-ray-induced transformation in C3H10T1/2 cells by interferon. *Cancer Res. 37*, 2714-2716.

Christensen, R. C., C. A. Tobias, and W. D. Taylor. 1972. Heavy-ion-induced single and double-strand breaks in X-174 replicative form DNA. *Int. J. Radiat. Biol. 22*, 457-477.

Cleaver, J. E. 1969. Xeroderma pigmentosum: A human disease in which an initial stage of DNA repair is defective. *Proc. Natl. Acad. Sci.* (U.S.A.) *63*. 428-433.

Cox, R., J. Thacker, D. T. Goodhead, and R. J. Munson. 1977. Mutation and inactivation of mammalian cells by various ionizing radiations. *Nature* (London) *267*, 425-427.

Cox, R., and W. K. Masson. 1978. Do radiation-induced thioquanine-resistant mutants of cultured mammalian cells arise by HGPRT gene mutation or X-chromosome rearrangement? *Nature* (London) *276*, 629-630.

Goodhead, D. T. 1983. Deductions from cellular studies of inactivation, mutagenesis, and transformation. In *Radiation Carcinogenesis: Epidemiology and Biological Significance*, pp. 369-385. Ed. by J. D. Boice, Jr., and J. F. Fraumeni, Jr. New York: Raven Press.

Gross, L. 1958.   Attempt to recover filterable agent from X-ray-induced leukemia.   *Acta Haematol. 19*, 353-361.

Han, A., and M. M. Elkind. 1979.   Transformation of mouse C3H10T1/2 cells by single and fractionated doses of X-rays and fission-spectrum neutrons.   *Cancer Res. 39*, 123-130.

Han, A., and M. M. Elkind. 1982.   Enhanced transformation of mouse 10T1/2 cells by 12-0-tetracanoyl-phorbol-13-acetate following exposure to X rays or to fission-spectrum neutrons.   *Cancee Res. 42*, 477-483.

Harisiadis, L., R. C. Miller, E. J. Hall, and C. Borek. 1978.   A vitamin A analogue inhibits radiation-induced oncogenic trans-formation.   *Nature* (London) *274*, 486-487.

Hart, R. W., and R. B. Setlow. 1975.   Direct evidence that pyrimidine dimers in DNA result in neoplastic transformation.   In *Molecular Mechanisms for Repair of DNA*, pp. 719-724.   Ed. by P. C. Hanawalt and R. B. Setlow.   New York: Plenum Press.

Kaplan, H. S. 1967.   On the natural history of the murine leukemias: Presidential address.   *Cancer Res. 27*, 1325-1340.

Kennedy, A. R., S. Mondal, C. Heidelberger, and J. B. Little. 1978.   Enchantment of X-ray transformation by 12-o-tetradecanoyl-phorbol-13-acetate in a cloned line of C3H mouse embryo cells.   *Cancer Res. 38*, 429-433.

Kennedy, A. R., G. Murphy, and J. B. Little. 1980.   Effect of time and duration of exposure to 12-0-tetradecanoyl-phorbol-13-ace-tate on X-ray transformation of C3H10T1/2 cells.   *Cancer Res. 40*, 1915-1920.

LeMotte, P. K., S. J. Adelstein, and J. B. Little. 1982.   Malignant transformation induced by incorporated radionuclides in Balb/3T3 mouse embryo fibroblasts.   *Proc. Natl. Acad. Sci. U.S.A. 79*. 7763-7767.

Little, J. B. 1977.   Radiation carcinogenesis in vitro: Implication for mechanisms.   In *Origins of Human Cancer* (B), pp. 923-939.   Ed. by H. H. Hiatt, J. D. Watson, and J. A. Winsten.   New York: Cold String Harbor Laboratory.

Lytle, C. D., K. B. Hellman, and N. C. Tells. 1970.   Enhancement of viral transformation by ultra-violet light.   *Int. J. Radiat. Biol. 18*, 297-300.

Maher, V. M., L. A. Rowan, K. C. Silinskas, S. A. Kateley, and J. J. McCormick. 1982.   Frequendy of U.V. induced neoplastic transformation of diploid human fibroblasts is higher in xeroderma pigmentosum cells than in normal cells.   *Proc. Natl. Adac. Sci. U.S.A. 79*, 2613-2617.

Miller, R. C., C. R. Geard, R. S. Osmak, M. Rutledge-Freeman, A. Ong, H. Mason, A. Napholz, N. Perez, L. Harisiadis, and C. Borek. 1981.   Modification of sister chromatid exchanges and radiation-induced transformation in rodent cells by the tumor promotor

12-0-tetradecanoyl-phorbol-13-acetate and two retinoids. *Cancer Res. 41*, 655–659.

Pollock, E., and G. Todaro. 1968. Radiation enhancement of SV40 transformation in 3T3 and human cells. *Nature* (London) *219*, 520–521.

Rapp, F. 1978. Herpes viruses, venereal disease, and cancer. *Amer. Sci. 66*, 670–674.

Ritter, M. A., J. E. Cleaver, and C. A. Tobias. 1977. High-LET radiation induced a large proportion of non-rejoining DNA breaks. *Nature* (London) *266*, 653–655.

Setlow, R. B., T. D. Regan, J. German, and W. L. Carries. 1969. Evidence that Xeroderma pigmentosum cells do not perform the first step in the repair of ultraviolet damage to their DNA. *Proc. Natl. Acad. Sci. U.S.A. 64*, 1035–1040.

Stoker, M. 1963. Effect of X-irradiation on susceptibility of cells to transformation by polyoma virus. *Nature* (London) *200*, 756–758.

Terzaghi, M., and J. B. Little. 1975. X-irradiation-induced transformation in a C3H mouse embryo-derived cell line. *Cancer Res. 36*, 1367–1374.

Walburg, H. R., Jr. 1974. Experimental radiation carcinogenesis. *Adv. Radiat. Biol. 4*, 209–254.

Wolff, S., J. Bodycote, G. H. Thomas, and J. E. Cleaver. 1975. Sister chromatide exchanges in xeroderma pigmentosum cells. *Genetics 81*, 349–355.

Yang, T. C. H., and C. A. Tobias. 1980. Radiation and cell transformation in vitro. *Adv. Biol. Med. Phys. 17*, 417–461.

Yang, T. C., and C. A. Tobias. 1982. Studies on the survival frequencies of irradiated mammalian cells with and without cancer cell morphology. In *Probability Models and Cancer*, pp. 189–210. Ed. by L. Le Cam and J. Neyman. Amsterdam: North-Holland Publishing Co.

Yang, T. C. H., C. A. Tobias, E. A. Blakeley, L. M. Craise, I. S. Madfes, C. Perez, and J. Howard. 1980. Enhancement effects of high-energy neon particles on the viral transformation of mouce C3H10T1/2 cells in vitro. *Radiat. Res. 81*, 208–223.

THE ROLES OF IONIZING RADIATION IN CELL TRANSFORMATION

CORNELIUS A. TOBIAS, NORMAN W. ALBRIGHT, and TRACY CHUI-HSU YANG
University of California, Berkeley

INTRODUCTION

Significant amounts of quantitative data are now available on a variety of responses of mammalian cells to viruses and to ionizing radiation. Ionizing radiations, including heavy ions, can cause lethal effects and at the same time can transform some surviving normal cells to cancer cells. When oncogenic viruses are applied to cells, transformation is enhanced by the additional application of ionizing radiations. Much of this information is summarized in the preceding paper by Yang and Tobias (1984).

We have also developed some mathematical models for the actions of radiations on living cells with the aim of better understanding the underlying molecular processes. The goal of this paper is to present a hypothesis that both lethal radiation action and cell transformation are results of similar processes: the production of

These studies were supported by the National Cancer Institute Grant CA-15185 and the Department of Energy Contract DE-AC03-76SF00098.

lesions in DNA followed by time-dependent repair. Lethal effects of radiation damage and radiation-induced cell transformation both appear to be the results of misrepair. The main feature of viral-induced cell transformation is the integration of viral oncogenes into the DNA of the host, and the role of radiations seems to be the introduction of lesions that become the sites of viral integration.

We have earlier described the repair-misrepair model, or the RMR-I, which is applicable for radiations of low LET, such as x rays and gamma rays (Tobias et al., 1980); RMR-II is described in the paper immediately following this one (Albright and Tobias, 1984). Here we introduce a mathematical modification fo the RMR model, RMR-III, which is intended to describe lethal effects caused by heavily ionizing tracks.

BIOLOGICAL OBSERVATIONS OF CELL TRANSFORMATION IN VITRO

Information from three kinds of model systems has helped to define the biological problem:

1. The occurrence of spontaneous cell transformation in mammalian cells, grown in monolayer.

2. Certain viruses, those carrying oncogenes, can add new genetic material to the genome of a host, causing cell transformation.

3. Bacterial plasmids constructed with specially coded DNA can also cause cell transformation. In plasmid form, one can obtain relatively large amounts of oncogenic DNA, and the plasmids have been used to transfect cells with specific segments of DNA.

Ionizing radiation has been shown to enhance transformation events in all three of these biological model systems. Some additional background information on each system is necessary.

There are certain classes of "normal" mammalian fibroblastic cells that can be grown in tissue culture medium on sterile glass or specially treated plastic dishes. It was observed more than twenty-five years ago that these normal cells do not multiply well in cell suspension, but prefer to adhere to the surfaces on which they are growing. The cells flatten out on the surface to form a monolayer; their number increases until the monolayer is confluent. When the cells come in contact with each other, cell division ceases. This phenomenon is called *contact-inhibition.*

Ambercrombie and Heaysman (1954) and Ambercrombie et al. (1957) notices that treatment with a noxious substance or environment can cause some cells to lose the property of contact inhibition. These treated cells gave rise to clones, that were not limited to mono- layers, but instead formed three-dimensional "transformed" clones containing cells that piled on top of each other. Working with cells taken from mice and hamsters, these investigators showed that many such transformed cells, when reintroduced into the bodies of the syn- geneic host from which they were originally obtained, produced can- cerous tumors; euploid cells that were allowed to grow in monolayers did not produce tumors when reintroduced into the host.

Since that time, quantitative assay systems have been developed for the formation of transformed foci from normal monolayers, and several additional criteria were added to define cell transformation. Borek and Sachs (1968) initially showed that ionizing radiation alone can transform cells. Later it was demonstrated that most, if not all, chemical carcinogens can also transform cells. Some cancer- causing viruses under certain conditions can also transform normal monolayer cells. The quantitative relationships obtained in our laboratory with such agents enabled us to begin development of mathematical models to describe the kinetics of this process.

We know today that normal and transformed cells are distinguished by certain characteristics: (1) Transformed cells develop three- dimensional colonies in three-dimensional agar medium; normal cells do not. (2) Normal cells are contact-inhibited on two-dimensional surfaces; transformed cells are not. (3) If certain compounds are added to the medium (e.g., retinoic acid and related substances), transformed cells can revert, and contact-inhibition is re-established. Additional criteria for transformation are detailed in Yang and Tobias (1980).

Cell transformation can be due to mutation or to changes in gene expression; evidence exists for both of these mechanisms. Todaro and Green (1966) hypothesized the existence of oncogenes, genes that can transform normal cells into cancerous ones. Several human oncogenes are known (Cooper, 1982); however, the mere presence of the oncogene in the genome of a mammalian cell does not guarantee that the transformation properties are phenotypically expressed. One of the important parameters that relates to expression of the transformation properties is the location of the oncogene in the genome.

Most of the viral work we report here relates to the use of simian virus, SV40, which was initially isolated from primates. SV40 consists of a single loop of supercoiled, double-stranded, circular DNA, 5840 base pairs long, which has been completely decoded (Fiers et al., 1978). In typical in vitro cell transformation experiments with rodents, an inactivated form of the virus is used, which can infect cells but does not permit the virus itself to proliferate and will not kill the host cell (Topp et al., 1980).

Work with the SV40 virus has shown that the *A* segment of the viral genome is 3218 base pairs long and can produce a protein called the *T antigen*, which has the crucial information for cell transformation (Reddy et al., 1978; Fiers et al., 1978). Botchan et al. (1976) demonstrated that in the typical process of in vitro cell transformation, several copies of the T antigen-producing-DNA sequence, most of which are not effective in transforming the host cell, can become incorporated into the host of the genome. This is one of the reasons why the notion has developed that there is a gene position effect: Apparently, the T region must be integrated into the host genome in certain specific sites and, perhaps, must be adjacent to appropriate promoter regions to be effective.

An important area of research is to discover the mechanism whereby the *A* region of double-stranded DNA becomes integrated into the host genome. This integrative process appears to be similar to the genetic recombination process, ubiquitous in nature, that normally includes

scission and recombination of double-stranded segments of DNA. The
normal recombination process usually occurs in the process of producing germ cells and appears to be strongly inhibited in the somatic
cells of mammals.

A dose of ionizing radiation, applied at the appropriate time,
can enhance the frequency of SV40-induced cell transformation.
The fraction of transformed cells can rise dramatically, sometimes
to an enhancement ratio of up to 50. (The *enhancement ratio* is the
number of viable cells transformed by the virus with radiation,
divided by those transformed by the virus alone.)

Cell transformation can also be accomplished without intact
viruses. Perez et al. (1984) grew the oncogene portion of the SV40
gene in bacterial plasmids together with a marker gene (thymidine
kinase +). The host cells were then exposed to segments of the
plasmid DNA. By the processes of transfection, some of the DNA
entered the cells and integrated with the cell's genome. The transfection process was significantly enhanced by exposing the host cells
to various doses of ionizing radiation. The incorporation of the
marker gene and of the oncogene was enhanced in a similar manner,
and the enhancement had a similar dose-effect relationship to that
obtained in actual viral infection; however, cell transformation was
expressed several times less frequently than the expression of the
marker gene. We believe these experiments demonstrated that the
DNA integration event is a key process in cell transformation and
that oncogene integration follows the same steps as the integration
of some other DNA segment.

THE REPAIR-MISREPAIR MODEL: GENERAL FEATURES

The RMR model attempts to provide a mathematical framework for cellular responses to injury in genetic material and in the extragenic
structures of cells. In the RMR model, it is important to note that
one is usually unable to observe radiation-induced molecular lesions in
individual, living cells. The loci of these lesions are so small (on

the order of one Angstrom, or a portion of a nucleotide) that we would have to use x rays to observe the lesions. According to the uncertainty principle, these x rays would in all likelihood further injure the genetic material in the course of observation.

It is important not to make deterministic statements about these initial lesions, such as "lesion X kills the cells" or "lesion Y is repaired." The theory should attempt to follow events as they happen in the cell. In the RMR model, this is done in the following manner: (1) Let the number of initial lesions be U_0 lesions, where U stands for *uncommitted*. (2) U lesions are recognized internally by the cell's informational apparatus. (3) These cells attempt enzymatically to repair U lesions to restore the normal properties of genetic material. (4) In the course of attempted repair, U lesions might be modified, enlarge, or diminished, and their numbers will usually decrease until they give rise to an observable expression of radiation injury.

We should be mindful here of the usual methods of observation available to radiobiologists, which fall into three categories:

1. If a population of irradiated cells (e.g., $> 10^6$ cells) is killed and chemically processed, we can obtain information on the mean number of certain lesions produced by radiation using an additional assumption that the chemical assay itself did not introduce new lesions. Thus, we can measure the mean numbers of single- and double-stranded scissions produced in DNA per cell, the number of nucleotide bases lost or impaired, and so on. For these variables, at low doses D of x rays, $U_0 = \delta D$, where δ is the lesion yield per cell per unit dose. The δ is usually constant and independent of D; however, the value of δ depends on the microstructure of the cell nucleus and on the quality of the radiation.

2. Changes in macroscopically expressed quantities, such as metabolism and protein synthesis, can also be measured for an irradiated cell population. Usually such measurements also represent mean values because measurements from a large number of cells are averaged.

3. By appropriate plating techniques, after a suitable time interval has elapsed, we can observe in a cell population the dose- and time-dependent probabilities for survival, mutation, and cell transformation as end results of a set of very complex cellular phenomena. The RMR model was constructed to calculate the mean number of a variety of U lesions in the course of

time-dependent repair processes. Combined with statistical
calculations, the probabilities of observable effects are
calculated.

The RMR-I model for cellular ionizing radiation effects with
no defined track structures is summarized in the Appendix. The RMR-I
model assumes a linear self-repair mechanism, in which a single U
lesion is repaired, and a quadratic process, in which a pair of DNA
lesions interact over the course of repair. Furthermore, we assume
that the repair is equivalent to restored DNA. We call this repair
eurepair when the restored DNA has identical structure and function to
the DNA that existed prior to irradiation. When the restoration produces
DNA radically different in structure or base sequence from that which
existed prior to irradiation, we call the result of the process *misrepair*.
Eurepair and misrepair are simplified assumptions, based on statistics,
of the actual processes that are probably very complex. To be practical,
it is necessary to use such simplifying assumptions.

THE PHYSICS OF TRACK STRUCTURE

In condensed media, many radiations (particularly electrons, neutrons,
and fast heavy ions) produce discrete tracks that consist of arrays
of ionization and excitation. The tracks extend along the path of
the primary particle producing the track and also radially away from
the central path. For a homogeneous medium, considerable knowledge
exists of the distribution of energy transfer along tracks. A
Monte Carlo statistical approach has been used by Paretzke et al.
(1978) and Turner et al. (1982). This approach assumes that the
initial energy deposition events are very fast (10^{-16} to 10^{-8} s)
and that secondary electrons (delta rays) are produced in collisions
along the primary particle path. The delta rays come to a stop after
a more or less tortuous path in the medium and produce further ionizations and excitations. The Paretzke approach has yielded spatial
distributions of primary and secondary ionization events.

Mozumder (1969), Chatterjee and Magee (1980), and Magee and
Chatterjee (1980) have realized that the lifetimes of primary

ionization products are very short. The primary ionizations are followed by diffusion controlled radical and radical-ion reactions.
A large fraction of the initial energy transfer, perhaps 80 to 90%,
ends up as thermal energy, diffusing rapidly away from the track
(Tobias et al., 1979). The diffusion of chemical products, such as
free radicals in aqueous living systems, is slower. To follow these
events, a distinction is made between the track core and the surrounding penumbra. The core is limited by the radius at which the primary
particle can still cause electronic excitation. The penumbra is much
larger, and its size is limited by the maximum range of the more
energetic delta rays.

   After about 10^{-7} s, macromolecular lesions exist in the irradiated cells. These lesions can be further modified by the presence
of radiation sensitizers or protectors (Klayman and Copeland, 1975;
Stradford, 1982), but after about 10^{-3} s, we can regard many of the
macromolecular U lesions in DNA and other cellular macromolecules
as having stabilized. The enzymatic apparatus of mammalian cells
appears to take about 10^{-3} s or longer to recognize lesions in its
fabric and to start repairing them.

   The number of U lesions per cell nucleus is much smaller than
the number of initial energy transfer events. In a single human
cell, we estimate perhaps 10^5 to 10^6 primary energy transfer events,
about 1000 to 4000 DNA single-strand break events per cell nucleus,
and perhaps up to 120 DNA double-strand scissions per cell from a dose
that can inhibit the reproductive integrity of 50% of the cells.
Most of the single-strand breaks eurepair; it is likely that many of
the U lesions that relate to the production of lethal effects are
misrepaired double-strand breaks.

   The yield of U lesions is a function of LET (linear energy
transfer); it also depends on the radial structure of tracks, which
is a function of the particle velocity and charge. The yield also
depends on the distribution of chromatin in the cell nucleus. It is
not likely that a track can produce a greater number of U lesions
than the number of times it crosses chromatin fibers. Microscopic
evaluation suggests that the structural distribution of chromatin is

uneven in the cell nucleus; this distribution is different in various physiological states.

THE TRACK SURVIVAL EQUATION (RMR-III)

Assume that a uniform cell population is exposed to a parallel stream of particles of linear energy transfer L and that the particles produce a distribution of U lesions along their tracks. We know that the energy transferred to a thin slab of matter follows the Landau-Vavilov distribution (Landau, 1944; Symon, 1948; Vavilov, 1957); however, we can simplify the treatment by assuming Poisson statistics. In the first part of this calculation, we assume that the mean number of lesions per particle is $n(t)$. At a given particle fluence F, let the mean number of particles that crosses a cell nucleus be $\overline{m}$. Since the particles each arrive independently, $\overline{m}$ represents the mean of a Poisson distribution,

$$\overline{m} = \sigma \cdot F. \tag{1}$$

If we neglect the radial dimensions of tracks, σ is the geometrical cross-section of the region represented by DNA in the cell nucleus. Following the RMR model, we assume that two kinds of repair processes take place in DNA. A linear self-repair process is characterized by rates λ_0 and λ. In quadratic processes, interactions between two different lesions take place in the course of repair to produce misrejoinings. If these quadratic repair events occur between lesions produced by the *same track* (intratrack repair), let their rate by k_I; when these occur between lesions produced by different tracks (intertrack repair), let their rate be k_{II}.

When $\overline{m}$ particles cross the cell nucleus at $t = 0$, then the time rate of change of the number of lesions due to repair can be written

$$\underbrace{\frac{d(mn)}{dt}}_{} = \underbrace{-\lambda_0 mn}_{\text{self-repair}} - \underbrace{k_I mn(n-1)}_{\substack{\text{intratrack} \\ \text{coop repair}}} - \underbrace{k_{II} m(m-1)n^2}_{\substack{\text{intertrack} \\ \text{coop repair}}}. \tag{2}$$

This equation is analogous to Eq. (2) of the Appendix, except that there are two cooperative terms, corresponding to the idea that cooperative repair occurs at a higher rate for lesions within the same track than between lesions of separate tracks. Since m is not a function of time, we obtain the differential equations for lesions along a single track:

$$\frac{dn}{dt} = -\lambda n - k(m)n^2 \tag{3}$$

where $\lambda = \lambda_0 - k_I$ and usually $\lambda \gg k_I$.

The term $k(m)$ is a function of the number of particles crossing tthe cell nucleus:

$$k(m) = k_I + (n - 1)k_{II}. \tag{4}$$

If $n(\infty) = 0$, then for the initial number of n_0 lesions per track, we obtain

$$n(t) = \frac{\lambda n_0 e^{-\lambda t}}{\lambda + n_0 k(m)(1 - e^{-\lambda t})}. \tag{5}$$

If all linear self-repair is eurepair, and all quadratic repair is lethal misrepair, we obtain for the survival probability $S(m)$:

$$S(m) = e^{-mn_0}\left[1 + \frac{n_0(1 - e^{-\lambda t})}{\varepsilon(m)}\right]^{m\varepsilon(m)} \tag{6}$$

where the repair ratio $\varepsilon(m)$ is given by

$$\varepsilon(m) = \frac{\lambda}{k_I + (m - 1)k_{II}} \tag{7}$$

when densely ionizing tracks are present, usually $k_I \gg k_{II}$. If a single particle track produces so many lesions and has so much intra-track interaction that they can kill a cell, and if intertrack inter-action is absent, then $k_{II} = 0$ and

$$\varepsilon(m) = \varepsilon(1) = \lambda/k_I = \varepsilon_I. \tag{8}$$

On the other hand, when we deal with loe-LET radiation, and there are no special intratrack effects, then $k_I = k_{II}$ and

$$m\varepsilon(m) = \lambda/k_{II} = \varepsilon_{II}. \tag{9}$$

In this case, for simpler notation, let the time factor be denoted as T, where $T = 1 - e^{-\lambda t}$; then

$$S(m) = e^{-\delta D} \cdot \left[1 + \frac{\delta D T}{\varepsilon_{II}} \right]^{\varepsilon_{II}}. \tag{10}$$

In this equation, we substituted for the number of lesions mn_0 the more conventional expression δD, where δ is the number of lesions produced per cell nucleus by a unit dose.

An important special case of Eq. (2) is $m = 1$, when only a single ionizing track crosses the cell nucleus. The single track survival curve is $S(1)$:

$$S(1) = e^{-n_0} \cdot \left[1 + \frac{n_0 T}{\varepsilon_I} \right]^{\varepsilon_I}. \tag{11}$$

If there is no intertrack interaction ($k_{II} = 0$), then the survival curve for m tracks is exponential:

$$S(m) = S(1)^m. \tag{12}$$

The slope of the exponential lesion curve, the first term in Eq. (6), is different from the slope of the actual survival curve, which is modified by the second term of Eq. (6). If there is no repair, or if all repair is lethal misrepair, $\varepsilon_I = 0$, and the first term of Eq. (6) expresses the probability of survival. The single track survival equation $S(1)$ is important for characterizing the biological effects of densely ionizing particulate radiations. For Eq. (6) to be useful, we should determine the dependence of n_0 on the linear energy transfer and on the track structure. These depend on the atomic number and velocity of the ionizing particles. We have made an experimental study of this problem using cell cultures of human Ataxia telangiectasia cells, the AT-2SF cell line, and also Thuan T-1 cells (Tobias et al., 1984). We find that at low values of the linear energy transfer L, n_0 is proportional to L, whereas at medium L values, n_0 increases approximately as $L^{2.2}$. The amount of structural configuration of DNA in the cell nucleus is also important. At very high L values, n_0 saturates due to the limited amount of DNA in the part of the particles. Figure 1 shows $S(1)$ as a function of L for T-1 cells exposed to neon ions.

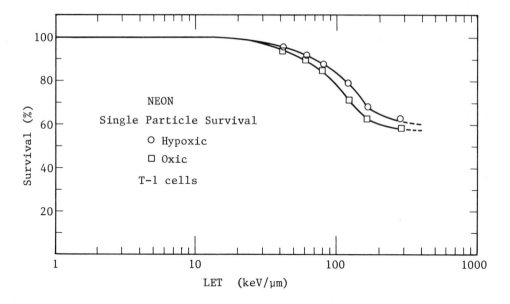

Figure 1. The probability of cell survival as a function of LET when a single neon particle passes through the nucleus of a human T-1 cell in culture.   (XBL 843-7618)

The particles arrive independently of each other, so that the probability $P(m)$ of m particles crossing the cell nucleus is

$$P(m) = \frac{\overline{m}^m}{m!}\, e^{-\overline{m}}. \tag{13}$$

If we use Eqs. (2), (6), (8), and (9), and Eq. (12) of the Appendix, the survival probability $S(m)$ with $T = 1$ is

$$S(\overline{m}) = e^{-\sigma F} \cdot \left[1 + S(1)\sigma F + S(2)\frac{\sigma^2 F^2}{!} + \cdots + S(m)\frac{(\sigma F)^m}{m!} + \cdots \right]. \tag{14}$$

Instead of Eq. (14), we usually employ a simpler, approxiamte RMR survival expression that is usually identical to Eq. (13) of the Appendix.   The coefficient of fidelity is the same as the single particle survival probability $\Phi = S(1)$.   Equation (13) of the Appendix is usually sufficiently accurate to represent radiobiological survival curves.

THE ROLE OF RADIATION IN DNA INTEGRATION

When radiation is used in conjunction with specific oncogenic viruses, it enhances the viral transformation process. Recent evidence has been obtained in our laboratory to support the hypothesis that a key step in cell transformation is the integration of oncogenic segments of DNA. These experiments also indicate that the role of DNA viruses is primarily ato assure propagation of the oncogenes independently of host cells and to provide a convenient entry of the transforming DNA into the cells of the host.

The question of how foreign DNA is integrated into a mammalian genome after its entry into the cells remains unknown. Links must be provided between both strands of the host DNA and of the viral DNA. We do not know at present whether the steps in this process involve initial linking of single-stranded DNA (strand invasion), whether severed double-stranded breaks rejoin, or whether the intermediate linking steps include other molecules, for example, DNA chains. For this discussion, we assume that the process involves rejoining completely scissioned double-stranded DNA. This approach seems fruitful in the RMR model, in which misrepair occurs when mismatched DNA endings join. Whereas in discussing RMR-III, we assumed that all misrepair is lethal, we now modify that assumption, admitting that a small fraction of misrepair events may not necessarily be lethal. The evidence for this is that chromosomal translocations are known to occur in cells that were exposed to deleterious agents but remained viable. Thus, the incorporation of viral DNA into mammalian cell DNA according to this model is a misrepair step. The incorporation of a linear segment of DNA would require at least two nonlethal misrepair steps. If there are more than two misrepair steps, the probability for a lethal effect increases.

Assume that in the course of integration, a quantity of r scissions per cell are made in the DNA by ionizing radiation. We can calculate the probability $Q(x, r - s|r)$ that a cell that sustained r lesions at $t = 0$ will, in the course of the repair process, develop x quadratic misrepairs and $(r - x)$ eurepairs. In the calculation, the repair process was viewed as a Markov chain, as represented in the Appendix. Without presenting details, we propose that the probability of dual misrepairs may be expressed as

For $x = 2$:

if $r \leqslant 3$, then $Q(2, r - 2 | r) = 0$;

if $r > 3$, then $Q(2, r - 2 | r) = Q(0, r | r) \left\{ \sum_{q=1}^{r-2} \frac{q}{r-q} \cdot \sum_{1=q+1}^{r-1} \frac{1}{r-1} \right\}.$ (15)

If one follows the reasoning according to Eq. (12) in the Appendix, the overall probability $M(x)$ that a fluence of F particles will produce cells with two misrepairs is

$$M(F, 2) = \sum_{r} e^{-\sigma F} \cdot \left[s(0 | r) Q(2, (r - 2) | r) \frac{(\sigma F)^r}{r!} \right].$$ (16)

Assume that viruses alone, in the absence of radiation, can transform a fraction ν of the isolated cells (usually $10^{-4} > \nu > 10^{-7}$). Consider further the subpopulation of cells that have suffered $x = 2$ misrepairs in the postirradiation period. This represents cells with chromosome deletions and symmetric of asymmetric translocations. Most of these cells die, but a small portion γ_s might survive (we extimate $10^{-1} > \gamma_s > 10^{-3}$). Let γ_ν be the probability that survivors express the transformation property. With these factors in mind we can write an expression for the transformation enhancement ratio ψ:

$$\psi = \frac{\nu + \gamma_s \gamma_\nu \dfrac{M(F, 2)}{S(F)}}{\nu}.$$ (17)

$S(F)$ is the overall probability of survival, including individual cells that survived DNA and chromosome rearrangements.

Our approach has tacitly assumed that the oncogene has been integrated into the host cell DNA between the two misrepairs. There are many missing details of the biological process that must be incorporated into the model at some future time, but Eq. (23), with appropriately adjusted constants, can be made to fit the experimental data. In Figure 2, a set of x-ray and neon-ion cell tranformation data (Yang et al., 1980) are superimposed on a set of curves

representing Eq. (23). Cell survival data following irradiation
by the same ions is given in Figure 3, where the continuous lines
represent the RMR survival curve, using the same values for the
coefficients as used in Figure 2.

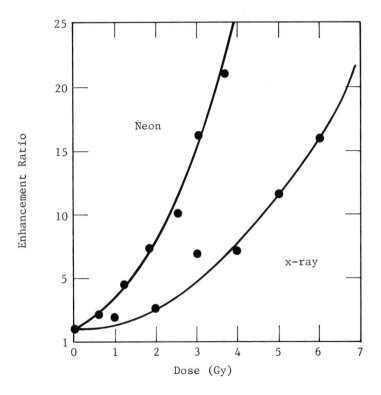

Figure 2. Data on the dose dependence of the enhancement ratio of
transformation of viable CeH10T1/2 cells by SV40 virus augmented by
exposure to x rays or neon ions of 1.4-cm residual range. Solid
lines represent calculations based on the RMR model as described
in the text. The RMR coefficients were determined to fit the
survival curves shown in Figure 3. Additional adjustable constants
γ_s and γ_v were used from Eq. (17). In the case of neon, the entire
effect appears to be due to intratrack lesion repair. (XBL 843-
7636)

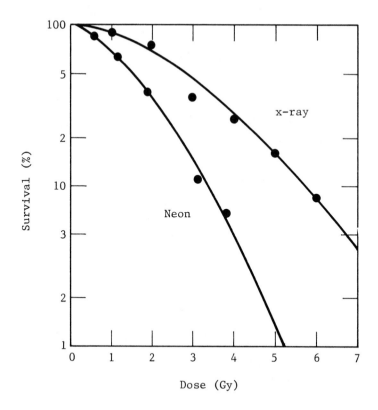

Figure 3. Survival curves of C3H10T1/2 cells followind administration of x rays or neon ions of 1.4-cm residual range. The solid lines are drawn by the use of the RMR model. (XBL 843-7637)

The calculations, as presented here, are obviously incomplete and are only useful if the rate-limiting steps in the integration process are the misrepair steps. Further experimental data are needed to clarify the mode of entry of the oncogenic virus and of plasmid DNA into the cell and its nucleus, and the biochemical steps leading to integration. The use of labeled plasmid DNA may be helpful in this regard. We are, in fact, engaged in a set of new experiments (Perez and Tobias, 1984) using plasmid recombination technology, which provides a powerful tool to answer these questions.

ACKNOWLEDGMENTS

The authors attended many of the seminar discussions, led by Jerzy Neyman, on the applications of statistics to biomedical science. Although our main field is not mathematical statistics, we have received inspiration and guidance from his innovative applications of mathematics to the biological sciences. His curiosity about the laws of nature knew no bounds, and his idealism and helpfulness were not dampened by age.

We gratefully acknowledge the editorial assistance of Ms. Mary P. Curtis and the assistance of Ms. Mardel Carnahan in typing this manuscript.

APPENDIX: DESCRIPTION OF THE RMR-I MODEL

Let the yield of a population of uncommited U lesions be proportional to dose D. At $t = 0$

$$U_0 = \delta D. \tag{1}$$

$U(t)$ represents the mean number of uncommited lesions per cell at time t.

The cells recognize the U lesions and attempt to repair them. The time rate of self-repair of individual lesions is λ. In the course of repair, cooperative processes with a rate k participate with pairs of U lesions.

The dose-dependent production of U lesions is not coupled to the time-dependent repair processes. The rate of repair is expressed by the RMR equation, which we assume is valid for the mean of the U distribution:

$$dU/dt = -\lambda U - kU^2. \tag{2}$$

Initially there are U_0 lesions, and $U(t)_{t \to \infty} = 0$. This leads to

$$U(t) = \frac{[U_0 e^{-\lambda t}]}{1 + [U_0(1 - e^{-\lambda t})/\varepsilon]} \tag{3}$$

where ε is the repair ratio: $\varepsilon = \lambda/k$.

The repair of each lesion may result either in *eurepair*, perfect re-establishment of DNA structure and coding, or in *misrepair*, where there are structural and/or coding defects. Examples of misrepair are the chromosome fragments and mis-rejoinings often found in cells exposed to radiation.

Equation (3) often describes the mean values of the quantity of DNA strand breaks that can be chemically assayed in the post-irradiation period.

The cumulative value of linear repairs $R_L(t)$ is

$$R_L(t) = \lambda \int_0^t U(t)dt. \tag{4}$$

The value of quadratic repairs $R_Q(t)$ is

$$R_Q(t) = k \int_0^t U^2(t)dt. \tag{5}$$

The initial distribution of the number of U lesions per cell caused by a homogeneous radiation field—x rays, for example—is assumed to be a Poisson distribution. As repair progresses, the distributions of U, R_L, and R_Q deviate from a Poisson distribution. A conservation condition,

$$U(t) + R_L(t) + R_Q(t) = U_0, \tag{6}$$

assures that in the course of time the sum of the distributions remains Poisson. From Eq. (6), at time t the mean lesions per cell are $U(t) + R_Q(t)$, because $U(t)$ represents uncommitted lesions and $R_Q(t)$ lethally misrepaired lesions.

If survival is determined at time t, and the condition of survival is that cells have no lesions at all, then the probability $S(t)$ of survival is

$$S(t) = \exp(-R_Q(t) - U(t)) = e^{-U_0}\left[1 + \frac{U_0(1 - e^{-\lambda t})}{\varepsilon}\right]^\varepsilon. \tag{7}$$

We know that the distribution of $R_Q(t) + U(t)$ is not Poisson, but in deriving Eq. (7) we assumed that the first term of the distribution is identical to the Poisson expression, an assumption that is verified below. The first term in Eq. (7) represents the probability that no U lesions are present in the cell initially; the second term represents the probability for eurepair. The time factor is also important: t is the time when the lethal effect is expressed; it is often the time when the irradiated cells divide.

The RMR process can also be described as a result of a statistical branching process. Let $P(J)$ be the probability that a cell has J lesions of the U type. Initially,

$$P(J) = \frac{U_0^J}{J!} e^{-U_0} \quad \text{where} \quad U_0 = \sum_0^\infty JP(J). \tag{8}$$

We are interested in how the probability $P(J)$ varies in time as the U lesions are repaired at a rate of $du/dt = \dot{U}$; differentiating Eq. (8) we obtain

$$\frac{dP(J)}{dt} = \dot{U} \left\{ \frac{U^{J-1}}{(J-1)!} e^{-U} - \frac{U^J}{J!} e^{-U} \right\}. \tag{9}$$

Substituting for $\dot{U}$ the right side of Eq. (1) and rearranging, we obtain

$$\frac{dP(J)}{dt} = \lambda[(J+1)P(J+1) - JP(J)]$$
$$+ k[(J+2)(J+1)P(J+2) - (J+1)JP(J+1)]. \tag{10}$$

The first term corresponds to linear self-repair, and the second term corresponds to quadratic misrepair. By analyzing the meaning of Eq. (10) for all J, we realize that it represents a branching process.

If a given cell has initially r lesions with probability $P(r)$, we can use Eq. (10) for describing the repair process. The rate of repair of lesions in that cell is described by a set of equations:

$$\frac{dP(r)}{dt} = -\lambda r P(r).$$

$$\frac{dP(r-1)}{dt} = \lambda r P(r) - \lambda(r-1)P(r-1) - r(r-1)P(r). \left.\rule{0pt}{5.5em}\right\} \quad (11)$$

$$\frac{dP(r-2)}{dt} = \lambda(r-1)P(r-1) - \lambda(r-2)P(r-2)$$
$$+ kr(r-1)P(r) - k(r-1)(r-2)P(r-2).$$

We can integrate Eqs. (11) and obtain for $t \to \infty$ the probability of all eurepair. In the process of integration, we assume that once a quadratic misrepair event developed (terms with k), the cell is slated to die and will not contribute further to the probability of survival, which implies that each of the successive repair steps must be linear eurepair. For cells with r initial lesions, the survival probability $S(0|r)$, expressed for the simple case in which λ/k is an integer is

$$S(0|r) = \frac{\lambda(\lambda - k)(\lambda - 2k) \dots (\lambda - (r-1)k)}{\lambda^r} \quad (12)$$

where $S(0|r)$ is the probability that a cell has performed in succession r eurepairs in the course of time. The total survival probability S, when $t \to \infty$, is

$$S = e^{-U_0} \cdot \sum_{r=0}^{\infty} \frac{S(0|r)U_0}{r!}. \quad (13)$$

Equation (13), with the value of $S(0|r)$ from Eq. (12), is identical to Eq. (7) if the repair term of Eq. (7) is represented by a Taylor expansion.

There are many applications of the RMR-I model (Tobias et al., 1980; Pirruccello and Tobias, 1980). Perhaps the most interesting applications relate to experiments in which the radiation is delivered in successive dose fractions, separated by time. The function $U(t)$ can also be termed as a measure of remnant lesions, representing the memory of the cell for previous radiation injuries. New U lesions can be added to remnant U lesions, and the RMR process as described here can predict the final outcome.

A frequent occurrence is that the linear self-repair does not have perfect fidelity, meaning that some linear repairs are misrepairs. We can then introduce a coefficient Φ for the fidelity of linear repair: $0 \leqslant \Phi \leqslant 1$. In this case, we introduce the eurepair ratio μ with $\mu = \Phi\epsilon$. Equation (7) then modified to

$$S = e^{-U_0} \cdot \left\{ 1 + \frac{U_0 \Phi (1 - e^{-\lambda t})}{\mu} \right\}^{\mu}. \tag{14}$$

Equation (14) describes very well the shape of most experimental cellular radiation survival curves.

Perhaps the most important contribution of the RMR-I model was the introduction of a method for quantitating eurepair and misrepair. Recently, several modifications have been suggested. For example, Curtis described the LPL model (1983). The next article by Albright and Tobias (1984) introduces RMR-II, a model based entirely on a Markov description. In the main body of this report, we extended the RMR model to describe radiation effects of heavy ion tracks. We have named this track version of the model: RMR-III.

REFERENCES

Albirght, N. W., and C. A. Tobias. (1984). Extension of the time-dependent repair-misrepair model of cell survival to high LET and multicomponent radiation. *Proceedings, Neyman-Kiefer Conference July 1983*. Belmont, Calif.: Wadsworth Publishing Co., in press.

Ambercrombie, M., and J. E. M. Heaysman. (1954). *Exp. Cell Res. 6.* 293–306.

Ambercrombie, M., J. E. M. Heaysman, and H. M. Karthauser. (1957). *Exp. Cell Res. 13,* 276.

Borek, C., and L. Sachs. (1968). In vitro cell transformation by x-irradiation. *Nature* (London) *210,* 276–278.

Botchan, M., W. Topp, and J. Sambrook. (1976). The arrangement of simian virus 40 sequences in the DNA of transformed cells, *Cell 9,* 269–287.

Chatterjee, A., and J. L. Magee. (1980). Energy transfer from heavy particles. In *Biological and Medical Research with Heavy Ions at the Bevalac, 1977–1980,* pp. 53–61. Ed. by M. C. Pirruccello and C. A. Tobias. Lawrence Berkeley Laboratory Report LBL-11220.

Cooper, G. M. (1982). Cellular transforming genes. *Science 218,* 801–806.

Curtis, S. B. (1983). Ideas on the unification of radiobiological theories. In *Proceedings, Eighth Symposium on Microdosimetry*, Report EUR-8395, pp. 527-536. Brussels, Belgium: Commission of the European Communities.

Fiers, W., R. Contreras, G. Haegerman, R. Rogiers, A. Van de Voorde, H. Van Heuversyn, J. Van Hereweghe, G. Volckaertand, and M. Ysebaer. (1978). Complete nucleotide sequence of SV40 DNA. *Nature* (London) *273*, 113-120.

Klayman, D. L., and E. S. Copeland. (1975). The design of anti-radiation agents. In *Drug Design*, Vol. 6, pp. 81-126. New York: Academic Press.

Landau, L. (1944). *J. Phys. USSR 8*, 201.

Magee, J. L., an A. Chatterjee. (1980). Radiation chemistry of heavy particles. In *Biological dn Medical Research with Accelerated Heavy Ions at the Vebalac, 1977-1980*, pp. 63-69. Ed. by M. C. Pirruccello and C. A. Tobias. Lawrence Berkeley Laboratory Report LBL-11220.

Mozumder, A. (1969). Advances in Radiation Chemistry, Vol. 1, pp. 1-102. Ed. by M. Burton and J. L. Magee. New York: Wiley (Interscience).

Paretzke, H. G., and M. J. Berger. (1978). Stopping power and energy degradation for electrons in water vapor. In *Proceedings, Sixth Symposium on Microdosimetry*, pp. 749-758. Ed. by J. Booz and H. G. Ebert. Brussels, Belgium: Commission of the European Communities.

Perez, C. F., M. R. Botchan, and C. A. Tobias. (1984). Ionizing radiation enhances the efficiency of transformation of marker genes transfected into cells *in vitro*. *Br. J. Cancer 49*, 314.

Perez, C. F., and C. A. Tobias. (1984). Ionizing and ultraviolet radiation enhances the efficiency of both eukaryotic-DNA and prokaryotic-DNA mediated gene transfer. *Radiat. Res.*, in press. (Abstract for Radiation Research Society Annual Meeting, Orlando, Florida, 25-29 March 1984.)

Pirruccello, M. C., and C. A. Tobias, eds. (1980). *Biological and Medical Research with Heavy Ions Accelerated at the Bevalac, 1977-1980*. Lawrence Berkeley Laboratory Report LBL-11220.

Reddy, V. B., B. Thimmappaya, R. Dhar, K. N. Subramanian, B. S. Zain, J. Pan, P. K. Ghosh, M. L. Clema, and S. M. Weissman. (1978). The genome of simian virus 40. *Science 200*, 494-502.

Stradford, I. J. (1982). Mechanisms of hypoxic cell radiosensitization and the development of new sensitizers. *Int. J. Radiat. Oncol. Biol. Phys. 8*, 391-398.

Symon, K. R. (1948). Ph.D. dissertation, Harvard University.

Tobias, C. A., A. Chatterjee, M. Malachowski, E. A. Blakely, and T. L. Hayes. Tracks in Condensed Systems. In *Proceedings of* the *Sixth International Congress of Radiation Research, May 1979, Tokyo, Japan*, pp. 156-165. Ed. by S. Okada, M. Mamura, T. Terashima, and H. Yamaguchi. Tokyo: Japanese Association for Radiation Research.

Tobias, C. A., E. A. Blakely, F. Q. H. Ngo, and T. C. H. Yang. (1980). The repair-misrepair model of cell survival. In *Radiation Biology in Cancer Research*, pp. 195-230. Ed. by R. E. Meyn and H. R. Withers. New York: Raven Press.

Tobias, C. A., E. A. Blakely, P. Y. Chang, L. Lommel, and R. Roots. (1984). Response of sensitive human ataxia and resistant T-1 cell lines to accelerated heavy ions. *Br. J. Cancer 49*, 175-185.

Todaro, G. J., and Green, H. (1966). *Virology 28*, 756-759.

Topp, W., D. Lane, and R. Pollack. (1980). Transformation by SV40 and polyoma virus. In *DNA Tumor Viruses*, pp. 205-296. Ed. by J. Tooze. New York: Cold Spring Harbor.

Turner, J. E., J. L. Magee, A. J. Wright, A. Chatterjee, R. N. Hamm, and R. H. Ritchie. (1983). Physical and chemical development of electron tracks in liquid water. *Radiat. Res. 96*, 437-449.

Turner, J., H. G. Paretzke, R. N. Hamm, H. A. Wright, and R. H. Ritchie. (1982). Comparative study of electron energy deposition and yields in water in the liquid and vapor phases. *Radiat. Res. 92*, 47-60.

Vavilov, P. V. (1957). *Zh. Experim. i. Theor. Fiz. 32*, 920. [English translation: *Soviet Phys. JEPT 5*, 749 (1957).]

Yang, T. C., and C. A. Tobias. (1980). Radiation and cell transformation in vitro. In *Advances in Biological and Mecical Physics*, Vol. 17, pp. 417-461. Ed. by T. L. Hayes. New York: Academic Press.

Yang, T. C., and C. A. Tobias. (1985). Mechanism of radiation-induced neoplastic cell transformation. In *Proceedings of the Berkeley Conference in Honor of Jerzy Neyman and Jack Kiefer*, Vol. I. Belmont, Calif.: Wadsworth Publishing Co., in press.

Yang, T. C., C. A. Tobias, E. A. Blakely, L. M. Craise, I. S. Madfes, C. Perez, and J. Howard. (1980). Enhancement effects of high-energy neon particles on the viral transformation of mouse C3H10T1/2 cells in vitro. *Radiat. Res. 81*, 208-223.

EXTENSION OF THE TIME-INDEPENDENT REPAIR-MISREPAIR MODEL OF CELL SURVIVAL TO HIGH-LET AND MULTICOMPONENT RADIATION

NORMAN W. ALBRIGHT and CORNELIUS A. TOBIAS

University of California, Berkeley

ABSTRACT: The repair-misrepair (RMR) model of cell survival is formulated as a sequence of discrete repair steps. The time-independent
RMR model is extended to high-LET and multicomponent beams, and the
probability of cell survival is calculated. The effects of the variances in the number of tracks through the nucleus, the type of ionizing particle, the LET, the track length, and the quantity of DNA are
estimated. The variance in the number of tracks raises the dose-log
survival curve and reduces its curvature. The variance in the LET
and the type of ionizing particle induces a variance in the number of
lesions per track which augments the effect of the variance in the
number of tracks. This increase can be as much as 70% for Bevalac
accelerated heavy-ion beams. The dose-log survival curve for a beam
with a range of LET values is less curved and lies above that for a
beam with a single LET that gives the same value for the mean number
of lesions per track. For the low-LET case the probability of survival is compared to that given by the original formulation of Tobias
et al. (1980). Both survival curves have the same initial and final
slopes; however, the change in slope occurs at lower doses for the
original curve. This difference is approximately equivalent to using
a smaller value for the parameter ε in the present formula than was
used in the original formulation.

This work was supported in part by the Office of Energy Research of
the Department of Energy under Contract No. DE-AC03-76SF00098, and by
the National Cancer Institute of the Department of Health and Human
Services under Research Grant CA-15184.

INTRODUCTION

The repair-misrepair (RMR) model of cell survival describes the
process by which cellular enzymatic mechanisms either eurepair or
misrepair lesions in the nucleus that have been produced by exposure
to ionizing radiation.

The basic assumptions of the RMR model are the following [1, 2]:

1. Radiation causes an uncertain number and distribution of criti-
    cal lesions in the nucleus of a cell.

2. The eventual fate of the cell is uncertain before repair has
    committed the lesions to be expressed in particular ways; there-
    fore, the lesions are initially uncommitted.

3. There exist enzymatic mechanisms that can recognize lesions and
    attempt their repair.

4. The lesions are either eurepaired, in which case the original
    molecular structure is restored, or misrepaired, in which case
    the repair is incomplete or an altered structure is produced.
    A eurepaired lesion is not expressed. A misrepaired lesion
    can be expressed as cell death, variant, mutuation, cell trans-
    formation, or no expression.

The lesions responsible for most cell death are assumed to be
double strand breaks in DNA molecules. Strand ends from different
breaks can be joined if the ends are sufficiently close. Two types
of repair processes are considered: linear repair, which involves
rejoining the strand ends of a single lesion, and quadratic repair,
which involves joining two of the four strand ends from two different lesions.

The RMR model is principally a model of the repair of lesions.
The original formulation of the RMR model is based on a differential
equation for the mean number of lesions per nucleus [1, 2]. The
present formulation centers on the probability of a sequence of
discrete repairs of the lesions in a single nucleus. These two
approaches are sufficiently different that, to avoid confusion. the
former will be called RMR-I and the latter RMR-II.

The various processes in the radiation biology of a cell occur
on widely different time scales. Processes occurring on a faster
time scale (e.g., radical-DNA interactions) are essentially complete

before those on a slower time scale (e.g., enzymatic repair processes) have progressed significantly. Therefore, events occurring on different time scales can be analyzed separately and sequentially. The key approximation that will be used here is based on the following assumptions: Irradiation and the production of lesions are complete before significant repair has occurred, and repair is complete before cell survival is tested. These assumptions allow the radiobiological processes to be divided into three distinct stages that can be analyzed separately: the production of lesions, the repair of lesions, and the subsequent survival or death of the cell. The formulation used in this paper will consider only the final result of irradiating mammalian cells in terms of cellular survival and therefore will not address the time evolution of the repair process.

The division of the radiobiological processes into separate time scales can be made also in the case of a dose rate that is sufficiently low that all the lesions produced by the passage of one particle will be repaired or misrepaired before the passage of a second particle produces new lesions. The case of an intermediate dose rate, for which the irradiation time and repair time are comparable, must be analyzed differently.

The probability that a cell will survive if it is irradiated with a dose D, $P(S|D)$, can be written as the product of four factors which is summed over the ranges of the intermediate random variables: (1) the probability $P(n|D)$ that n tracks will cross the nucleus if the average does is D, (2) the probability $P(k|n)$ that k lesions will be produced if n tracks cross the nucleus, (3) the probability $P(m|k)$ that m lethal misrepairs will be made if the cell has k lesions, and (4) the probability $P(S|m)$ that the cell will survive if it has m lethal misrepairs.

$$P(S|D) = \sum_{n=0}^{\infty} \sum_{k=0}^{\infty} \sum_{m=0}^{k} P(S|m)P(m|k)P(k|n)P(n|D). \tag{1}$$

This factoring is a consequence of the assumptions about the time scales. $P(S|m)$ and $P(m|k)$ can be calculated separately only if the repair processes are complete and no uncommitted lesions remain when survival is tested. $P(m|k)$ and $P(k|n)$ can be calculated separately only if the production of lesions is complete before repair has begun.

Each of the four probability factors defined above can depend on: (1) the physical state of the cell (for example, the temperature to which the cell is exposed and the shape and dimensions of the nucleus at the time of irradiation); (2) the chemical state of the cell (the oxic state, the presence and concentration of radioprotectors or radiosensitizers); (3) the biological state of the cell (the cell line, genetic deficiencies, the cell age, and the number of chromosomes and their distribution through the nucleus); (4) the properties of the radiation (the percentage of each type of ionizing particle of a multicomponent beam and their energy distributions); and (5) other parameters such as the dose rate. The calculations in this paper assume that each of these parameters has the same value for every cell. To keep the notation as simple as possible, these constant parameters will be suppressed.

It is convenient to characterize an ionizing particle by its linear energy transfer (LET) rather than by its energy. The LET is the rate of energy loss by one particle per unit of track length through the cell. The LET relates the dose to the number of ionizing particles passing through a cell. The dose is defined as the energy deposited in the cells by all of the particles per unit of cell mass.

In this paper, each of the four probability factors of Eq. (1) will be calculated, beginning with the number of particle tracks from a specific dose and the resultant production of lesions in DNA. The probability of no lethal misrepairs will then be calculated as a Markov process of lesion repair. The four probability factors will be combined to give the probability of cell survival for a specific dose, and the resulting expression will be approximated. The special cases of low dose rate and survival curves without shoulders will be discussed. Next, a comparison with RMR-I will be made. Finally, the effects on cell survival of the variances in the number of tracks through the nucleus, the type of ionizing particle, the LET, the track length, and the quantity of DNA will be discussed.

NUMBER OF TRACKS

A random variable, such as the number of tracks through a nucleus, will be denoted by a boldface upper case letter $\mathbf{N}$, the mean value of this random variable will be denoted by the corresponding upper case letter N, a specific value of the random variable by the corresponding lower case letter n, and the standard deviation of the random variable by σ with the lower case letter as a subscript σ_n. Let $P(n|D)$ denote the probability of n tracks through the nucleus given that the dose was D. An alternative notation for this probability is $P(\mathbf{N} = n|D)$. The former notation will be preferred because of its brevity, and the latter will be used only where required for clarity.

The number of tracks through a nucleus has a Poisson distribution that depends on the dose D, the track average LET of the beam L, the density of the nucleus ρ, and its cross-sectional area A, which might depend on the cell age or other factors. Then

$$P(n|D) = e^{-\omega D} (\omega D)^n /n! \tag{2}$$

where the mean number of tracks is $N = \omega D$, and $\omega = \rho A/L$.

PRODUCTION OF LESIONS

Synchronized cells, all of which are in the same physical, biological, and chemical state, are irradiated with a dose D by a multicomponent beam (e.g., a beam of accelerated heavy ions and the fragments produced by the passage of these ions through a water column). The number of lesions in a cell resulting from the irradiation is a random variable. This randomness has several sources, quantities that vary from one cell to the next or from track to track, such as (1) the number of tracks through the nucleus n, (2) the type of ionizing particle that made the ith track z_i, (3) the LET of the ith

ionizing particle ℓ_i, (4) the length of the ith track x_i, (5) the
quantity of DNA in the nucleus q, and (6) the physical and chemical
steps in the production of lesions by a track with specific values
for z_i, ℓ_i, x_i, and q.

The Distribution of Ionizing Particles and Their LETs

The distribution of the types of ionizing particles and their LETs
must be determined experimentally for a particular beam. Let $f(z, \ell)$
denote the joint probability and probability density that an ionizing
particle will be of type z and will have LET ℓ. Then $\int f(z, \ell)\,d\ell$
is the fraction of the beam that is particle type z, and the track
average LET of the beam is

$$L = \sum_z \int f(z, \ell)\,\ell\,d\ell. \tag{3}$$

The Distribution of Track Lengths Due to Nuclear Shape

During the interphase between mitotic events, the DNA is distributed
throughout the nucleus. Therefore the mean amount of DNA that is
exposed to the free radicals produced by a track is proportional to
the track length. The distribution of track lengths depends on the
shape of the nucleus and its orientation to the beam. In track-
segment exposures to charged particle beams, the cells are grown in
monolayer on petri dishes. The attachment of the cells to the dish
results in the cell nucleus being partially flattened during the
interphase between mitotic events. A cross-section of the nucleus
in a plane parallel to the bottom of the dish is approximately circular, but a cross-section in a plane perpendicular to the bottom
has a somewhat elliptical shape. The nuclear shape can be approximated by an ellipsoid generated by revolving an ellipse about its
minor axis, whose length is b. The cells are irradiated by a beam
perpendicular to the bottom of the dish and therefore parallel to
the minor axis of the ellipsoid. The tracks through the nucleus are
approximately parallel to the beam. Let $f(x)$ denote the probability
density of the length of one track. The range of x is $0 \leqslant x \leqslant b$,

and its density is

$$f(x) = 2x/b^2.$$ (4)

The mean and variance of the track lengths are $X = 2b/3$ and $\sigma_x^2 = X^2/8$. Note that the distribution of track lengths depends on the length of the minor axis but not on the length of the major axis.

The Distribution of the Quantity of DNA

If DNA is the locus for lethal lesions, then the mean number of lesions produced by a track will be proportional to the quantity of DNA near the track q. Let $P(q)$ denote the distribution of **Q**. For in vitro experiments with synchronized cells of established cell lines, q varies from cell to cell if the cells are aneuploid. For simplicity, let the units of q be the number of chromosomes, and as an example, consider Tl cells. Todd counted the number of chromosomes in a sample of 32 Tl cells and found that q had a range of 52 to 77, a mean (Q) equal to 63.5, and a standard deviation, σ_q, of 6.67 [3]. For this sample,

$$\sigma_q^2/Q^2 \approx 0.011,$$ (5)

which is quite small.

The Distribution of the Number of Lesions Produced by One Track with Specific Values for z, ℓ, x, and q

In the absence of a detailed theory of the physical and chemical steps in the production of lesions, a simple probabilistic model for the number of lesions can be based on three assumptions: (1) Each lesion is produced by a single track (intertrack effects are negligible). (2) The lesions from a single track are produced nearly independently and approximately follow a Poisson distribution. Let $w(z, \ell, x, q)$ denote the mean number of lesions produced by one track with specific values for the type of ionizing particle LET, track length, and quantity of DNA, and let W denote the average of w over these variables; W is the mean number of lesions per track. (3) w is proportional to the track length and the number of target loci for the lesions, $w(z, \ell, x, q) = c(z, \ell)xq$ [4]. The coefficient c depends also on the volume and the physical, biological, and chemical state of the cell. Arguments for these assumptions are given in Appendix I.

The Distribution of the Number of Lesions

The probability that k lesions wil be made if n tracks cross the nucleus can be calculated from these assumptions. Let $P(k|z_1, \ell_1, x_1, \ldots, z_n, \ell_n, x_n, q)$ denote the probability that k lesions will be made if n tracks of particle types $z_1 \ldots z_n$ with LETs $\ell_1 \ldots \ell_n$ and track lengths $x_1 \ldots x_n$ cross a nucleus containing a quantity of DNA q. Then

$$P(k|z_1 \ldots x_n, q) = \left(\sum_i c_i x_i q \right)^k \exp\left(- \sum_i c\, x\, q \right)/k! \tag{6}$$

where $c_i = c(z_i, \ell_i)$. The probability that k lesions will be made if n tracks cross the nucleus is the average over $z_1 \ldots x_n$, q of $P(k|z_1 \ldots x_n, q)$.

$$P(k|n) = \sum_q \sum_{z_1} \cdots \sum_{z_n} \int d\ell_1 \cdots \int d\ell_n \int dx_1 \cdots \int dx_n f(z_1, \ell_1) f(x_1)$$
$$\cdots f(z_n, \ell_n) f(x_n) P(q) P(k|z_1 \ldots x_n, q). \tag{7}$$

It is shown in a later section that the variation in the quantity of DNA from cell to cell can be neglected, and $P(k|z_1 \ldots x_n, q)$ can be evaluated at $q = Q$. The neglect of the variation in the quantity of DNA simplifies the calculation of the probability that k lesions will be made if n tracks cross the nucleus, $P(k|n)$. This calculation can be performed by use of the probability generating function (pgf) for $P(k|n)$, $g(s|n)$, which is defined by

$$g(s|n) = \sum_{k=0}^{\infty} s^k P(k|n). \tag{8}$$

$P(k|n)$ can be obtained from the kth derivative of $g(s|n)$, which can be calculated by substituting Eq. (7) into Eq. (8).

$$g(s|n) = \gamma(s)^n, \tag{9}$$

where

$$\gamma(s) = \sum_z \int d\ell \int dx\, f(z, \ell) f(x) \exp[(s - 1)w(z, \ell, x, Q)]. \tag{10}$$

If the distribution of w is approximately normal, then γ can be calculated analytically.

$$\gamma(s) = \exp[(s - 1)W + (s - 1)^2 \sigma_w^2/2]. \tag{11}$$

Equation (11) gives the simplest representation of a multicomponent beam with a range of LET values. A single component beam with a single LET is represented by $\gamma = \exp[(s - 1)W]$.

In Eq. (1), $P(k|n)$ is multiplied by $P(n|D)$, and the result is summed over n. This sum yields $P(k|D)$, the probability that k lesions will be made in cells receiving a dose D.

$$P(k|D) = \sum_{n=0}^{\infty} P(k|n)P(n|D). \tag{12}$$

The simple form of Eq. (9) enables the pgf for $P(k|D)$, that is, $g(s|D)$, to be calculated as a function of $\gamma(s)$. Multiplying Eq. (12) by s^k and summing over k gives

$$g(s|D) = \sum_{n=0}^{\infty} g(s|n)P(n|D). \tag{13}$$

Substituting Eqs. (2) and (9) into (13) gives

$$g(s|D) = \exp[(\gamma(s) - 1)\omega D]. \tag{14}$$

$P(k|D)$ is given in terms of the kth derivative of this pgf.

$$P(k|D) = \left[\frac{1}{k!}\left(\frac{d}{ds}\right)^k g(s|D)\right]_{s=0}. \tag{15}$$

CELL SURVIVAL

Linear misrepair can produce chromosome and chromatid aberrations such as altered DNA sequences, deletion of small segments of DNA, and stabilization of breaks. Quadratic misrepair can produce chromosome and chromatid aberrations with two misrejoinings. The effect of an aberration can be either no expression, variant, mutation, cell transformation, or cell death. The possible ways that aberrations can lead to reproductive cell death include deficient synthesis of DNA, synthesis of abnormal DNA or deletion of small segments of DNA, and unequal

segregation of DNA at mitosis including formation of an anaphase
bridge.

It will be assumed here that all misrepairs can be divided into
two classes: viable and lethal. These classes denote that a cell
can survive an arbitrary number of viable misrepairs but will be
killed by a single lethal misrepair. Let m denote the number of
lethal misrepairs in a cell. The probability of cell survival from
m lethal misrepairs is

$$P(S|m) = 0 \tag{16}$$

for $m \geqslant 1$, and $P(S|M = 0) = 1$.

REPAIR OF LESIONS

For the case considered here, the possible states of an individual
nucleus at any time can be described by the number of unrepaired,
lethally misrepaired, and viably repaired lesions. These states
can be divided into three categories: a survival state in which all
the original lesions have been viably repaired, lethal states in
which one or more lethal misrepairs have occurred, and uncommitted
states in which all lesions have not been repaired but no lethal
misrepairs have occurred. The fraction of cells that survive is
equal to the fraction that have no lesions plus the fraction whose
lesions are all viably repaired, as in the sequence of repairs
$U \rightarrow U \rightarrow U \rightarrow S$ shown in Figure 1, where U = uncommitted states and
S = the survival state.

The probability that no lethal misrepairs will be made if a
cell has k lesions, $P(M = 0|k)$, is the product of probabilities that
each repair in a sequence of repairs will be a viable repair. Let
$P(VR|k)$ denote the probability that the next repair will be viable
if a cell has k unrepaired lesions.

$$P(M = 0|k) = P(VR|k)P(VR|k - 1) \cdots P(VR|1). \tag{17}$$

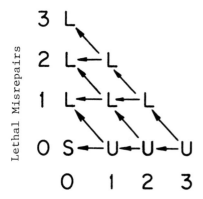

Unrepaired Lesions

Figure 1. Possible states of a nucleus at any time after irradiation. L = lethal states, U = uncommited states, and S = the survival rate.

It is assumed that the probabilities of linear and quadratic repairs are proportional to the number of distinct ways in which each repair can occur. (A quadratic repair joins two of the four ends from a pair of lesions. The remaining two strand ends will be considered a single residual lesion.) Specifically, it is assumed that if a cell has k lesions, then the probability of linear repair in a short time interval Δt is $\lambda k \Delta t$. And the probability of quadratic repair is $\kappa k(k - 1)\Delta t$. Consequently, the probability that the next repair will be a linear one if a cell has k unrepaired lesions is

$$\frac{\lambda k}{\lambda k + \kappa k(k - 1)}$$

and the probability that it iwll be a quadratic repair is

$$\frac{\kappa k(k - 1)}{\lambda k + \kappa k(k - 1)} .$$

This assumes that the coefficient for the probability of quadratic repair κ is the same for the interactions between lesions on the same track and between lesions on different tracks. This assumption was used in the original formulation of the RMR model [1, 2]. The possibility that the rates of intratrack and intertrack quadratic repair differ is considered by Tobias in another paper [5]. Let φ be the fraction of linear repairs that are viable and δ be the fraction of quadratic repairs that are viable. Then the probability that the next repair will be viable if a cell has k unrepaired lesions is

$$P(VR|k) = \frac{\varphi \lambda k + \kappa k(k-1)}{\lambda k + \kappa k(k-1)}. \tag{18}$$

Define ε as the ratio $\varepsilon = \lambda/\kappa$, then $P(VR|k)$ can be written

$$P(VR|k) = \frac{\varphi \varepsilon + \delta(k-1)}{\varepsilon + k - 1} \tag{19}$$

$P(M = 0|k)$ is the product of factors of the type given by Eq. (19).

$$P(M = 0|k) = \frac{\varphi \varepsilon(\varphi \varepsilon + \delta) \cdots (\varphi \varepsilon + \delta(k-1))}{\varepsilon(\varepsilon + 1) \cdots (\varepsilon + k - 1)} \tag{20}$$

for $k \geqslant 1$, and, of course $P(M = 0|K = 0) = 1$.

Fits of survival curves to experimental data indicate that ε is about 20 or 30 [6]. Consequently, $P(M = 0|k)$ can be approximated accurately using Stirling's formula for the factorial [7]. This gives

$$P(M = 0|k) = \varphi^k f(x) \tag{21}$$

where

$$f(k) = \left(\frac{\varepsilon}{\varepsilon + k}\right)^{\varepsilon + k - 1/2} \left(\frac{\varphi \varepsilon + \delta k}{\varphi \varepsilon}\right)^{\varphi \varepsilon/\delta + k - 1/2}. \tag{22}$$

For $\varepsilon = 20$, the error in the approximation divided by the true value of f is less than 0.0002 at $k = 1$, 0.0013 at $k = 10$, and 0.0021 at $k = 20$.

PROBABILITY OF CELL SURVIVAL

If all misrepairs are either viable or lethal, then a cell will survive if it has no lethal misrepairs and not otherwise. For this case, Eq. (1) for the probability of survival for cells irradiated with a dose D reduces to

$$P(S|D) = \sum_{k=0}^{\infty} P(M = 0|k)P(k|D) \tag{23}$$

where the first factor is given by Eq. (20) and the second by
Eq. (15).

$$P(S|D) = \sum_{k=0}^{\infty} f(k)\frac{\varphi^k}{k!}\left[\left(\frac{d}{ds}\right)^k g(s|D)\right]_{s=0}. \tag{24}$$

It is shown in Appendix II that Eq. (24) can be approximated by

$$P(S|D) = \exp[(\gamma - 1)\omega D](1 + \vartheta\alpha D/\varepsilon)^{-\varepsilon}(1 + \delta\vartheta\alpha D/\varphi\varepsilon)^{\varphi\varepsilon/\delta} \tag{25}$$

for $\delta \neq 0$, and by

$$P(S|D) = \exp[(\gamma - 1)\omega D + \vartheta\alpha D](1 + \vartheta\alpha D/\varepsilon)^{-\varepsilon} \tag{26}$$

for $\delta = 0$ or $\delta \ll \varphi$; where $\alpha = \omega\gamma'$, $\gamma' = d\gamma/ds$, both γ and γ' are
evaluated at $s = \vartheta$, and ϑ is the solution of the pair of equations

$$\alpha = \omega\gamma'(\vartheta)$$

$$\vartheta = \frac{2\varphi}{(1 - \delta\alpha D/\varepsilon) + \sqrt{(1 - \delta\alpha D/\varepsilon)^2 + 4\varphi\alpha D/\varepsilon}}. \tag{27}$$

Equations (27) can be solved by iteration using W as an initial esti-
mate of $\gamma'(\vartheta)$. The parameter ϑ decreases monotonically with increasing dose. It equals φ at $D = 0$ and is asymptotic to δ as $D \to \infty$. The
quantity $\vartheta(D)$ is approximately the fraction of repairs that are
viable for dose D.

Equations (25) and (26) provide the extension of the time-
independent RMR model to high-LET and multicomponent beams with a
range of LET values. These features are contained in the functions
$\gamma(\vartheta)$ and $\vartheta(D)$. The dose-log survival curve given by Eq. (25) bends
downward continuously. The slope equals $-(1 - \gamma(\varphi))\omega$ at $D = 0$ and is
asymptotic to $-(1 - \gamma(\delta))\omega$ as $D \to \infty$. For low LET these slopes are
$-(1 - \varphi)\alpha$ at $D = 0$ and $-(1 - \delta)\alpha$ as $D \to \infty$. The limiting slopes correspond to the linear and quadratic repair processes, respectively.
Linear repair predominates if the number of initial lesions is small,
and $(1 - \varphi)$ is the fraction of linear repairs that are lethal misrepairs.
Quadratic repair predominates if the number of initial lesions is large,
and $(1 - \delta)$ is the fraction of quadratic repairs that are lethal misrepairs. Although the slope of the dose-log survival curve has an
asymptotic value at large doses, the curve itself does not have an
asymptote, and so it has no extrapolation number.

For the case of low-LET irradiation and $\delta = 0$, Eq. (26) reduces to

$$P(S|D) = \exp[2\vartheta\alpha D - \alpha D](1 + \vartheta\alpha D/\varepsilon)^{-\varepsilon}. \tag{28}$$

Moreover, since $\gamma' \approx W$ for this case, $\alpha = \omega W$, which does not depend on ϑ; consequently, ϑ is obtained without iteration.

$$\vartheta = \frac{2\varphi}{1 + \sqrt{1 + 4\varphi\alpha D/\varepsilon}}.$$

Some qualitative properties of $P(S|D)$ are evident in the two families of survival curves displayed in Figures 2 and 3 for the case $\delta = 0$ and low LET. Survival curves for larger values of ε or φ lie above those for smaller values of these parameters. Also, $P(S|D)$ is more sensitive to changes in φ than to changes in ε.

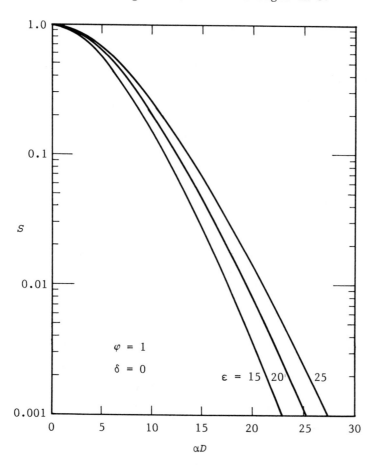

Figure 2. Survival curves of the time-independent RMR model for the case $\delta = 0$, $\varphi = 1$, $\varepsilon = 15$, 10, and 25, and low-LET radiation.

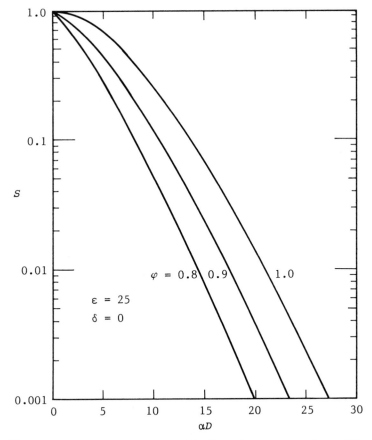

Figure 3. Survival curves of the time-independent RMR model for the case $\delta = 0$, $\varphi = 0.8$, 0.9, and 1.0, $\varepsilon = 25$, and low-LET radiation.

LOW DOSE RATE

The probability of cell survival for the case of very low dose rate can be written in terms of the same four probability factors used in Eq. (1) for the high dose rate case. This limiting case corresponds to the natural background radiation. Let $P(S|n)$ denote the probability that a cell will survive if it has n tracks through the nucleus.

$$P(S|n) = \sum_{k=0}^{\infty} \sum_{m=0}^{k} P(S|m)P(m|k)P(k|n). \tag{30}$$

Then the probability that a cell will survive if it is irradiated with a dose D can be written

$$P(S|D) = \sum_{n=0}^{\infty} P(S|n)P(n|D). \tag{31}$$

Equating the slope of $P(S \; D)$ at $D = 0$ given by Eq. (31) to that given by Eq. (25) yields the probability that a cell will survive if it has one track through the nucleus.

$$P(S|\text{N} = 1) = \gamma(\varphi). \tag{32}$$

If the dose rate is sufficiently low, then the lesions produced by one track will have had time to repair before a second track produces more lesions. The probability that a cell will survive n tracks with complete repair between the occurrence of successive tracks is $P(S|\text{N} = 1)^n$. For this case the sum over n can be done analytically. The probability of survival for cells receiving a dose D at very low dose rate is

$$P(S|D) = e^{(\gamma(\varphi) - 1)\omega D}. \tag{33}$$

If the dose rate is too high for complete repair between successive tracks, then quadratic repairs will be a larger fraction of the repairs that occur. If, in addition, quadratic repair is viable less often than linear repair, then the probability that a cell will survive n tracks through the nucleus is less than $P(S|\text{N} = 1)^n$.

$$0 \leqslant P(S|n) \leqslant P(S|\text{N} = 1)^n. \tag{34}$$

From which it follows that, for any dose rate,

$$e^{-\omega D} \leqslant P(S|D) \leqslant e^{(\gamma(\varphi) - 1)\omega D}. \tag{35}$$

At sufficiently high LET, $\gamma(\varphi)$ decreases, and the region between these two bounds narrows. The relative viability of quadratic and linear repairs is considered in the next section.

SURVIVAL CURVES WITHOUT SHOULDERS

For the special case $\kappa = 0$, that is, no quadratic repair, $P(\mathbf{M} = 0|k)$
reduces to φ^k, and the probability of cell survival is given by
Eq. (33). The log of survival is a linear function of the dose for
this case. If quadratic repair does not occur, then the survival
curve will not have a shoulder; however, even if quadratic repair
does occur, the shoulder may be absent, as the next special case
demonstrates.

For the special case $\delta = \varphi$, $P(\mathbf{M} = 0|k)$ equals φ^k, and the
probability of cell survival again is given by Eq. (33). The graph
of the log of survival versus dose is a straight line. If δ is less
than φ, that is, if quadratic repair is viable less often than linear
repair, then the graph of the log of survival curves down; and if δ
is greater than φ, it curves up. Repair-deficient cells correspond
to the case $\delta = \varphi = 0$ [8]. If such cells are irradiated by a single-
component, single-LET beam, then $\gamma(0) = \exp(-W)$, and $W \approx w(z, \ell, X, Q)$,
because variations in track length and quantity of DNA are negligible.
This provides a simple way to measure the mean number of lesions per
track as a function of the beam species z and the LET ℓ. The probability of survival for this case is

$$P(S|D) = \exp[(e^{-W} - 1)\omega D]. \tag{36}$$

The mean number of lesions per track W can be found from the slope of
the dose-log survival line, since ω is a known function of two quantities that can be measured directly, the LET, and the nuclear cross-
sectional area.

Even if quadratic repair occurs and is viable less often than
linear repair, the survival curve will lack a shoulder if there is no
interaction between the lesions produced by different tracks as was
shown by the low dose rate case. In the RMR model, the presence of
a shoulder requires three factors: that quadratic repair occur, that
it occur between lesions produced by different tracks, and that it
is viable less often than linear repair.

COMPARISON WITH THE ORIGINAL FORMULATION

The original formulatin of the RMR model, RMR-I, is based on two
assumptions: (1) that the average rate of repair is given approximately by

$$dU/dt = -\lambda U - \kappa U^2 \tag{37}$$

where $U(t)$ is the average number of unrepaired lesions at time t
after irradiation, and (2) that the probability of a cell having zero
lethal misrepairs is given by e^{-M}, where M denotes the average number
of lethal misrepairs per cell. The average rate of lethal misrepair
is

$$dM/dt = \lambda(1 - \varphi)U + (1 - \delta)U^2. \tag{38}$$

The equations for U and M are easily integrated. The resulting
probability of survival for cells receiving a dose D is

$$P(S|D) = e^{-(1-\delta)U_0}(1 + U_0/\varepsilon)^{(\varphi-\delta)\varepsilon} \tag{39}$$

where $U_0 = \alpha D$ is the average number of initial lesions.

   RMR-I is applicable to the case of low-LET radiation because the
effect of fluctuations in the number of tracks through the nucleus
has not been included. Therefore, Eq. (39) for the case $\delta = 0$ is
comparable to Eq. (28) of RMR-II.

   For low LET, the probabilities of cell survival given by RMR-I
and RMR-II generate qualitatively similar dose-log survival curves.
Both curves have the same initial and final slopes; however, the
RMR-I survival curve undergoes more of the change in slope at lower
doses than does the RMR-II curve. The slope of the RMR-I curve is
halfway between the initial and final slopes at the dose $\alpha D = \varepsilon$.
The RMR-II survival curve reaches this halfway point at $\alpha D = 2\varepsilon$. So
the probability of survival given by RMR-I decreases with increasing
dose more rapidly than that given by RMR-II for the same value of ε.
However, if different values of ε are used, the two curves can remain
quite close over the range of survival levels that is important for
experiments. A numerical comparison shows that the probability of
survival given by RMR-I with a value of ε equal to ε_0 remains within

±20% of that given by RMR-II using a value of ε equal to $\varepsilon_0/1.4$ for survival levels in the range 0.001 to 1.0.

EFFECTS OF THE VARIANCES

The size and effect of the variance associated with each factor in Eq. (7) will be calculated in the next few paragraphs. The mean number of lesions produced by dose D is $K = WN$, where $N = \omega D$ is the mean number of tracks. The variance of the total number of lesions for dose D is

$$\sigma_k^2 = K + K^2 \sigma_q^2/Q^2 + KW(1 + \sigma_\omega^2/W^2) \tag{40}$$

where

$$(1 + \sigma_\omega^2/W^2) = (1 + \sigma_c^2/C^2)(1 + \sigma_x^2/X^2)(1 + \sigma_q^2/Q^2). \tag{41}$$

The first term on the right-hand side of Eq. (40) is the variance that would be present even if there were no variance in the number of tracks, the type of ionizing particle, the LET, the track length, and the quantity of DNA. It is due to the assumption that, for fixed values of these variables, the production of lesions follows a Poisson distribution. The variance in the quantity of DNA contributes twice to the variance in k: the second term in Eq. (40) and part of the third term. The factor KW of the third term is due to the variance in the number of tracks. This factor is augmented by the variance in $w(\mathbf{Z}, \mathbf{L}, \mathbf{X}, \mathbf{Q})$.

If the variance in $\mathbf{Q}$ is sufficiently small and if $W \ll 1$ (low LET), then the variance in the number of lesions approximately equals the first term in Eq. (40), and the probability $P(k|n)$ is approximately given by a Poisson distribution. This approximation holds regardless of the size of σ_ω; for $W \ll 1$ the distribution of the number of lesions is a function only of W.

Goodhead estimates that the average number of "lethal lesions" (lethal misrepairs in the RMR model) per track in T1 cells is roughly 0.01 for radiation with a LET of 10 keV/μm, 0.1 at 50 keV/μm, and 1.0 at 200 keV/μm [9]. The mean numbers of initial lesions per track

W are larger than these estimates, because some initial lesions are viably repaired. So, for a LET somewhere between 50 and 200 keV/μm, W is near 1, the third term in Eq. (40) becomes comparable to the first term, and the distribution of the number of lesions ceases to be Poisson. The combination of fewer tracks and more lesions per track is the origin of the overkill phenomenon; if a nucleus has no tracks, then it will have many lesions and probably dies. Overkill reduces the curvature of the dose-log survival curve.

The effects of the variances in the quantity of DNA and the number of lesions per track on the probability of cell survival can be shown by a simplified model. (These estimates are not based on the RMR model; they are used only to show the magnitudes of the effects.) For this simplified model let the probability that the repair of a lesion will be eurepair be approximated by a decreasing function of dose $\psi(D)$, then the probability of cell survival from k lesions is ψ^k. Let $w(\mathbf{Z}, \mathbf{L}, \mathbf{X}, \mathbf{Q})$ and $\mathbf{Q}$ have normal distributions. Then the probability of cell survival from dose D for the low-LET case can be expanded in powers of W.

$$\log P(S|D) \approx -(1 - \psi)K + \frac{K^2}{2}(1 - \psi)^2 \sigma_q^2/Q^2$$
$$+ \frac{KW}{2}(1 - \psi)^2(1 + \sigma_w^2/W^2). \tag{42}$$

Since K is linear in dose, Eq. (42) shows that the effect on survival of the variance in the quantity of DNA (the K^2 term) increases with dose. The result is an upward bending contribution to the dose-log survival curve. However, since this is a small effect, it could be masked by a larger downward bending term due to the declining fraction of eurepair. For a dose corresponding to a survival level of 0.001 and for the value of σ_q estimated above for synchronized T1 cells, the K^2 term in Eq. (42) equals 0.038, which will be neglected. (The variation of $\mathbf{Q}$ with cell age has a major effect on the variation of the probability of survival with cell age. However, for synchronized cells the variation in $\mathbf{Q}$ due to aneuploid spread has so small an effect that it will be neglected.)

Since W increases with LET, Eq. (42) shows that the effect on survival of the variance in the number of tracks (the KW term) becomes important for values of W near to or greater than 1, and that this effect is augmented by the variance in the number of lesions per track. The KW term has two effects: It raises the dose-log survival curve, and it reduces its curvature. The first effect is evident from Eq. (42). The second effect can be shown by calculating the slope of the dose-log survival curve at small and large doses. To illustrate this calculation assume that $\psi(D)$ declines from the value φ at $D = 0$ to 0 as $D \rightarrow \infty$, and ignore the K^2 term. Then the change in the slope between $D = 0$ and $D = \infty$ is

$$\Delta slope = \alpha\varphi[1 - (1 - \varphi/2)W(1 + \sigma_w^2/W^2)] \tag{43}$$

where $\alpha = K/D$. As W increases, the change in the slope decreases; that is, the curvature of the dose-log survival curve is reduced. Equations (42) and (43) show that the variance in the type of ionizing particle, LET, track length, and quantity of DNA raise the survival curve and reduce its curvature by contributing to the variance in $w(\mathbf{Z}, \mathbf{L}, \mathbf{X}, \mathbf{Q})$. The survival curve for a beam with a range of LET values is less curved and lies above that for a beam with a single LET that gives the same value for W. The size of each of the contributions to σ_w will be estimated in the next two paragraphs.

The variance in the quantity of DNA augments the curvature-reducing term in Eq. (42) by only 1.6%, which is negligible. The variance in the track length σ_x^2 increases this term by 12%. An effect of this size is just barely detectable because the curvature can be estimated to about 10% from survival data. Furthermore, the variance in track length is a constant effect; it does not vary with any experimental parameter such as oxic state, beam composition, or cell age in the interphase between mitotic events. This effect cannot explain any observable variation, and so it will be ignored.

The variance in the LET and type of ionizing particle has a major effect on the probability of survival, and this effect varies with the residual range of the beam. Irradiation in the plateau of the Bragg ionization curve of an accelerated monoenergetic heavy-ion beam is virtually free of contaminating particles from fragmentation evente; therefore, there is almost no variance in the type of ionizing particle or LET.   In contrast, irradiation near the Bragg ionization peak uses particles that have passed through a water column whose thickness is nearly equal to the range.   Depending on the original ion species, up to 50% of this beam can be fragments.   The LET of individual ions ranges from very small values to a maximum which, depending on the atomic number, can be several orders of magnitude greater.   The relative variance in LET, σ_ℓ^2/L^2, can be roughly a third [6, 10].   The data from Goodhead for the LET range 10-200 keV/μm can be fit approximately by $w = 0.01(\ell/10)^{1.5}$, which gives $\partial w/\partial \ell = 1.5w/\ell$. Combining these numbers gives a rough estimate for $\sigma_w^2/W^2 \approx 1.5^2 \sigma_\ell^2/L^2$ ≈ 0.7.   The variance in LET augments the effect of the variance in the number of tracks by roughly 70% in the Bragg peak.

CONCLUSION

The repair-misrepair model of cell survival has been formulated as a sequence of discrete repair steps, and the probability of a viable sequence of repairs has been calculated.   The time-independent RMR model has been extended to high LET and multicomponent beams.   The probability of cell survival has been derived as an infinite sum and reduced to a simplified expression.

Magnitudes have been estimated for the effects of the variances in the number of tracks through the nucleus, the type of ionizing particle, the LET, the track length due to the shape of the nucleus, and the quantity of DNA due to aneuploid spread.   The variance in the quantity of DNA for synchronized cells is negligible; it causes only a 4% difference in the survival of T1 cells at the 0.001 survival level.   The variance in the track length for irradiation in the interphase between mitotic events has a small effect that does not vary with any experimental parameter.   It augments by only 12% the curvature-reducing effect of the variance in the number of tracks, and so it can be neglected.

The variance in the number of tracks has two major effects on the probability of survival for high-LET radiation: It raises the dose-log survival curve and reduces its curvature. The variance in the LET and type of ionizing particle induces a variance in the number of lesions per track which augments the effects of the variance in the number of tracks. This increase can be as much as 70% for Bevelac accelerated heavy-ion beams. Furthermore, experiments in which the residual range is varied combine the effects of increasing average LET and increasing variance in the distribution of LETs. Consequently, if survival curve parameters are estimated from such experiments, up to 40% (= 70%/170%) of the change in these parameters with LET can be due to the variance in the LET and the type of ionizing particle. The dose-log survival curve for a beam with a range of LET values is less curved and lies above that for a beam with a single LET that gives the same value for the mean number of lesions per track.

In the RMR model, the presence of a shoulder on the dose-log survival curve requires three factors: that quadratic repair occurs, that it occurs between lesions produced by different tracks, and that it is viable less often than linear repair.

The slope of the dose-log survival curve of the time-independent RMR model equals $-(1 - \gamma(\varphi))\omega$ at $D = 0$, reflecting the preponderance of linear repair at low doses; and it is asymptotic to $-(1 - \gamma(\delta))\omega$ as $D \to \infty$, reflecting the preponderance of quadratic repairs a high doses. Although the slope of the curve has an asymptotic value at high doses, the curve itself does not have an asymptote, and so it has no extrapolation number.

For low-LET radiation exposures, the probability of cell survival given by the original formulation of the RMR model, RMR-I, and that given by the present formulation, RMR-II, generate qualitatively similar survival curves. Both curves have the same initial and final slopes; however, the transition from initial to final slope occurs at lower doses for the RMR-I curve. So the probability of survival given by RMR-I decreases with dose more rapidly than that given by RMR-II for the same value of ε. However, if different values of ε are used, the two survival curves can remain quite close for survival levels in the range 0.001 to 1.0. Specifically, experimental data that was best fit by an RMR-II survival curve with a value of ε equal to ε_0 will be best fit by an RMR-II survival curve with a value of ε equal to $\varepsilon_0/1.4$.

ACKNOWLEDGMENT

The authors thank E. A. Blakely for numerous helpful discussions of this paper.

APPENDIX I

Arguments for the three assumptions of the simple probabilistic model for the production of lesions are given in this appendix, where assumptions are denoted by 1, 2, or 3, and arguments by a or b.

For a beam with an energy on the order of 500 MeV per nucleon, about half the energy transfer is in glancing collisions with electrons (spurs) that create the core of the track, and the other half in knock-on electrons that compose the penumbra [11, 12]. These secondaries produce further ion pairs, which produce free radicals, in particular OH and atomic H, through a sequence of chemical reactions. One OH radical or one H atom can attack a DNA sugar by hydrogen abstraction from a carbon and initiate the sequence of chemical steps that lead to a single-strand break (SSB). Other reactions may give SSBs also. These reactions compete with recombination and with reactions that do not lead to strand breaks. If two SSBs occur sufficiently close together and on opposite strands, the result is a double-strand break (DSB), which is assumed to be the potentially lethal lesion produced by the radiation.

The SSBs are confined to a narrow region around the core and the tracks of the knock-on electrons, whose extent is determined by the distance the radicals can diffuse before chemical reactions eliminate them. It is unlikely that the regions of SSBs from two tracks will overlap significantly. Thus, the probability that two SSBs from different tracks will be close enough to form a DSB is much less than the corresponding probability for two SSBs from the same track. Consequently, (1) intertrack effects are neglected in this analysis.

A single track produces a large number of radicals. The production of one DSB consumes only two radicals, which is a negligible reduction in the number of radicals available for the production of other DSBs. Therefore, (2a) for a single track the DSBs are produced nearly independently.

The remaining part of the second assumption is that the number of DSBs is given by a Poisson distribution. The argument for this is more detailed. For each DNA molecule, label one of the strands as the first strand and the other as the second. Let subscript i index the sugars of all the first strands, and suppose there are N of them. Let q_i be the probability of a DSB at the ith sugar, which depends on the initial radical concentration and the subsequent reactions. Since the DSBs are produced independently, the probability of no DSB is

$$P(0) = (1 - q_1) \cdots (1 - q_N).$$ (A1)

The probability of one DSB is

$$P(1) = \sum_i q_i (1 - q_1) \cdots (1 - q_N)/(1 - q_i) = P(0)W$$ (A2)

where

$$W = \sum_i q_i/(1 - q_i).$$ (A3)

The probability of two DSBs is

$$P(2) = \sum_{i \neq j} q_i q_j (1 - q_1) \cdots (1 - q_N)/2(1 - q_i)(1 - q_j)$$

$$= \frac{1}{2} P(0) \left[\sum_i q_i (1 - q_i) \right]^2 - \frac{1}{2} P(0) \sum_i q_i^2/(1 - q_i)^2.$$ (A4)

Since there are a large number of sugars in the region of high radical concentration, the second term in Eq. (A4) can be neglected compared with the first. Thus,

$$P(2) \approx P(0)W^2/2. \tag{A5}$$

A continuation of this argument shows that the probability of n DSBs is

$$P(n) \approx P(0)W^n/n! \tag{A6}$$

for n small compared with the number of sugars in the region of high radical concentration. Thus, (2b) the number of DSBs for a single track is given approximately by a Poisson distribution.

The mean amount of energy deposited in the cell is proportional to the length of the primary track; so the number of ion pairs, the number of free radicals, and consequently (3a) the mean number of DSBs also are proportional to the track length.

The rate at which radicals attack sugars to initiate SSBs is proportional to the quantity of DNA in the region of high radical concentration. However, only a small fraction of the radicals produce SSBs; so the radical-DNA reaction is not saturated, and the mean number of SSBs produced is proportional to the quantity of DNA. The number of sugars in the region of high radical concentration also is proportional to the quantity of DNA; so the fraction of sugars that have a SSB is independent of the quantity of DNA. Therefore, for a given SSB on the first strand, the probability of a SSB on the second strand that is close enough to form a DSB is independent of the quantity of DNA. The mean number of DSBs is proportional to the mean number of SSBs on the first strands times the probability of a close SSB on the second strands, and thus (3b) it is proportional to the quantity of DNA.

APPENDIX II

To derive Eq. (25) define

$$F(k) = f(k)(\varphi/\vartheta)^k \tag{A7}$$

and

$$G(k) = \frac{\vartheta^k}{k!}\left[\left(\frac{d}{ds}\right)^k g(s\,D)\right]_{s=0} \tag{A8}$$

then

$$P(S|D) = \sum_{k=0}^{\infty} F(k)G(k). \tag{A9}$$

F and G each have a single peak for some value of k. By the proper choice of ϑ, these peaks can be made to coincide. G is much more sharply peaked than is F; so Eq. (A9) can be approximated by

$$P(S|D) \approx F(\xi) \sum_{k=0}^{\infty} G(k) \tag{A10}$$

where ξ denotes the G-weighted mean value of k, that is,

$$\xi = \sum_{k=0}^{\infty} kG(k) \bigg/ \sum_{k=0}^{\infty} G(k) \tag{A11}$$

and ϑ is chosen so that

$$F'(\xi) = 0. \tag{A12}$$

Each of the G-sums can be done exactly and give

$$\sum_{k=0}^{\infty} G(k) = g(\vartheta|D) \tag{A13}$$

and

$$\xi = \gamma'\vartheta\omega D \tag{A14}$$

where γ' is evaluated at $s = \vartheta$. The right-hand side of Eq. (A10) is the first term in the expansion of $F(k)$ in Eq. (A9) in a power series about the value $k = \xi$. The error in Eq. (A10) can be calculated from the first term in the expansion that is neglected, which is the second derivative of F. The error as a fraction of $P(S|D)$ is approximately

$$\frac{0.5\xi}{\varepsilon + \xi}$$

The error is zero at $D = 0$ and reaches 0.19 at the dose $\varphi\alpha D = \varepsilon$.

Equation (A12) can be solved for ϑ and yields

$$\vartheta = \frac{\varphi\varepsilon + \delta\xi}{\varepsilon + \xi} \exp\left[\frac{0.5\xi}{\varepsilon + \xi} - \frac{0.5\delta}{\varphi\varepsilon + \delta\xi}\right]. \tag{A15}$$

Since ε is about 20 or 30, neglect of the exponential in Eq. (A15) gives an error of less than 0.025. When the value of ξ given by Eq. (A14) is substituted into Eq. (A15), the result is a quadratic equation in ϑ, whose solution is given by Eq. (27). When the expression for ξ is substituted into Eq. (A10), the result is Eq. (25).

REFERENCES

1. C. A. Tobias, E. A. Blakely, F. Q. H. Ngo, and A. Chatterjee. 1978. Repair-misrepair (RMR) model for the effects of single and fractionated dose of heavy accelerated ions. *Radiat. Res.* *74*, 589-590 (abstract).

2. C. A. Tobias, E. A. Blakely, F. Q. H. Ngo, and T. C. H. Yang. 1980. The repair-misrepair model of cell survival. In *Radiation Biology and Cancer Research*, pp. 195-230. (R. E. Meyn and H. R. Withers, eds.). New York: Raven Press.

3. P. W. Todd. 1964. Ph.D. Dissertation, University of California, Berkeley, p. 11.

4. E. L. Lloyd, M. A. Gemmell, C. B. Henning, D. S. Gemmell, and B. J. Zabransky. 1979. Cell survival following multiple-track alpha particle irradiation. *Int. J. Radiat. Biol. 35*, 23-31.

5. C. A. Tobias and T. C. Yang. 1984. The roles of ionizing radiation in cell transformation. *Proceedings of the Neyman-Kiefer Conference, July 1983, Berkeley, California.* (L. Le Cam, ed.).

6. E. A. Blakely, C. A. Tobias, T. C. H. Yang, K. C. Smith, and J. T. Lyman. 1979. Inactivation of human kidney cells by high-energy monoenergetic heavy-ion beams. *Radiat. Res. 80*, 122-160.

7. C. D. Hodgman, ed. 1959. *Standard Mathematical Tables*, 12th ed., p. 387. Cleveland: Chemical Rubber Publishing Co.

8. C. A. Tobias, E. A. Blakely, P. Y. Chang, L. Lommel, and R. Roots. 1984. Response of sensitive human ataxia and resistant T-1 cell lines to accelerated heavy ions. *Br. J. Cancer 49* (Suppl. VI), 175-185.

9. D. T. Goodhead, R. J. Munson, J. Thacker, and R. Cox. 1980. Mutation and inactivation of cultured mammalian cells exposed to beams of heavy ions IV: Biophysical interpretation. *Int. J. Radiat. Biol. 37*, 135-167.

10. J. Llacer, C. A. Tobias, W. R. Holley, and T. Kanai. 1984. On-line characterization of heavy ion beams with semiconductor detectors. *Med. Phys.*, in press.

11. J. L. Magee and A. Chatterjee. 1980. Radiation chemistry of heavy-particle tracks. 1. General considerations. *J. Phys. Chem. 84*, 3529-3536.

12. A. Chatterjee and J. L. Magee. 1980. Radiation chemistry of heavy-particle tracks. 2. Fricke dosimeter system. *J. Phys. Chem. 84*, 3537-3543.